Arithmetic Circuits for DSP Applications

Arithmetic Circuits for DSP Applications

Edited by Pramod Kumar Meher and Thanos Stouraitis

IEEE PRESS

WILEY

Published by John Wiley & Sons, Inc., Hoboken, New Jersey.
Published simultaneously in Canada.

For general information on our other products and services or for technical support, please contact our Customer Care Department within the United States at (800) 762-2974, outside the United States at (317) 572-3993 or fax (317) 572-4002.

Wiley also publishes its books in a variety of electronic formats.Some content that appears in print may not be available in electronic formats.For more information about Wiley products, visit our web site at www.wiley.com.

Library of Congress Cataloging-in-Publication Data is available.

ISBN: 978-1-119-20677-4

Printed in the United States of America.

10 9 8 7 6 5 4 3 2 1

Contents

Preface

Digital Signal Processing (DSP) has become increasingly indispensable in most of our daily activities. It enables modern communications, image and video creation, storage, processing and transmission, medical instrumentations and diagnostics, automobile functioning, military surveillance and target tracking operations, satellite and aerospace control, DNA analysis, the Internet, energy smart grids, a host of digital consumer products, and myriad other applications.

DSP's algorithms are heavily computation-intensive and many applications are real time, thus imposing very stringent performance requirements on the used devices. Depending on the application, these devices have to be optimized for speed, energy, power consumption, and/or area. Running these applications on the CPUs of general-purpose processors, as was mostly done up until the 1990s, usually does not meet their demanding requirements. It is often the case that dedicated processors, with customized arithmetic units, have to be designed by computer engineers well versed in computer arithmetic options.

Although computer arithmetic is one of the first fields of computer science and engineering, its growth has not stopped over the decades. It has often been demonstrated that it plays a most important role in the performance of digital systems, as it is closely related to the data representation schemes. Especially for DSP systems, its algorithms have a number of interesting characteristics, which can be exploited in the design of the arithmetic circuits, so that they can be implemented more efficiently, in terms of computation time, chip area, and power consumption. As we move well into the era of embedded systems and the Internet of everything, the arithmetic circuits, and particularly the ones for DSP applications, will retain and expand their significance of system on chip designs.

This book aims to give a comprehensive coverage of fundamental concepts and advantages for various classes of arithmetic circuits, their digital implementation schemes, performance considerations, and applications. It comes in the form of an edited volume, as leading researchers in various important classes of arithmetic circuits have presented the relevant state of the art, possible applications, and scope for future research.

Chapter 1 is devoted to circuits used in implementing basic operations, like additions and multiplications. The authors then move on to discussing the realization of more complex operations, such as sums of products, squaring, square roots, and complex multiplications. At the end of this chapter, they present circuits for special functions, as well as polynomial and piecewise polynomial approximations of nonlinear functions.

Multiplication is a common but costly arithmetic operation, in terms of silicon area, computation time, and power consumption. Many real-time applications utilize algorithms involving multiplications by constants. Chapter 2 is devoted to the presentation of the basic techniques for shift-add circuits for multiplications by constants, along with an overview of advanced optimization schemes and a brief sketch of application domains.

Computation of inner products of two vectors, particularly when one of the vectors is fixed and known *a priori*, is frequently encountered in DSP applications. Chapter 3 focuses on the use of distributed arithmetic (DA) as a technique for the multiplier-less implementation of inner products and their application in a host of different DSP operations and transforms. As DA is primarily a lookup table technique, it is also appropriate for FPGA implementations.

Chapter 4 continues with the discussion of lookup table (LUT)-based techniques, especially for the implementation of operations such as multiplications and divisions by a constant, exponentiations of constants, inversion, evaluation of squares and square roots of integers, as well as sinusoidal functions, sigmoid function, logarithms, antilogarithms. The chapter provides an overview of the key concepts, important developments, implementations, and applications of some popular LUT-based systems.

The calculation of trigonometric, hyperbolic, and logarithmic functions, real and complex multiplications, division, square root, solution of linear systems, eigenvalue estimation, singular value decomposition, QR factorization, and many other functions through the use of simple shift-add operations is the focus of Chapter 5. It presents the COordinate Rotation DIgital Computer (CORDIC) technique, which has been utilized for applications in diverse areas, such as signal and image processing, communication systems, and robotics and 3D graphics, in addition to general scientific and technical computations.

Chapter 6 considers the use of the Residue Number Systems (RNS) as a means of data representation and its impact on the design of arithmetic circuits. The various design considerations, such as choice of modulus sets, algorithms for converting binary numbers to RNS and vice versa, multiplication, scaling, sign detection, and base extension operations are reviewed. Special attention is given to power-of-two-related modulus sets, in view of their computational advantages. Some applications of RNS in signal processing, communication systems, and cryptography are presented, in order to highlight the possible advantages as compared to conventional arithmetic-based implementations.

The representation of data as logarithms has been long used, as a means to simplify particular arithmetic operations and utilize improved numerical properties, even before the modern digital computer era. The Logarithmic Number System (LNS) is a formalization of the logarithmic representation of the data in a digital system and it can be conceived as a generalization of the floating-point representation. It is the focus of Chapter 7, which discusses the implementations of basic operations in logarithmic arithmetic circuits, presents forward and inverse conversion schemes, the implementation of complex arithmetic, and finally discusses the impact of logarithmic arithmetic on power dissipation and presents several applications amenable to LNS processing.

Chapter 8 deals with the implementation of arithmetic when other redundant number systems (RDNS) are used for data representation. For a radix-r RDNS, each digit can assume more than γ values. The extra digit values allow for constant time and carry-free addition and subtraction. In this chapter, RDNS is formally defined, its carry-free property is analyzed, and the algorithms and circuits for the corresponding add, subtract, and special functions are provided.

The primary readership for this edited volume will be senior undergraduate and graduate students in computer science, computer engineering, and electrical and electronics engineering. It may be used as a textbook for courses on embedded systems, signal processing, and IC design, as well as a reference book. Practicing engineers or researchers in academia will also find it very handy and useful. The prerequisites required for a complete understanding of the book include fundamentals of digital logic and a basic ability in mathematics.

The editors would like to thank all chapter authors for their valuable contributions.

Nanyang Technological University
Singapore

Pramod K. Meher

Khalifa University UAE
University of Patras Greece

Thanos Stouraitis

About The Editors

Pramod Kumar Meher received the B.Sc. (Honours) and M.Sc. degrees in physics, and the Ph.D. degree in science from Sambalpur University, India, in 1976, 1978, and 1996, respectively. Currently, he is a Senior Research Scientist with Nanyang Technological University, Singapore. Previously, he was a Professor of Computer Applications with Utkal University, India, from 1997 to 2002, and a Reader in electronics with Berhampur University, India, from 1993 to 1997. His research interest includes design of dedicated and reconfigurable architectures for computation-intensive algorithms pertaining to signal, image and video processing, communication, bioinformatics, and intelligent computing. He has contributed more than 200 technical papers to various reputed journals and conference proceedings. Dr Meher has served as a speaker for the Distinguished Lecturer Program (DLP) of IEEE Circuits Systems Society during 2011 and 2012, Associate Editor of the *IEEE Transactions on Circuits And Systems-II: Express Briefs* during 2008–2011, the *IEEE Transactions on Circuits And Systems-I: Regular Papers* during 2012–2013, and the *IEEE Transactions on Very Large Scale Integration (VLSI) Systems* during 2009–2014. Currently, he is serving as Associate Editor for the *IEEE Transactions on Circuits and Systems for Video Technology* and *Journal of Circuits, Systems, and Signal Processing*. Dr Meher is a Senior Member of IEEE and Fellow of the Institution of Electronics and Telecommunication Engineers, India. He was the recipient of the Samanta Chandrasekhar Award for excellence in research in engineering and technology for 1999.

Thanos Stouraitis, an IEEE Fellow for his "contributions in digital signal processing architectures and computer arithmetic," is currently Professor and Chair of the ECE department of Khalifa University, UAE (on leave from the University of Patras, Greece). He has served on the faculties of University of Florida, Ohio State University, Polytechnic University (NYU), and University of British Columbia. His research interests include cryptography, computer arithmetic, signal/image processing systems, and optimal processor design. He has coauthored over 180 technical papers, holds one patent on DSP processor design, and has authored several book chapters and books. He has been serving as an Editor or/and Guest Editor for numerous technical journals. He served as General Chair, TPC Chair, and Symposium Chair for many international conferences, like ISCAS, SiPS, and ICECS. He received the IEEE Circuits and Systems Society Guillemin-Cauer Award. He has served IEEE in many ways, including as CAS Society President, 2012–2013.

1

Basic Arithmetic Circuits

Oscar Gustafsson and Lars Wanhammar

Division of Computer Engineering, Linköping University, Linköping, Sweden

1.1 Introduction

General-purpose DSP processors, application-specific processors [1], and algorithm-specific processors are used to implement different types of DSP systems or subsystems. General-purpose DSP processors are programmable and therefore, provide maximum flexibility and reusability. They are typically used in applications involving complex and irregular algorithms while application-specific processors provide lower unit cost and higher performance for a specific application, particularly when the volume of production is high. The highest performance and lowest unit cost is obtained by using algorithm-specific processors. The drawback is the restricted or even lack of flexibility, and very often the nonrecurring engineering (NRE) cost could be very high.

The throughput requirement in most real-time DSP applications is generally fixed, and there is no advantage of an implementation with throughput than that design to minimize the chip area, and power consumption. Now in a CMOS implementation with higher throughput than required, it is possible to reduce the power consumption by lowering the supply voltage and operating the system at lower frequency [2].

1.2 Addition and Subtraction

The operation of adding two or more numbers is the most fundamental arithmetic operation, since most other operations in one or another way are based

Arithmetic Circuits for DSP Applications, First Edition. Edited by Pramod Kumar Meher and Thanos Stouraitis.
© 2017 by The Institute of Electrical and Electronics Engineers, Inc. Published 2017 by John Wiley & Sons, Inc.

 **Figure 1.1** Addition of two binary numbers.

on addition. The operands of concern here are either two's-complement or unsigned representation.

Most DSP applications use fractional arithmetic instead of integer arithmetic [3]. The sum of two w-bit numbers is a $(w + 1)$-bit number while the product of two w-bit binary numbers is a $2w$-bit number. In many cases and always in recursive algorithms the resulting number needs to be quantized to a w-bit number. Hence, the question is which bits of the result are to be retained. In fractional arithmetic, the input operands as well as the result are interpreted as being in the range $[0, 1]$, that is,

$$x = \sum_{i=1}^{w} x_i 2^{-i} \tag{1.1}$$

for unsigned numbers and in the range $[-1, 1]$, that is,

$$x = -x_0 \sum_{i=1}^{w} x_i 2^{-i} \tag{1.2}$$

for signed numbers in two's-complement representation. For convenience, we let w denote the number of fractional bits and one additional bit is used for a signed representation.

We use the graphic representation shown in Figure 1.1 to represent the operands and the sum bits with the most significant bit to the left.

1.2.1 Ripple-Carry Addition

Ripple-carry addition is illustrated in Figure 1.2. A ripple-carry adder performs addition of two numbers; adds the bits of the same significance and the carry-bit from the previous stage sequentially using a full adder (FA), and propagates the carry-bit to the next stage. Obviously, the addition takes w addition cycles, where duration of each clock cycle is the time required by an FA to complete the addition of three bits. This type of adder can add both unsigned and two's-complement numbers.

The major drawback with the ripple-carry adder is that the worst-case delay is proportional to the word length. Also, typically, the ripple-carry adder produces many glitches since the full adder cells need to wait for the correct carry input. This situation is improved if the delay for the carry bit is smaller than that of the sum bit [4].

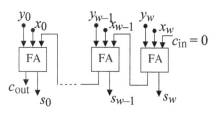

Figure 1.2 Ripple-carry adder.

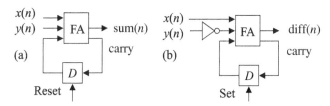

Figure 1.3 Bit-serial (a) adder and (b) subtractor.

Alternatively, all pairs of bits of the same significance can be added simultaneously and then the carries are added using some efficient scheme. Many additional schemes have been proposed. For more details, we refer to, for example, Reference [5].

It is also possible to perform addition in constant time using redundant number systems such as signed-digit or carry-save representations [6]. An alternative is to use residue number systems (RNS), which split the carry-chain into several shorter ones [7].

1.2.2 Bit-Serial Addition and Subtraction

Bit-serial addition and subtraction of numbers in a two's-complement representation can be performed with the circuits shown in Figure 1.3. A pair of input bits of operands X and Y are fed to the circuit in the least-significant-bit-first order, and the carry generated during at any given cycle is returned as input of the circuit during the next clock cycle. Since the carries are saved from one bit position to the next, the circuits of Figures. 1.3a and b are called carry-save adder and carry-save subtractor, respectively. At the start of the addition, the D flip-flop is reset (set) for the adder (subtractor), respectively.

Figure 1.4 shows two latency models for bit-serial adders. The leftmost is suitable for static CMOS and the rightmost for dynamic (clocked) logic styles. The time to add two w-bit numbers is w or $w + 1$ clock cycles. However, the throughput rate of computation would not be affected by this additional latency of one clock cycle since the two successive sum values computed in an interval is w cycles.

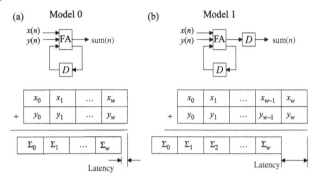

Figure 1.4 Latency models for a bit-serial adder.

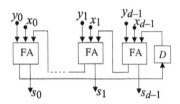

Figure 1.5 Digit-serial adder with digit-size d obtained from unfolding a bit-serial adder.

1.2.3 Digit-Serial Addition and Subtraction

In case of digit-serial processing [8–11] a group of bits (called a digit) of input words are processed at a time. From speed and power consumption points of view, it is advantageous to process several bits at a time. The number of bits processed in a clock cycle is referred to as the digit size.

Figure 1.5 shows a digital-serial adder, where d is the digit size. The D-flip-flop transfers the output carry. In case of subtraction of a two's-complement number, the negative value is instead added by inverting the bits and setting the carry flip-flop during the addition of the least significant digit.

Most of the principles used for bit-serial arithmetic can easily be extended to digit-serial arithmetic.

The bit-serial and digit-serial arithmetic circuits require less chip area and therefore their equivalent switched capacitance and leakage current are relatively low compared with word-level circuits [3,12].

1.3 Multiplication

Multiplication of two numbers can be realized into the following two main steps:

1. Partial product generation where partial product is the result of multiplication of a bit of the multiplier with the multiplicand.
2. Accumulation of the partial products.

In the following subsections, we discuss various techniques to simplify and speed-up the summation of the partial products.

1.3.1 Partial Product Generation

In unsigned binary number representation, bit-wise multiplications can be written as

$$Z = XY = \sum_{i=1}^{w} x_i 2^{-i} \sum_{j=1}^{w} y_j 2^{-j} = \sum_{i=1}^{w} \sum_{j=1}^{w} x_i y_j 2^{-i-j} \tag{1.3}$$

This leads to a partial product array as shown in Figure 1.6. The partial product generation can be readily realized using AND gates.

For two's-complement representation, the result is very similar to that in Figure 1.6, except that the sign-bit causes some of the bits to have negative weight. This can be seen from

$$Z = XY$$
$$= \left(-x_0 + \sum_{i=1}^{w} x_i 2^{-i} \right) \left(-y_0 + \sum_{j=1}^{w} y_j 2^{-j} \right)$$
$$= x_0 y_0 - x_0 \sum_{j=1}^{w} y_j 2^{-j} - y_0 \sum_{i=1}^{w} x_i 2^{-i} + \sum_{i=1}^{w} \sum_{j=1}^{w} x_i y_j 2^{-i-j} \tag{1.4}$$

The corresponding partial product matrix is shown in Figure 1.7.

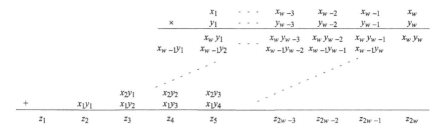

Figure 1.6 Partial products for unsigned binary numbers.

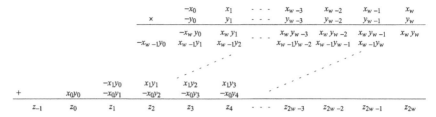

Figure 1.7 Partial products for two's-complement numbers.

1.3.2 Avoiding Sign-Extension (the Baugh and Wooley Method)

Addition of any two numbers in two's-complement representation requires that the word lengths are equal. Hence, it is necessary to sign-extend the MSB of partial products in Figure 1.7 to obtain the same word length for all rows.

To avoid this the computation resulting from sign-extension, it is possible to perform the summation from top to bottom and perform sign-extension of the partial results to match the next row to be added. However, if we want to add the partial products in an arbitrary order using a multioperand adder, the following technique, proposed by Baugh and Wooley [13], can be used.

Note that for a negative partial product, we have $-p = \overline{p} - 1$. Hence, we can replace all partial products with negative weight with an inverted version. Then, we need to subtract a constant value from the result. Since there will be several constants, one from each negative partial product, we can sum these up and form a single compensation vector to be added. When this is applied we get the partial product array as shown in Figure 1.8 as

$$
Z = x_0 y_0 + \sum_{j=1}^{w} \overline{x_0 y_j} 2^{-j} - \sum_{j=1}^{w} 2^{-j} + \sum_{i=1}^{w} \overline{y_0 x_i} 2^{-i} - \sum_{i=1}^{w} 2^{-i}
$$

$$
+ \sum_{i=1}^{w} \sum_{j=1}^{w} x_i y_j 2^{-i-j}
$$

$$
= x_0 y_0 + \sum_{j=1}^{w} \overline{x_0 y_j} 2^{-j} + \sum_{i=1}^{w} \overline{y_0 x_i} 2^{-i} + \sum_{i=1}^{w} \sum_{j=1}^{w} x_i y_j 2^{-i-j}
$$

$$
- \left(2 - 2^{-w+1} \right) \tag{1.5}
$$

1.3.3 Reducing the Number of Partial Products

There are several methods for reducing the number of partial products. A technique to reduce the number of partial products using small ROM-based

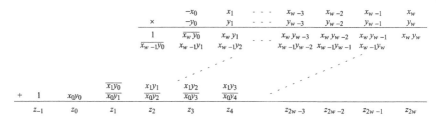

Figure 1.8 Partial products for two's-complement numbers without sign-extension.

Table 1.1 Rules for the radix-4 modified Booth encoding.

$x_{2k}x_{2k+1}x_{2k+2}$	r_k	$d_{2k}d_{2k+1}$
000	0	00
001	1	01
010	1	01
011	2	10
100	−2	$\bar{1}0$
101	−1	$0\bar{1}$
110	−1	$0\bar{1}$
111	0	00

multipliers is found in Reference [14]. Other methods are based on number representation or encoding of one of the operands. It is possible to reduce the number of nonzero positions by using a signed-digit representation, for example, canonical signed digit (a CSD) representation to obtain a minimum number of nonzeros. However, the drawback is that the conversion from two's-complement to CSD involves carry-propagation [15]. Furthermore, the worst case is that half of the positions are nonzero, and, hence, one would still need to design the multiplier to deal with this case.

Instead, it is possible to derive a signed-digit representation that is not necessarily minimal, but has at most half of the positions being non-zero. This is referred to modified Booth encoding [16,17] and is often described as being a radix-4 signed-digit representation where the recoded digits $r_k \in \{-2, -1, 0, 1, 2\}$.

The logic rules for performing the modified Booth encoding are based on the idea of finding strings of ones and replace them as $011...11 = 100...0\bar{1}$, illustrated in Table 1.1. From this, one can see that there is at most one nonzero digit in each pair of digits $\left(d_{2k}d_{2k+1}\right)$.

Now, to perform the multiplication, we must be able to possibly negate and multiply the operand with 0, 1, or 2. As discussed earlier, the negation is typically

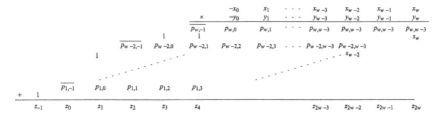

Figure 1.9 Partial products for radix-4-modified Booth encoding.

performed by inverting the bits and add a one in the column corresponding to the LSB position. The partial product array for a multiplier using the modified Booth encoding is shown in Figure 1.9. From this, it can be seen that each row now is one bit longer and that the least significant position contains two bits, where the additional bit is used for negation.

1.3.4 Reducing the Number of Columns

The result of multiplication is usually quantized to be represented with fewer bits. To reduce the complexity of the multiplication of such cases it has been proposed to perform the quantization at the partial product generation stage and partial product summation stage [18]. This is commonly referred to as fixed-width multiplication referring to the fact that (most of) the partial product rows have the same width. Simply truncating the partial products will result in a rather large error. Several methods have therefore been proposed to reduce the error by introducing compensating terms [19,20].

1.3.5 Accumulation Structures

The problem of summing up the partial products can be solved in the following three general ways:

1. Sequential accumulation, where a row or a column of partial products are accumulated in each cycle.
2. Array accumulation, which gives a regular structure.
3. Tree accumulation, which gives the smallest logic depth but in general an irregular structure.

1.3.5.1 Add-and-Shift Accumulation

In multipliers based on add-and-shift, the partial products are successively accumulated in w cycles for w-bit operands. During each cycle a new partial product is added with the current sum of partial products divided by two. In most number systems dividing by two can be implemented by a shift of the bits.

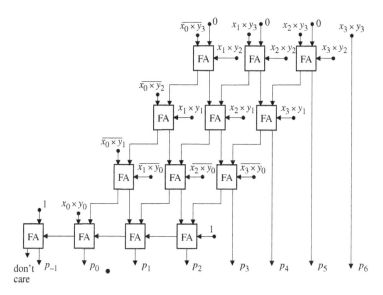

Figure 1.10 Baugh–Wooley array multiplier.

The multiplication time becomes large if two's-complement representation and ripple-carry adders are used. Redundant number representation is therefore used to alleviate the carry propagation problem. We will defer the discussion of add-and-shift multipliers based on bit-serial and digit-serial realizations to Section 1.3.6.

1.3.5.2 Array Accumulation

Array multipliers use an array of almost identical cells for generation and accumulation of the partial products. Figure 1.10 shows a realization of the Baugh–Wooley array multiplier [13], leading to a multiplication time proportional to $2w$.

The array multiplier is a highly regular structure resulting in short wire lengths between the logic circuits, which is important for high-speed design in nanometer processes where wiring delay gives a significant contribution to the overall delay. However, in a process where cell delay dominates wire delay, the logic depth of the design is more important than regularity. In the array multiplier the logic depth is $\mathcal{O}(w)$, where w is the input word length. In the adder tree multiplier, which is discussed in Section 1.3.5.3, the depth is $\mathcal{O}(\log(w))$. Even for short word lengths, this leads to a significant shorter delay.

1.3.5.3 Tree Accumulation

The array structure provides a regular structure, but at the same time the delay grows linearly with the word length. All the partial product generation methods in Figures 1.6–1.9 provide a number of partial products that can be accumulated in arbitrary order.

The first approach is to use as many full adders as possible to reduce as many partial products as possible. Then, we add as many half adders as possible to minimize the number of levels and try to shorten the word length for the vector-merging adder. This kind of approach is followed in the Wallace tree proposed in Reference [21]. The main drawback of this approach is the excessive use of half adders. Dadda [22] instead proposed that full adders and half adders should only be used if required to obtain a number of partial products equal to a value in the Dadda series. The value of position n in the Dadda series is the maximum number of partial products that can be reduced using n levels of full adders. The Dadda series starts {3, 4, 6, 9, 13, 19, … }. The benefit of this is that the number of half adders is significantly reduced while still obtaining a minimum number of levels.

However, the length of the vector-merging adder increases in case of Dadda reduction trees. A compromise between these two approaches is the 'reduced area' heuristic [23], which is similar to the Wallace tree, since as many full adders as possible are introduced in each level. Half adders are on the other hand only introduced if they are required to reach a number of partial products corresponding to the Dadda series, or if exactly two partial products have the same least significant weight. In this way a minimum number of stages is obtained, while at the same time both the length of the vector-merging adder and the number of half adders are kept small.

To illustrate the operation of the reduction tree approaches we use dot diagrams, where each dot corresponds to a bit of the partial product to be added. Bits with the same weight are placed in the same column and bits in adjacent columns are either of one position higher or one position lower weight, with higher weights toward left. The bits are manipulated by using either full or half adders. The operation of these are illustrated in Figure 1.11.

The reduction schemes are illustrated for an unsigned 6 × 6-bit multiplication in Figure 1.12. The complexity results are summarized in Table 1.2. It should be noted that the positioning of the results in the next level is resulting partial products done based on ease of illustration. From a functional point of view, this step is arbitrary, but it is possible to optimize the timing by carefully utilizing different delays of the sum and carry outputs of the adder cells [24]. Furthermore, it is possible to reduce the power consumption by optimizing the interconnect ordering [25].

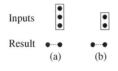

Figure 1.11 Operation on bits in a dot diagram with (a) full adder and (b) half adder.

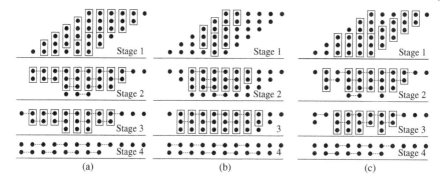

Figure 1.12 Reduction trees for a 6 × 6-bit multiplier: (a) Wallace, (b) Dadda, and (c) Reduced area.

Table 1.2 Complexity of the three reduction trees in Figure 1.12.

Tree structure	Full adders	Half adders	VMA length
Wallace	16	13	8
Dadda	15	5	10
Reduced area	18	5	7

The reduction trees in Figure 1.12 do not provide any regularity. This means that the routing is complicated and may become the limiting factor in an implementation. Reduction structures that provide a more regular routing, but still a small number of stages, include the overturned-stairs reduction tree [26] and the HPM tree [27].

1.3.5.4 Vector-Merging Adder

The role of the vector-merging adder is to add the outputs of the reduction tree. In general, any carry-propagation adder can be used. However, the signals corresponding to different bits of input to the vector-merging adders are typically available after different delays after the arrival of input to the multiplier. It is possible to derive carry-propagation adders that utilize this different signal arrival times to optimize the adder delay [28].

1.3.6 Serial/Parallel Multiplication

Serial/parallel multipliers are based on the add-and-shift principle where the multiplicand, x, arrive serially while the multiplier, a is available in bit-parallel format. Several forms of serial/parallel multipliers have been proposed [29].

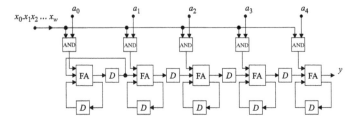

Figure 1.13 Serial/parallel multiplier based on carry-save adders. The box containing "AND" refers to AND gate throughout this chapter.

They differ mainly by the order in which bit-products are generated and added and in the way subtraction is handled.

The generation of partial products are handled by the AND gates shown in Figure 1.13. The first partial products correspond to the first row of partial products in Figure 1.6. Thus, in the first time slot, we add the partial products, $a \times x_w$ to the initially cleared accumulator. Next, the D flip-flops are clocked and the sum-bits from the FAs are shifted by one bit to the right; each carry-bit is saved to be used in the next clock cycle: the sign-bit is copied; and the LSB of the product is produced as output bit. These operations correspond to dividing the accumulator contents by 2. Note that the value in the accumulator is in redundant form and that carry propagation is not needed.

In the following clock cycle, the next bit of x is used to form the next partial product and added to the value in the accumulator, and the value in the accumulator is again divided by 2. This process continues for w clock cycles, until the sign bit x_0 is reached, whereupon a subtraction must be done instead of an addition.

During the first w clock cycles, the least significant part of the product is computed and the most significant is stored in the D flip-flops. In the next $w_c - 1$ clock cycles, we apply zeros to the input so that the most significant part of the product is shifted out of the multiplier. Hence, the multiplication requires $w + w_c - 1$ clock cycles. Two successive multiplications must therefore be separated by $w + w_c - 1$ clock cycles. Note that the accumulation of the partial products is performed using a redundant representation, which do not require carry propagation. The redundant value stored in the accumulator in carry save format is converted to a nonredundant representation in the last FA.

An alternative and better solution is to copy the sign-bit in the first multiplier stage as shown in Figure 1.14. The first stage, corresponding to the sign-bit in the coefficient, is replaced by a subtractor. In fact, only an array of half-adders is needed since one of the input bits to the 1-bit adders is zero.

The subtraction of the bit-products required for the sign-bit in the serial/parallel multiplier can be avoided by extending the input by $w_c - 1$ copies of the sign-bit. After w clock cycles the most significant part of the product is

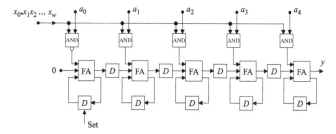

Figure 1.14 Modified serial/parallel multiplier.

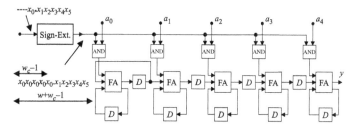

Figure 1.15 Serial/parallel multiplier with sign extension-circuit.

stored in the D flip-flops. In the next w_c clock cycles the sign bit of x is applied to the multiplier's input. This is accomplished by the sign extension-circuit shown in Figure 1.15. The sign extension-circuit consists of a latch that transmits all bits up to the sign-bit and thereafter latches the sign-bit. For simplicity, we assume that $w = 5$ bits and $w_c = 5$ bits.

The product is

$$y = a \times x = a \times \left(-x_0 + \sum_{k=1}^{5} x_k 2^{-k} \right) \tag{1.6}$$

but the multiplier computes

$$y = a \left(x_0 2^4 + x_0 2^3 + x_0 2^2 + x_0 2^1 + x_0 2^0 + \sum_{k=1}^{5} x_k 2^{-k} \right)$$

$$= a \left(x_0 2^4 + x_0 2^3 + x_0 2^2 + x_0 2^1 + x_0 2^1 - x_0 2^0 + \sum_{k=1}^{5} x_k 2^{-k} \right)$$

$$= a x_0 \left(2^4 + 2^3 + 2^2 + 2^1 + 2^1 \right) + a \times x$$

$$= a x_0 2^5 + a \times x \tag{1.7}$$

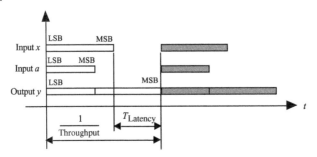

Figure 1.16 Latency and throughput for a multiplication $y = a \times x$ using bit-serial arithmetic with least significant bit (LSB) first.

The first term above contributes an error in the desired product. However, there will not be any error in the $w + w_c - 1$ least-significant bits since the error term only contributes to the bit positions with higher significance.

A bit-serial multiplication takes at least $w + w_c - 1$ clock cycles. However, two successive multiplications can partially be overlapped to increase the throughput [3]. These serial/parallel multipliers using this technique can be designed to perform one multiplication every $\max\{w, w_c\}$ clock cycles. Latency and throughput for a multiplication using bit-serial arithmetic with least significant bit (LSB) first is illustrated in Figure 1.16.

A major advantage of bit-serial over bit-parallel arithmetic is that it significantly reduces chip area. This is done in two ways. First, it eliminates the need of wide buses and simplifies the wire routing. Second, by using small processing elements, the chip itself becomes smaller and requires shorter wiring. A small chip can support higher clock frequencies and is therefore faster.

Bit-serial multiplication can be done either by processing the least significant or the most significant bit first. The former is the most common since the latter is more complicated and requires the use of so-called redundant arithmetic.

The latency of a multiplication is equal to the number of fractional bits in the coefficient. For example, a multiplication with a coefficient $w_c = (1.0011)_{2C}$ will have a latency corresponding to four clock cycles. A bit-serial addition or subtraction has in principle zero latency while a multiplication by an integer may have zero or negative latency. But, the latency in a recursive loop is always positive, since the operations must be performed by causal PEs. In practice, the latency may be somewhat longer, depending on the type of logic that is used to realize the arithmetic operations, as will be discussed shortly.

Here, we define two latency models for bit-serial arithmetic. Two latency models for a bit-serial adder are shown in Figure 1.4. In model 0, which can be used to model a static CMOS logic style without pipelining of the gates, the latency is equal to the gate delay at a full adder. In model 1, which can be used to model a dynamic CMOS logic style, or a static CMOS logic style

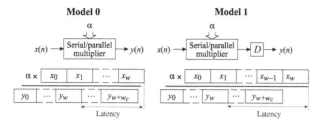

Figure 1.17 Latency models for a serial/parallel multiplier.

with pipelining at the gate level, the full adder, followed by a D flip-flop, causes the latency to become one clock cycle. Model 1 generally results in faster bit-serial implementations, due to the shorter logic paths between the flip-flops in successive operations [30].

For low-complexity implementation of a serial/parallel multiplier, bit-serial adders may be used. The corresponding latency models for a serial/parallel multiplier are shown in Figure 1.17. Denoting the number of fractional bits of the coefficient w_c, the latencies become w_c for latency model 0, and $w_c + 1$ for latency model 1.

A digit-serial multiplier, which accumulate several bits in each stage, can be obtained either via unfolding of a bit-serial multiplier or via folding of a bit-parallel multiplier [10]. The execution time is $\lceil w/d \rceil$ clock cycles for a digit-serial adder. For a serial/parallel multiplier, it is $\lceil w/d \rceil + \lceil w_a/d \rceil$ clock cycles, where w and w_a are the data and coefficient word lengths, respectively.

1.4 Sum-of-Products Circuits

Multimedia and communication applications involve real-time audio and video/image processing which very often require sum-of-products (SOP) computation. The SOP of two N-point vectors is given by

$$y = \sum_{i=1}^{N} a_i x_i \tag{1.8}$$

SOP is a common operations in most DSP applications, for example, IIR and FIR filters and correlation. Sum-of-product is also referred to as inner products of two vectors. Most digital signal processors are optimized to perform multiply-accumulate operations.

When N is large or varies, it is not a constant, the SOP given by Eq. (1.8) is computed sequentially using a multiply-accumulate circuit (MAC) by repeating the MAC operation shown in Figure 1.18 N times.

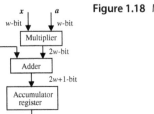

Figure 1.18 Multiplier-accumulator.

Typically, the multiplier is realized by modified radix-4 Booth's algorithm (MBA) [16,31], a carry-save adder (CSA) tree to add the partial products, and a final adder, similar to the multipliers discussed in a previous section. The delay of a Wallace tree is proportional to $\mathcal{O}(\log(N))$, where N is the number of inputs. However, the throughput is limited by the long critical path for multiplication. (4:2) or (7:3) compressors may therefore be used to reduce the number of outputs in each step. Typically, the accumulations are combined with the adder tree that compresses partial products, for example, it is possible to use two separate $N/2$-bit adders instead of one N-bit adder to accumulate the N-bit MAC result.

It is instructive to study the multiplier-accumulator in more detail. Typically, in the first stage some form of Booth encoding is used to reduce the number of partial products. The adder and multiplier are usually realized by carry-select or carry-save adders, as throughput is of utmost importance. In addition, pipelining is often used to increase the throughput further.

The adder must be operated $N \times 1$ times, which will be expensive if we use a conventional techniques with carry propagation. However, using carry-save adders (CSA), we eliminate the need for carry propagation, except of the last addition. In the intermediate steps we use a redundant number representation with separate sum and carry words. A CSA is a 3:2 compressor, that is, a full adder. Hence, the intermediate results are represented by a sum and carry word.

In the final step, however, we most often need to use a nonredundant representation for the further processing of the SOP. This step involves a vector-merging adder, which may be realized using any of the previously discussed fast adders.

The multiplier typically consists of a partial-product generation stage and an adder-tree. Often Booth encoding is used to reduce the number of partial products. A Wallace carry-save adder-tree is typically used for long word lengths, while overturned-stairs may be better for shorter word length due to the more regular wire routing.

The resulting realization is shown in Figure 1.19, where the dashed box indicates that parts that use a redundant carry-free representation. Note that the output bits (sum and carry words) of the accumulator register can be

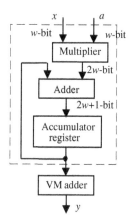

Figure 1.19 Carry-save multiplier-accumulator.

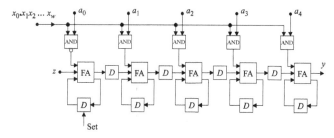

Figure 1.20 Serial/parallel multiplier-accumulator.

partially merged into the adder-tree of the multiplier. It is favourable to insert the sum and carry bits as early as possible in the tree in order to fully exploit the compressors. It may even be favourable to use 4:2 compressors.

1.4.1 SOP Computation

Alternatively, SOP can be realized by serial/parallel multiplier with an additional input that allows computations of the type $y = a \times x + z$ (is shown in Figure 1.20). The extra input allows a value z to be added at the same level of significance as x. A multiplier-accumulator is obtained if the output y is truncated/rounded to the same word length as x and added to the subsequent multiplication. A full precision multiplier-accumulator is obtained if the part of y that is truncated is saved and used to set the sum D flip-flops instead of resetting them at the start of a multiplication.

1.4.2 Linear-Phase FIR Filters

Multiplier-accumulator processing elements are suitable for implementing linear-phase FIR filters, where Eq. (1.8) can be rewritten like

$$y = \sum_{i=1}^{N/2} a_i \left(x_i \pm x_{N-i} \right) \tag{1.9}$$

where we have assumed that N is even. This requires pre-addition of the two delayed input samples. This addition can be incorporated into the adder-tree of the multiplier.

FIR filters realized in the transposed direct form can be implemented by replacing the accumulator register with a register bank that stores the values in the delay elements.

1.4.3 Polynomial Evaluation (Horner's Method)

A related case to SOP is the evaluation of a polynomial, or a series expansion, using Horner's method[1]. An expression of the form

$$y = \sum_{i=1}^{N} a_i x^i \tag{1.10}$$

is rewritten

$$y = \left(\dots \left(\left(\left(0 + a_N \right) x + a_{N-1} \right) x + a_{N-2} \right) x + \dots + a_1 \right) x \tag{1.11}$$

Figure 1.21 shows how the generic MAC processor can be modified to evaluate Eq. (1.11). Also in this case can the addition be incorporated into the adder-tree of the multiplier.

1.4.4 Multiple-Wordlength SOP

Different applications could have different requirements on accuracy and dynamic range. For example, it is desirable to use the same hardware to process both music and images, but it may be inefficient if the image processing is dominant, since the former require a word length in the range 16–24 bits, while for the latter it is often sufficient with eight bits. However, it is possible to segment a carry-save multiply-accumulate processor with a long word length. For example, a 64×64-bit SOP processor may be segmented to simultaneously perform either two 32×32-bit, four 16×16-bit, or eight 8×8-bit SOP

---

1 William Horner, 1819. In fact, this method was derived by Isaac Newton in 1711.

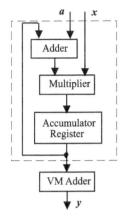

Figure 1.21 Carry-save realization of Horner's method.

products, However, this requires that some additions circuitry to be inserted into the adder-tree to separate input and outputs of different segments.

1.5 Squaring

Squaring is a special case of multiplication, which is relevant since the complexity can be significantly reduced compared to a general multiplication. Squaring finds applications in, for example polynomial evaluation through Estrin's method, see Section 1.7.2, integer exponentiation, and in certain DSP algorithms such as spectrum analysis.

1.5.1 Parallel Squarers

The partial product array for a six-bit squarer using unsigned multiplication as in Figure 1.6 is shown in Figure 1.22. It can be noted that each partial product $x_i x_j$ appears twice when $i \neq j$. Hence, it is possible to replace these two partial products with one partial product in the column with next higher significance. This results in the folded partial product array illustrated in Figure 1.23. Here, it is also utilized that $x_i x_i = x_i$, so no gates are needed to compute those partial products.

It can be noted from Figure 1.23 that the middle column will always contain the most partial products. To reduce the maximum number of partial products it is possible to apply the identity from Reference [32]

$$x_i + x_i x_{i+1} = 2x_i x_{i+1} + \overline{x_i} x_{i+1} \tag{1.12}$$

This results in the partial product array shown in Figure 1.24.

Multiplication: $\times$ $x_1\ x_2\ x_3\ x_4\ x_5\ x_6$ / $x_1\ x_2\ x_3\ x_4\ x_5\ x_6$

	z_1	z_2	z_3	z_4	z_5	z_6	z_7	z_8	z_9	z_{10}	z_{11}	z_{12}
							x_6x_1	x_6x_2	x_6x_3	x_6x_4	x_6x_5	x_6x_6
						x_5x_1	x_5x_2	x_5x_3	x_5x_4	x_5x_5	x_5x_6	
					x_4x_1	x_4x_2	x_4x_3	x_4x_4	x_4x_5	x_4x_6		
				x_3x_1	x_3x_2	x_3x_3	x_3x_4	x_3x_5	x_3x_6			
			x_2x_1	x_2x_2	x_2x_3	x_2x_4	x_2x_5	x_2x_6				
+		x_1x_1	x_1x_2	x_1x_3	x_1x_4	x_1x_5	x_1x_6					
	z_1	z_2	z_3	z_4	z_5	z_6	z_7	z_8	z_9	z_{10}	z_{11}	z_{12}

Figure 1.22 Partial products for a six-bit squarer.

Multiplication: $\times$ $x_1\ x_2\ x_3\ x_4\ x_5\ x_6$ / $x_1\ x_2\ x_3\ x_4\ x_5\ x_6$

	z_1	z_2	z_3	z_4	z_5	z_6	z_7	z_8	z_9	z_{10}	z_{11}	z_{12}
		x_1x_2	x_1x_3	x_2x_3	x_2x_4	x_3x_4	x_3x_5	x_4x_5	x_4x_6	x_5x_6		x_6
		x_1		x_1x_4	x_1x_5	x_2x_5	x_2x_6	x_3x_6		x_5		
				x_2		x_1x_6		x_4				
+						x_3						
	z_1	z_2	z_3	z_4	z_5	z_6	z_7	z_8	z_9	z_{10}	0	x_6

Figure 1.23 Folded partial products for a six-bit squarer.

Multiplication: $\times$ $x_1\ x_2\ x_3\ x_4\ x_5\ x_6$ / $x_1\ x_2\ x_3\ x_4\ x_5\ x_6$

	z_1	z_2	z_3	z_4	z_5	z_6	z_7	z_8	z_9	z_{10}	z_{11}	z_{12}
	x_1x_2	$\overline{x}_1x_2$	x_1x_3	$\overline{x}_2x_3$	x_2x_4	$\overline{x}_3x_4$	x_3x_5	$\overline{x}_4x_5$	x_4x_6	$\overline{x}_5x_6$		x_6
		x_2x_3	x_1x_4	x_1x_5	x_2x_5	x_2x_6	x_3x_6	x_5x_6				
				x_3x_4	x_1x_6	x_4x_5						
+												
	z_1	z_2	z_3	z_4	z_5	z_6	z_7	z_8	z_9	$\overline{x}_5x_6$	0	x_6

Figure 1.24 Partial products for a six-bit squarer after applying Eq. (1.12) to reduce the maximum number of partial products in a column.

Multiplication: $\times$ $x_1\ x_2\ x_3\ x_4\ x_5\ x_6$ / $x_1\ x_2\ x_3\ x_4\ x_5\ x_6$

	z_1	z_2	z_3	z_4	z_5	z_6	z_7	z_8	z_9	z_{10}	z_{11}	z_{12}
	x_1x_2	$b_{1,2,3}$	$a_{1,2,3}$	$b_{2,3,4}$	$a_{2,3,4}$	$b_{3,4,5}$	$a_{3,4,5}$	$b_{4,5,6}$	$a_{4,5,6}$	$\overline{x}_5x_6$		x_6
			x_1x_4	x_1x_5	x_2x_5	x_2x_6	x_3x_6					
					x_1x_6							
+												
	z_1	z_2	z_3	z_4	z_5	z_6	z_7	z_8	$a_{4,5,6}$	$\overline{x}_5x_6$	0	x_6

Figure 1.25 Partial products for a six-bit squarer after applying Eq. (1.13) to reduce the number of partial products.

A method for reducing the number of partial products at the expense of slightly more complicated partial product generation was suggested in Reference [33]. Here, it is noticed that the partial products $\overline{x}_i x_{i+1}$, $x_i x_{i+2}$, and $x_{i+1}x_{i+2}$ can never be all at the same time, and, hence, the following identity can be applied:

$$2\overline{x}_i x_{i+1} + x_i x_{i+2} + x_{i+1}x_{i+2} = 2b_{i,i+1,i+2} + a_{i,i+1,i+2} \qquad (1.13)$$

where the expressions for $b_{i,i+1,i+2}$ and $a_{i,i+1,i+2}$ are straightforward to derive. The resulting partial product array is shown in Figure 1.25. Both methods discussed above can be adapted to be used for signed squarers.

1.5.2 Serial Squarers

To derive a suitable algorithm sequentially computing the square x^2 of a number x, we will first discuss the special case where x is a fractional, unsigned binary number as in Eq. (1.1) [34,35]. Now, let the function $f(x)$ represent the square of the number x, that is,

$$f(x) = x^2 \tag{1.14}$$

The computation of $f(x)$ can be carried out in w iterations by repeatedly squaring the sum of the most significant bit of a number and the other bits of that number. In the first step, $f(x)$ is decomposed into the square of the most significant bit of x with a rest term and a remaining term.

$$
\begin{aligned}
f_1 &= f\left(\sum_{k=1}^{w} x_k 2^{-k}\right) = \left(\sum_{k=1}^{w} x_k 2^{-k}\right)^2 \\
&= \left(x_1 2^{-1} + \sum_{k=2}^{w} x_k 2^{-k}\right)^2 \\
&= x_1 2^{-2} + x_1 2^0 \sum_{k=2}^{w} x_k 2^{-k} + \underbrace{f\left(\sum_{k=2}^{w} x_k 2^{-k}\right)}_{f_2}
\end{aligned}
\tag{1.15}
$$

In the next step, $f_2 = f(x - x_1)$ is decomposed in the same manner into the square of the most significant bit of $x - x_1$ with sum of other terms, resulting finally with a remaining square $f_3 = f(x - x_1 - x_2)$. The scheme is repeated as long as there are bits left to process. Examining this scheme we find that in order to input a bit-serial word x with the least significant bit first, we have to reverse the order of the iterations in the scheme above. The iterative algorithm then can be written as

$$f_j = \Delta_j + f_{j+1} \tag{1.16}$$

where

$$j = w, \, w - 1, \, \ldots, \, 1 \tag{1.17}$$

$$\Delta_j = 2^{-2j} x_j + 2^{1-j} x_j \sum_{k=j+1}^{w} x_k 2^{-k} \tag{1.18}$$

In each iteration j we accumulate the previous term f_{j+1} and input the next bit x_j. If $x_j = 1$, then we add the square of the bit weight and the weights of the bits that have arrived prior to bit x_j shifted left $(1 - j)$ positions. Examination

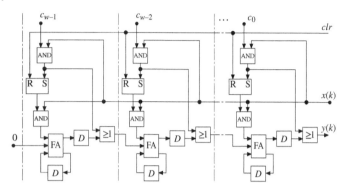

Figure 1.26 Serial squarer for unsigned numbers.

of the bit weights accumulated in each iteration reveals that the sum converges toward the correct square with at least one bit in each step, going from the least-significant bit toward more significant bits in the result.

An implementation of the algorithm above is shown in Figure 1.26. It uses a shift-accumulator to shift the accumulated sum to the right after each iteration. Thus, left-shifting of the stored x_i's are avoided and the addition of the squared bit weight of the incoming x_j is reduced to a shift to the left in each iteration. The implementation consists of w regular bit-slices, which make it suitable for hardware implementation.

The operation of the squarer in Figure 1.26 is as follows: All D flip-flops and SR flip-flops are assumed to be cleared before the first iteration. In the first iteration, the control signal c_0 is high while the remaining control signals are low. This allows the first bit $x(0) = x_w$ to pass the AND gate on top of the rightmost bit-slice. The value x_w is then stored in the SR flip-flop of the same bit-slice for later use. Also, x_w is added to the accumulated sum via the OR gate in the same bit-slice. The least significant output bit $y(0) = y_{2w}$ then becomes available at the output after pure gate delays. Then a shift to the left follows.

The following iterations are carried out in the same manner, with the input bits $x(i) = xw - i$ in sequence along with one of the control signals c_i high, respectively. During the iterations, the result bits $y(i) = y_{2w-i}$ will become available at the output. Then, $x(i)$ has to be zero-extended w locations to access the bits stored in the accumulator. The last control signal clr is used to clear the SR flip-flops before the squaring of next operand can take place.

To adapt the squarer to the case of a two's-complement number as in Eq. (1.2), we sign-extend a two's-complement number to at least $2w - 1$ bits and do not clear the SR flip-flops until the next computation is to take place. Then,

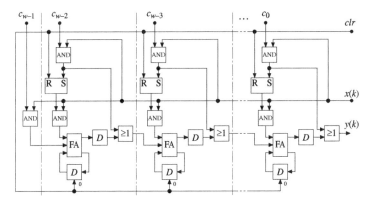

Figure 1.27 Serial squarer for two's-complement numbers.

the squarer computes

$$y = \sum_{j=1}^{w} x_0 2^{j+1} \sum_{k=1}^{w} x_k 2^{-k} + \left(\sum_{k=0}^{w} x_k 2^{-k} \right)^2$$

$$= x_0 2^{w+1} \sum_{k=1}^{w} x_k 2^{-k} + \underbrace{\left(-x_0 + \sum_{k=1}^{w} x_k 2^{-k} \right)^2}_{x^2} \quad (1.19)$$

Here, we can see that the result will contain an error in the accumulated sum. But, since this error exists only in bits of higher significance than the desired result, this error will not affect the desired output result. Further, if we sign extend the input signal with more than $2w - 1$ bits, the error will be scaled accordingly, and stay within the squarer. An implementation of a squarer for two's-complement numbers is shown in Figure 1.27. The only drawback compared to the squarer in Figure 1.26, is that we now need to clear the error in the accumulator before the next squaring can take place.

1.5.3 Multiplication Through Squaring

It is possible to multiply two numbers by using only squarers. This primarily finds applications in memory-based computations as storing the square of a number require significantly less memory compared to storing the product of two numbers. The product of two numbers, a and b, can be computed as [36]

$$a \times b = \frac{(a+b)^2 - (a-b)^2}{4} \quad (1.20)$$

Alternative expressions are also available.

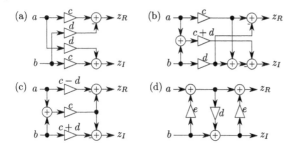

Figure 1.28 Complex multipliers based on (a) Eq. (1.21), (b) Eq. (1.22), (c) Eq. (1.23), and (d) lifting for reversible transforms.

1.6 Complex Multiplication

Multiplication of complex numbers is a time and area consuming arithmetic operation. It is required in many applications, for example, in a fast Fourier transform (FFT) processor for use in an orthogonal frequency-division multiplex (OFDM) system. Multiplication of complex numbers can also be considered into two cases: when both the multiplier and multiplicand are variable or when only the multiplicand is variable.

1.6.1 Complex Multiplication Using Real Multipliers

Consider the complex multiplication $z = k \times x$, where x is the multiplicand, $x = a + jb$ and k is the multiplier, $k = c + jd$. We have

$$z = (c + jd)(a + jb) = (ca - db) + j(da + cb) \qquad (1.21)$$

A direct realization of Eq. (1.21) requires four multipliers and two adders/subtractors as shown in Figure 1.28a.

We may rewrite Eq. (1.21) as follows:

$$z_R = ca - db$$
$$z_I = (c + d)(a + b) - ca - db \qquad (1.22)$$

which requires only three real multiplications and five real additions. An alternative version is

$$z_R = c(a + b) - (c + d)b$$
$$z_I = c(a + b) - (c - d)a \qquad (1.23)$$

which also requires only three real multiplications and five real additions. In fact, there exist several such expressions with only three multiplications [37].

In many applications, for example, FFT, the multiplier, $c + jd$, is constant and $c \pm d$ can be precomputed. This reduces a complex multiplication to only three real multiplications and three/four real additions, as shown in Figure 1.28b and c, for Eqs (1.22) and (1.23), respectively. Obviously, multiplications by ± 1 and $\pm j$ are easily implemented. In addition, further savings can be made for multipliers of the form $\pm c \pm jc$ since the complex multiplication reduces to two real additions of a and b with different signs and two real multiplications by c. This case applies when $\cos(n\pi/4) \pm j\sin(n\pi/4)$ and $n =$ odd. This simplification allows an eight-point DFT to be realized by using only four real multiplications. Hence, a complex multiplication requires three real multiplications and some additions. A more efficient technique is based on distributed arithmetic where only two units are required, which from chip area, throughput, and power consumption points of view are comparable to two real multipliers [3].

1.6.2 Lifting-Based Complex Multipliers

In many transforms and other applications, the complex multiplication is a rotation with an angle α, that is, $k = c + jd = \cos(\alpha) + j\sin(\alpha)$. First, it should be noted that when $\cos(\alpha)$ and $\sin(\alpha)$ are represented in a binary representation, the magnitude of the complex number can never be exactly one (unless $\alpha = n\pi/2$ rad). This in turn means that the inverse rotation can never be exact, there will always be a gain and/or angle error. To get around this, a lifting-based complex multiplier can be used [38]. The lifting-based complex multiplication is written in matrix form as

$$
\begin{aligned}
\begin{bmatrix} z_R \\ z_I \end{bmatrix} &= \begin{bmatrix} c & -d \\ d & c \end{bmatrix} \begin{bmatrix} a \\ b \end{bmatrix} \\
&= \begin{bmatrix} \cos(\alpha) & -\sin(\alpha) \\ \sin(\alpha) & \cos(\alpha) \end{bmatrix} \begin{bmatrix} a \\ b \end{bmatrix} \\
&= \begin{bmatrix} 1 & \frac{\cos(\alpha)-1}{\sin(\alpha)} \\ 0 & 1 \end{bmatrix} \begin{bmatrix} 1 & 0 \\ \sin(\alpha) & 1 \end{bmatrix} \begin{bmatrix} 1 & \frac{\cos(\alpha)-1}{\sin(\alpha)} \\ 0 & 1 \end{bmatrix} \begin{bmatrix} a \\ b \end{bmatrix}
\end{aligned}
\tag{1.24}
$$

$$
= \begin{bmatrix} 1 & e \\ 0 & 1 \end{bmatrix} \begin{bmatrix} 1 & 0 \\ d & 1 \end{bmatrix} \begin{bmatrix} 1 & e \\ 0 & 1 \end{bmatrix} \begin{bmatrix} a \\ b \end{bmatrix}
\tag{1.25}
$$

where

$$
e = \frac{\cos(\alpha) - 1}{\sin(\alpha)} = \frac{c - 1}{d}
\tag{1.26}
$$

The inverse rotation can similarly be written as

$$
\begin{bmatrix} \cos(\alpha) & -\sin(\alpha) \\ \sin(\alpha) & \cos(\alpha) \end{bmatrix}^{-1} = \left(\begin{bmatrix} 1 & e \\ 0 & 1 \end{bmatrix} \begin{bmatrix} 1 & 0 \\ d & 1 \end{bmatrix} \begin{bmatrix} 1 & e \\ 0 & 1 \end{bmatrix} \right)^{-1}
$$

$$
= \begin{bmatrix} 1 & e \\ 0 & 1 \end{bmatrix}^{-1} \begin{bmatrix} 1 & 0 \\ d & 1 \end{bmatrix}^{-1} \begin{bmatrix} 1 & e \\ 0 & 1 \end{bmatrix}^{-1}
$$

$$
= \begin{bmatrix} 1 & -e \\ 0 & 1 \end{bmatrix} \begin{bmatrix} 1 & 0 \\ -d & 1 \end{bmatrix} \begin{bmatrix} 1 & -e \\ 0 & 1 \end{bmatrix} \tag{1.27}
$$

Now, independent of what the values d and e are after quantization, the forward and reverse rotations will always cancel exactly. However, it should be noted that evaluating the three matrix multiplications leads to

$$
\begin{bmatrix} 1 & e \\ 0 & 1 \end{bmatrix} \begin{bmatrix} 1 & 0 \\ d & 1 \end{bmatrix} \begin{bmatrix} 1 & e \\ 0 & 1 \end{bmatrix} = \begin{bmatrix} 1 + de & e(de + 2) \\ d & 1 + de \end{bmatrix} \tag{1.28}
$$

and that the sin (α) terms of the resulting rotation are different, although similar since $d \approx -e(de + 2)$.

The lifting-based complex multiplier can be efficiently realized using three multiplications and three additions as shown in Figure 1.28d. However, a potential drawback from an implementation perspective is that all operations are sequential.

1.7 Special Functions

The need of computing non-linear functions arises in many different applications. The straightforward method of approximating an elementary function is of course to just store the values in a look-up table typically leads to large tables, even though the resulting area from standard cell synthesis grows slower than the number of memory bits [39]. Instead, it is of interest to find ways to approximate elementary functions using a trade-off between arithmetic operations and look-up tables. In addition, the CORDIC algorithm, which is discussed in a separate chapter, is an efficient approach for computing certain functions. For a more thorough explanation of these and other methods, the readers may refer to Reference [40] and other chapters of this book.

1.7.1 Square Root Computation

Computing the square root is commonly performed in one of two different iterative ways, by computing one digit per iteration or by iterative refinement of a temporary result. The methods used are similar to division algorithms because the computation of square root can be seen as dividing the radicand, x, with the square root, $z = \sqrt{x}$, as, $z = x/z$.

1.7.1.1 Digit-Wise Iterative Square Root Computation

While only the binary (radix-2) case is considered here, generalizations to higher radices can be performed. Assuming $0 < X < 1$ and using a "reminder" after the ith iteration, r_i, initialized to the radicand, $r_0 = X$, and a partially computed square root

$$Z_i = \sum_{k=1}^{i} z_k 2^{-k} \tag{1.29}$$

one digit of the result, z_i, is determined in iteration i. After each iteration, we have

$$X = Z_i Z_i + 2^{-i} r_i \tag{1.30}$$

Clearly, if $2^{-i} r_i$ is smaller than $2^{-i+1} r_{i-1}$, the result Z_i will be more accurate than Z_{i-1}. In general, the iteration for square root extraction is

$$r_i = 2r_{i-1} - z_i \left(2Z_{i-1} + z_i 2^{-i}\right) = 2r_{i-1} - 2Z_{i-1}z_i - z_i^2 2^{-i} \tag{1.31}$$

Schemes similar to restoring, nonrestoring, and SRT division can be defined for selecting the next digit [41]. For the quotient digit selection scheme similar to SRT division, the square root is usually restricted to $1/2 \leq z < 1$, which corresponds to $1/4 \leq x < 1$. The selection rule is then

$$z_i = \begin{cases} 1, & \frac{1}{2} \leq 2r_{i-1} \\ 0, & -\frac{1}{2} \leq 2r_{i-1} < \frac{1}{2} \\ -1, & 2r_{i-1} < -\frac{1}{2} \end{cases} \tag{1.32}$$

Meaning that only the first few bits of $2r_{i-1}$ must be inspected to determine the correct digit of the result. Note that the normalization, $1/4 \leq x < 1$, always gives $z_1 = 1$ and, hence, $r_1 = 2x - \frac{1}{2}$.

1.7.1.2 Iterative Refinement Square Root Computation

The Newton–Raphson method for solving equation systems can be applied for square root computations. The Newton–Raphson recurrence is

$$x_{i+1} = x_i - \frac{f\left(x_i\right)}{f'\left(x_i\right)} \tag{1.33}$$

and the equation to solve in this case is $f(z) = z^2 - x = 0$. This gives the recurrence

$$z_{i+1} = \frac{1}{2}\left(z_i + \frac{x}{z_i}\right) \tag{1.34}$$

Hence, each iteration requires a division, an addition, and a bit-shift.

Instead, it turns out that it is more efficient to compute the reciprocal square root $\rho = 1/\sqrt{x}$. Solving the equation $f(\rho) = 1/\rho^2 - z$, the recurrence for this computation is

$$\rho_{i+1} = \rho_i\left(\frac{1}{2}\left(3 - \rho_i^2 x\right)\right) \tag{1.35}$$

Each iteration now requires a square, two multiplications, subtraction from a constant, a bit-shift. Although there are more computations involved, the operations are simpler to implement compared to a division. The first estimate, ρ_0, can be read from a table. Then, for each iteration, the error is reduced quadratically.

The square root can then easily be determined from the reciprocal square root as $z = x\rho$.

1.7.2 Polynomial and Piecewise Polynomial Approximations

It is possible to derive a polynomial $p(x)$ that approximates a function $f(x)$ by performing a Taylor expansion for a given point d as

$$p(x) = \sum_{k=0}^{\infty} \frac{f^{(k)}(d)}{k!}(x - d)^k \tag{1.36}$$

When the polynomial is restricted to a certain number of terms it is often better to optimize the polynomial coefficients, as there are some accuracy to be gained. To determine the best coefficients is an approximation problem where typically there are more constraints (number of points for the approximation) than variables (polynomial order). This problem can be solved for a minimax solution using, for example, Remez's exchange algorithm or linear programming. For a least square solution the standard method to solve overdetermined systems can be applied. The result will be a polynomial of order N

$$p(x) = \sum_{k=0}^{N} a_k x^k \tag{1.37}$$

The polynomial approximations can be efficiently and accurately evaluated using Horner's method, discussed in Section 1.4.3. Hence, there is no need to compute any powers of X explicitly and a minimum number of arithmetic operations are used.

The drawback of Horner's method is that the computation is inherently sequential. An alternative is to use Estrin's method [40], where x^2 is used in a tree structure for increasing the parallelism and reducing the critical path. Estrin's method for polynomial evaluation can be written as

$$p(x) = \left(a_3 x + a_2\right) x^2 + \left(a_1 x + a_0\right) \tag{1.38}$$

for a third-order polynomial. For a seventh-order polynomial, it becomes

$$p(x) = \left(\left(a_3 x + a_2\right) x^2 + \left(a_1 x + a_0\right)\right) x^4 + \left(\left(a_3 x + a_2\right) x^2 + \left(a_1 x + a_0\right)\right) \tag{1.39}$$

As can be seen, Estrin's method also maps well to MAC-operations.

The required polynomial order depends very much on the actual function that is approximated [40]. An approach to obtain a higher resolution despite using a lower polynomial order is to use different polynomials for different ranges. This is referred to as piecewise polynomials. An L segment Nth-order piecewise polynomial with segment breakpoints $x_l, l = 0, 1, \ldots, L - 1$ can be written as

$$p(x) = \sum_{k=0}^{N} a_{k,l} \left(x - x_l\right)^k, \quad x_l \leq x \leq x_{l+1} \tag{1.40}$$

From an implementation point of view, it is often practical to have 2^k uniform segments and let the k most significant bits determine the segmentation, as these directly forms the segment number l. However, it can be shown that in general the total complexity is reduced for nonuniform segments. An illustration of a piecewise polynomial approximation is shown in Figure 1.29 where uniform segments and a parallel implementation of Horner's method is used for the polynomial evaluation.

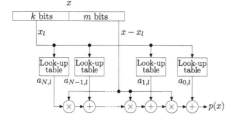

Figure 1.29 Piecewise polynomial approximation using uniform segmentation based on the k most significant bits.

References

1 D. Liu, *Embedded DSP Processor Design: Application Specific Instruction Set Processors*, Morgan Kaufmann, 2008.

2 A. P. Chandrakasan and R. W. Brodersen, Minimizing power consumption in digital CMOS circuits. *Proc. IEEE*, vol. 83, no. 4, 1995, pp. 498–523.

3 L. Wanhammar, *DSP Integrated Circuits*, Academic Press, 1999.

4 K. Johansson, O. Gustafsson, and L. Wanhammar, Power estimation for ripple-carry adders with correlated input data, in *Proc. Int. Workshop Power Timing Modeling Optimization Simulation*, Springer, January 2004, pp. 662–674.

5 R. Zimmermann, Binary adder architectures for cell-based VLSI and their synthesis, Ph.D. dissertation, Swiss Federal Institue of Technology (ETH), 1998.

6 T. G. Noll, Carry-save architectures for high-speed digital signal processing. *J. VLSI Signal Process. Syst.*, vol. 3, no. 1–2, 1991, pp. 121–140.

7 A. Omondi and B. Premkumar, *Residue Number Systems: Theory and Implementation.* Imperial College Press, 2007.

8 Y.-N. Chang, J. H. Satyanarayana, and K.K. Parhi, Systematic design of high-speed and low-power digit-serial multipliers. *IEEE Trans. Circuits Syst. II*, vol. 45, no. 12, 1998, pp. 1585–1596.

9 K. Johansson, O. Gustafsson, and L. Wanhammar, Multiple constant multiplication for digit-serial implementation of low power FIR filters. *WSEAS Trans. Circuits Syst.*, vol. 5, no. 7, 2006, pp. 1001–1008.

10 K. K. Parhi, A systematic approach for design of digit-serial signal processing architectures. *IEEE Trans. Circuits Syst.*, vol. 38, no. 4, 1991, pp. 358–375.

11 H. Suzuki, Y.-N. Chang, and K. K. Parhi, Performance tradeoffs in digit-serial DSP systems, in *Proc. Asilomar Conf. Signals Syst. Comput.*, vol. 2, November 1998, pp. 1225–1229.

12 P. Nilsson, Arithmetic and architectural design to reduce leakage in nano-scale digital circuits, in *Proc. Europ. Conf. Circuit Theory Design*, August 2007, pp. 372–375.

13 C. R. Baugh and B. A. Wooley, A two's complement parallel array multiplication algorithm. *IEEE Trans. Comput.*, vol. C-22, no. 12, 1973, pp. 1045–1047.

14 B. C. Paul, S. Fujita, and M. Okajima, ROM-based logic (RBL) design: a low-power 16 bit multiplier. *IEEE J. Solid-State Circuits*, vol. 44, no. 11, 2009, pp. 2935–2942.

15 M. Faust, O. Gustafsson, and C.-H. Chang, Fast and VLSI efficient binary-to-CSD encoder using bypass signal. *Electron. Lett.*, vol. 47, no. 1, 2011, pp. 18–20.

16 O. L. Macsorley, High-speed arithmetic in binary computers. *Proc. IRE*, vol. 49, no. 1, 1961, pp. 67–91.

17 Y. H. Seo and D. W. Kim, A new VLSI architecture of parallel multiplier–accumulator based on radix-2 modified Booth algorithm. *IEEE Trans. VLSI Syst.*, vol. 18, no. 2, 2010, pp. 201–208.

18 Y. C. Lim, Single-precision multiplier with reduced circuit complexity for signal processing applications. *IEEE Trans. Comput.*, vol. 41, no. 10, 1992, pp. 1333–1336.

19 J. P. Wang, S. R. Kuang, and S. C. Liang, High-accuracy fixed-width modified Booth multipliers for lossy applications. *IEEE Trans. VLSI Syst.*, vol. 19, no. 1, 2011, pp. 52–60.

20 D. D. Caro, N. Petra, A. G. M. Strollo, F. Tessitore, and E. Napoli, Fixed-width multipliers and multipliers-accumulators with min-max approximation error. *IEEE Trans. Circuits Syst. I*, vol. 60, no. 9, 2013, pp. 2375–2388.

21 C. S. Wallace, A suggestion for a fast multiplier. *IEEE Trans. Electron. Comput.*, vol. EC-13, no. 1, 1964, pp. 14–17.

22 L. Dadda, Some schemes for parallel multipliers. *Alta Frequenza*, vol. 34, no. 5, 1965, pp. 349–356.

23 K. C. Bickerstaff, M. J. Schulte, and E. E. Swartzlander, Parallel reduced area multipliers. *J. VLSI Signal Process. Syst.*, vol. 9, no. 3, 1995, pp. 181–191.

24 P. F. Stelling, C. U. Martel, V. G. Oklobdzija, and R. Ravi, Optimal circuits for parallel multipliers. *IEEE Trans. Comput.*, vol. 47, no. 3, 1998, pp. 273–285.

25 S. T. Oskuii, P. G. Kjeldsberg, and O. Gustafsson, Power optimized partial product reduction interconnect ordering in parallel multipliers, in *Proc. Norchip.*, November 2007, pp. 1–6.

26 Z. J. Mou and F. Jutand, overturned-stairs adder trees and multiplier design. *IEEE Trans. Comput.*, vol. 41, no. 8, 1992, pp. 940–948.

27 H. Eriksson, P. Larsson-Edefors, M. Sheeran, M. Sjalander, D. Johansson, and M. Scholin, Multiplier reduction tree with logarithmic logic depth and regular connectivity, in *Proc. IEEE Int. Symp. Circuits Syst.*, May 2006, pp. 4–8.

28 P. F. Stelling and V. G. Oklobdzija, Design strategies for optimal hybrid final adders in a parallel multiplier. *J. VLSI Signal Process. Syst.*, vol. 14, no. 3, 1996, pp. 321–331.

29 R. Lyon, Two's complement pipeline multipliers. *IEEE Trans. Commun.*, vol. 24, 1976, no. 4, 1976, pp. 418–425.

30 O. Gustafsson and L. Wanhammar, Optimal logic level pipelining for digit-serial implementation of maximally fast recursive digital filters, in *Proc. National Conf. Radio Sciences*, 2002.

31 A. R. Cooper, Parallel architecture modified Booth multiplier. *IEE Proc. G Circuits Devices Syst.*, vol. 135, no. 3, 1988, pp. 125–128.

32 R. K. Kolagotla, W. R. Griesbach, and H. R. Srinivas, VLSI implementation of 350 MHz 0.35 μm 8 bit merged squarer. *Electron. Lett.*, vol. 34, 1998, no. 1, 1998, pp. 47–48.

33 K.-J. Cho and J.-G. Chung, Parallel squarer design using pre-calculated sums of partial products. *Electron. Lett.*, vol. 43, no. 25, 2007, pp. 1414–1416.

34 P. Ienne and M. A. Viredaz, Bit-serial multipliers and squarers. *IEEE Trans. Comput.*, vol. 43, 1994, no. 12, 1994, pp. 1445–1450.

35 M. Vesterbacka, K. Palmkvist, and L. Wanhammar, Serial squarers and serial/serial multipliers, in *Proc. National Conf. Radio Sciences*, 1996.

36 E. L. Johnson, A digital quarter square multiplier. *IEEE Trans. Comput.* vol. 29, no. 3, 1980, pp. 258–261.

37 A. Wenzler and E. Luder, New structures for complex multipliers and their noise analysis, in *Proc. IEEE Int. Symp. Circuits Syst.*, vol. 2, April 1995, pp. 1432–1435.

38 S. Oraintara, Y. J. Chen, and T. Q. Nguyen, Integer fast Fourier transform. *IEEE Trans. Signal Process.*, vol. 50, 2002, no. 3, 2002, pp. 607–618.

39 O. Gustafsson and K. Johansson, An empirical study on standard cell synthesis of elementary function lookup tables, in *Proc. Asilomar Conf. Signals Syst. Comput.*, 2008, pp. 1810–1813.

40 J.-M. Muller, *Elementary Functions*. Springer, 2006.

41 M. D. Ercegovac and T. Lang, *Digital Arithmetic*, Elsevier, 2004.

2

Shift-Add Circuits for Constant Multiplications

Pramod Kumar Meher[1], C.-H. Chang[2], Oscar Gustafsson[3], A.P. Vinod[2], and M. Faust[4]

[1] *Independent Hardware Consultant*

[2] *Nanyang Technological University, Singapore, Singapore*

[3] *Linköping University, Linköping, Sweden*

[4] *mfnet gmbh, Switzerland*

2.1 Introduction

Several applications involving signal, image, and video processing, graphics and robotics, and so on, very often require large number of multiplications, which consume a major part of computation time, silicon area, and energy. But, for-tunately, in most of those multiplications, one of the multiplicands is a known constant, which leads to a great simplification of their implementation. Two interesting situations are frequently encountered. In one case, several different variable multiplicands are multiplied with a constant. Since the bit/digit-pattern of the constant is known *a priori*, the logic circuits for those multiplications can accordingly be optimized to achieve reduction in area and time complex-ity. This situation is categorized as single constant multiplication (SCM). The multiplications in the direct form finite impulse response (FIR) filters involves an SCM for each filter tap [1]. In another situation, several different constants are multiplied with a single variable multiplicand, which could be implemented with less hardware by sharing intermediate results among the multiplications. The optimization of hardware and time under this situation is handled by the multiple constant multiplication (MCM) techniques. The transpose form FIR filter is the most common application of MCM techniques.

Arithmetic Circuits for DSP Applications, First Edition. Edited by Pramod Kumar Meher and Thanos Stouraitis.
© 2017 by The Institute of Electrical and Electronics Engineers, Inc. Published 2017 by John Wiley & Sons, Inc.

In fixed-point binary representation, constants can always be treated as integers followed by a power-of-2 scaling, and multiplication by a known constant can be realized by a series of predefined shift-and-add operations [2–4]. A straightforward implementation of multiplication $p = cx$ of a variable x by a constant c involves $(n - 1)$ additions accompanied by bit-shift operations, if the fixed-point binary representation of c consists of n 1s. For example, if $c = 79$, we can write c as $[01001111]_2$ to obtain

$$79x = (x \ll 6) + (x \ll 3) + (x \ll 2) + (x \ll 1) + x \tag{2.1}$$

where $x \ll l$ represents the left-shift of x by l bit positions. Using signed digit (SD) representation, involving $-1, 0$, and 1, we can represent 79 as $[0101000\bar{1}]_{SD}$, where $\bar{1} = -1$. Based on this representation, the multiplication can be written as

$$79x = (x \ll 6) + (x \ll 4) - x \tag{2.2}$$

Note that according to Eq. (2.1), the multiplication involves four additions but according to Eq. (2.2), it involves one addition and one subtraction. Early evidence of SD representation for efficient realization of arithmetic operations in digital circuits dates back to 1961 [5] and SD representation has been used to improve the efficiency of shift-and-add implementation of constant multiplications later [6–8]. The understanding of different representations of constants is very important to arrive at an efficient shift-and-add configuration for constant multiplication, and forms the foundation of some of the techniques in this area of research.

If a single input operand is multiplied by more than one constant, a set of common subexpressions (CSs) or intermediate results (also called fundamentals[1]) could be identified not only for implementing the multiplication with a single constant but also to be shared among all other constants. The objective of the MCM scheme is to replace all the multipliers by an optimized shift-and-add/subtract network, where the digital circuit for each subexpression produces an intermediate result to be shared by different multiplications to minimize the overall implementation cost. Two popular cost metrics are used to evaluate the quality of the MCM solution: the number of logic operators (LOs) and logic depth (LD). Since a subtractor could be implemented with the similar architecture and cost as an adder, each of the adders and subtractors is counted as one LO, while fixed shifts can be implemented without cost as they are just a matter of wiring. The maximum number of LOs in the critical path of

---

1 The term "fundamental" is widely used in adder-graph approaches, which refers to a multiple of the input value that is used as an intermediate result in the shift-and-add network in the multiplication block. Note that a fundamental is not necessarily a subexpression. But the term subexpression, which is used in the CSE algorithms, can be said to be a fundamental. These terms, however, could be used synonymously in the context of the algorithms.

the shift-and-add network is called LD. The intricate problem of minimization of number of additions/subtractions for the implementation of fixed FIR filter, has attracted the attention of researchers since the early eighties [9]. Many heuristic algorithms have been suggested during the three decades of research for near optimal solution of this problem in programmable machines as well as in dedicated hardware for different applications. All these approaches can be grouped in many different ways, but can be placed in two main categories: Those are (i) adder-graph (AG) approach or graph dependent (GD) approach and (ii) common subexpression elimination (CSE) approach. The main objective of this survey is to present a brief description of the basic methods of optimization of shift-and-add configuration for constant multiplications, along with the major advances in the design techniques and possible limitations.

From the analysis of 200 industrial examples taken mainly from DSP, communication, graphics, and control applications, it is shown in References [10,11] that more than 60% of them have more than 20% of operations that are multiplications of constants. Constant multiplication optimization can also be applied readily to more generalized applications involving matrix multiplication and inner-product computation, where one of the matrices or vectors is constant, and could be extended to several applications in graphics, image/video processing, and robotic applications. We have devoted one full section of the survey on different applications of constant multiplications that could benefit from the techniques and possibly new applications could also be envisaged. Since deciding if the shift-and-add network has the minimum number of adders for a set of constant multiplications is NP-complete [12,13], one cannot guarantee the optimality of the solution found in reasonable time for large multiplication blocks. Some efforts have, however, been made to develop optimization schemes and optimal algorithms, which could be used for shift-and-add implementation for a small set of constants in MCM blocks. Considering the importance of optimal implementation, we have discussed those in another section. It should be noted that while the focus of this survey is on implementation in dedicated hardware, similar techniques are used in software compilers, where the technique is known as (weak) strength reduction [14], replacing expensive operations (multiplications) with less expensive (shifts, additions, and subtractions).

The remainder of the survey is organized as follows. In Section 2, the main aspects of representation of constants pertaining to constant multiplications are reviewed. In Sections 3 and 4, the basic approaches and key contributions on different design techniques of SCM and MCM are presented and discussed. The optimization schemes and optimal algorithms for shift-and-add implementation of MCM are discussed in Section 5. Various application perspectives of constant multiplication techniques are briefly sketched in Section 6. The pitfalls and scope for future work are presented in Section 7, and conclusions are presented in Section 8.

2.2 Representation of Constants

Computer arithmetics commonly use two's complement or sign-magnitude represented numbers in either fixed-point or floating-point representation. As fixed-point systems are far less complex to implement than their floating-point counterparts, the ASIC or FPGA implementations of DSP and related applications with limited dynamic ranges are based on fixed-point approximation to reduce the area, delay, and power complexity. In two's complement representation, subtraction can be implemented by a simple modification of adder circuit and no sign detection of the operands is required, which is an advantage over sign-magnitude representation. An integer C is described as an N-bit two's complement number as

$$C = -a_{N-1} \times 2^{N-1} + \sum_{i=0}^{N-2} a_i \times 2^i, \text{ where } a_i \in \{0, 1\} \tag{2.3}$$

In case of unsigned binary numbers, a_{N-1} is weighted positively and can be grouped into the sum. The use of unsigned binary or two's complement representation allows the variable operand in a constant multiplication to be shifted according to the positions of every nonzero bit of the constant and summed up to produce the product.

Since the hardware cost to replace an adder by a subtractor is not significant, the signed-digit (SD) representation using a symmetric digit set, that is, $a = (a_{n-1}, \ldots, a_2, a_1, a_0)$ with $a_i \in \{0, \pm 1, \pm 2, \ldots, \pm(r-1)\}$ is introduced in Reference [5]. For the ease of hardware implementation, binary signed digit set with $r = 2$, $a_i \in \{-1, 0, 1\}$ is considered.[2] An integer constant C can be written in SD form

$$C = \sum_i b_i \times 2^i, \text{ for } b_i \in \{\bar{1}, 0, 1\} \tag{2.4}$$

The SD representation is not unique, as $[1] = [1\bar{1}]$ and $[\bar{1}] = [\bar{1}1]$ that allows any $[1]$ or $[\bar{1}]$ to be replaced by its equivalent digit pair repeatedly as long as it has a zero adjacent to its left. If there is no bit width restriction, the leading nonzero digit can be replaced indefinitely [8]. This nonuniqueness poses additional complexity to a design algorithm to pick the right form of representation to minimize the number of adders for implementing the constant multiplication. The minimal signed digit (MSD) [6] representation is derived by restricting the SD representations to those with the minimum number of nonzero digits.

---

2 Binary SD is normally referred to as SD, unless otherwise specified.

Table 2.1 Binary, MSD, and CSD representations of 815.

Binary	MSD	CSD
1100101111	$1100110001\bar{1}$	$10\bar{1}010\bar{1}000\bar{1}$
	$1101010\bar{1}000\bar{1}$	
	$10\bar{1}00110001\bar{1}$	
	$10\bar{1}010\bar{1}000\bar{1}$	
	$100\bar{1}1\bar{1}0\bar{1}000\bar{1}$	

Since the bit patterns $[10\bar{1}]$ and $[\bar{1}01]$ have the same nonzero digit count as $[11]$ and $[\bar{1}\bar{1}]$, and correspond to the same integers 3 and -3, respectively, MSD representation can have more than one unique representation. By introducing the requirement that $b_i \times b_{i-1} = 0$, $\forall i \in \{1, 2, \ldots, N\}$, where b_i is the nth digit of the MSD representation of an N-bit number, the only way to represent 3 is $[10\bar{1}]$. This subset of SD representations is unique and is called the canonical signed digit (CSD) [2]. Both CSD and MSD might require $N + 1$ digits to represent an N-bit binary number, as can be seen on the example of 7 which is $[111]$ in binary and $[100\bar{1}]$ in CSD and MSD, but they always use the same minimum number of nonzero digits. The variations of MSD representation and equivalent binary and CSD representations of an integer 815 is illustrated in Table 2.1.

The evaluation of $P = 815 \times x$ can be performed by using the binary representation, the MSD, or the CSD representation of 815, respectively, from Table 2.1, as

$$(x \ll 9) + (x \ll 8) + (x \ll 5) + (x \ll 3) + (x \ll 2) + (x \ll 1) + x,$$

$$(x \ll 9) + (x \ll 8) + (x \ll 6) - (x \ll 4) - x, \text{ and}$$

$$(x \ll 10) - (x \ll 8) + (x \ll 6) - (x \ll 4) - x$$

It can be observed from above that the multiplication based on the unsigned binary representation of constant 815 involves six additions, while the multiplications based on MSD and CSD representations involve four addition/subtraction operations.

A detailed analysis of the CSD representation is given in References [2,15]. It is shown that $(S(C))_{\text{CSD}} \leq \left\lceil \frac{N+1}{2} \right\rceil$, where $(S(C))_{\text{CSD}}$ is the number of nonzero digits in the CSD or MSD representation of a number. This limits the maximum number of nonzero digits to about $1/2$ of the word-size of the binary representation, as a consequence of the rule that no two adjacent digits can be nonzero in CSD. Using this expression, the asymptotical bound of $(N/3 + 1/9)$ for the

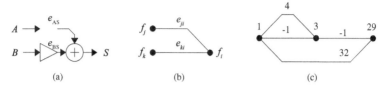

(a)   (b)   (c)

Figure 2.1 Arithmetic implementation of a shift-and-add operation (a), corresponding graph representation (b), and a numerical example (c).

average number of nonzero digits for an N-bit CSD number has been derived in References [2,15,16]. In other words, it indicates an average reduction of 33% of the LO count by CSD/MSD over the direct binary representation in a straightforward[3] implementation of multiplication [3]. Algorithms for finding the CSD representation for any given number can be found in References [4,17,18].

A CSD/MSD number contains the minimal number of nonzero digits, and no further reduction is possible for the straightforward implementation of constant multiplier based on a shift-and-add/subtract network. The optimization of the resulting shift-and-add network can only reduce the implementation cost, but not the nonzero digit count. From the timing perspective, particularly to have higher clock frequency, the LD of shift-and-add network is important. The minimum possible LD can be derived as

$$\text{LD}_{\min} = \max_{i} \left(\lceil \log_2 S(C_i) \rceil \right) \tag{2.5}$$

where $S(C_i)$ denotes the minimum number of nonzero digits for the constant C_i in CSD/MSD representation.

Using the minimal number of nonzero digits of CSD for constant representation is advantageous, but in some cases searching among all MSD representations [7] or among all SD representations bounded by a maximum number of additional nonzero digits [8] can produce better solutions.

Another possibility, which is suggested by several algorithms in the literature, is to reduce the number of adders for constant multiplication without fixing the number representation for the constant. The resulting shift-and-add network for a constant is represented as a directed acyclic graph. A single shift-and-add operation is shown in Figure 2.1a where e_{AS} and e_{BS} denote the power-of-two

---

3 The use of a balanced adder tree to add the products of every weighted nonzero digit of the constant and the multiplicand could be considered as straightforward implementation of multiplication.

multipliers (left shifts) of the variables A and B, respectively. The graph can be simplified to Figure 2.1b by assuming a left-to-right signal flow direction. Each vertex, f, represents an integer multiple of the input variable. The common input variable is implicitly defined and the output of the vertex is annotated only by a constant integer. Each edge from vertex f_a to vertex f_b, e_{ab}, of the graph is assigned a value of $\pm 2^m$, where m is a non-negative integer indicating the number of bits to be left shifted. The sign of the edgevalue determines if the output of the left vertex is to be added or subtracted at the vertex to its right. This operation can be expressed as

$$f_i = e_{ji} \times f_j + e_{ki} \times f_k \tag{2.6}$$

For a complete shift-and-add network representation, the leftmost vertex with an in-degree of zero represents the input variable and is always annotated by a fixed integer value of 1. Other vertices have an in-degree of two and each can be viewed as a two-operand carry propagation adder (CPA). The value of any vertex can be taken as an external output, irrespective of its number of output edges, which denote the fan-out of the adder it represented.

An example of a simple shift-and-add network is shown in Figure 2.1c. Let x be the input variable, the leftmost vertex has an output of $1x$. It has three outgoing edges. The two output edges with values 4 and -1 feed the middle vertex. Therefore, the middle vertex has an output value of $1x \times 4 + 1x \times (-1) = 3x$, which is equivalent to $x \ll 2 - x$. The middle vertex has an outgoing edge to the last vertex with edge value -1. The other input edge to the last vertex comes from the output edge of the first vertex with value 32. Therefore, the output value of the right vertex is $29x$ as $3x \times (-1) + 1x \times 32 = 29x$.

In hardware implementation, fixed shift is considered free since it can be realized by routing of signals during circuit layout. Furthermore, all even numbers have at least one 0 at the least significant position and can therefore be divided (right shifted) until they are odd. By reducing the output value of each vertex until it is odd, a minimized graph of a shift-and-add network contains only vertices with unique odd values. Such odd values are commonly referred to as odd fundamentals, as it is a fundamental value in the composition of a minimal shift-and-add network. The graph in Figure 2.1c only contains vertices with odd fundamentals and it represents the shift-and-add network for the multiplication of a variable by 29, 58, or any even constant given by 29×2^n, where the integer $n \geq 0$. The LO cost to implement an adder graph is equal to the number of vertices excluding the input vertex. Therefore, the graph in Figure 2.1c has a cost of 2. Based on the LO cost, a fundamental that requires only one vertex to implement is called a cost-1 fundamental. Examples for such fundamentals are all constants with values $C = 2^n \pm 1$.

There exist many different graphs for a single or multiple constant multiplication. The manipulations and approaches to find the graph with the lowest cost will be discussed further along with the respective algorithms and the search for the set of odd fundamentals for a FIR filter is shown in Table 2.2.

2.3 Single Constant Multiplication

Several techniques have been suggested in the literature, to simplify the shift-and-add network in SCM circuits. The main emphasis has been on minimizing the LO count, and in a few cases, however, some importance has been given to the LD. Although it is difficult to categorize these techniques into exclusively distinct classes, we can view them in three different design approaches such as (i) direct simplification from a given number representation, (ii) simplification by redundant SD representation, and (iii) simplification by adder graph.

2.3.1 Direct Simplification from a Given Number Representation

Bernstein [19] has some earliest contributions on single constant multiplication. He studied SCM from the programming perspective and implemented constant multiplication using addition, subtraction, and shift operation. Apart from unsigned binary implementations, Bernstein has used the fact that factorization of a number can reduce the number of additions and subtractions involved in a constant multiplication. A simple example of that is the multiplication of a variable with $119 = [1110111]_{BIN}$. Having six nonzero digits in binary, five additions are required, but only three adders are required if it is factored into $7 \times 17 = [111]_{BIN} \times [10001]_{BIN}$. If subtraction is allowable, the cost is further reduced to only one adder and one subtractor as $7 = [100\bar{1}]$. Factoring is advantageous if the factor is of the form $2^c \pm 1$, as such factor can be realized with only an adder or a subtractor. Other forms of factoring could lead to more adders than the direct binary implementation. Lefevre [20] analyzed the algorithm of Bernstein and proposed an algorithm for simplifying the constant multiplication in a general purpose digital signal processor. His algorithm searches for identical bit patterns called common subexpressions (CSs), in the radix-4 Booth encoded constant, which is often equal to its CSD representation. To improve the change of finding CSs, $[10\bar{1}]$ can be replaced with $[011]$, and similarly $[\bar{1}01]$ can be replaced with $[0\bar{1}\bar{1}]$ without increasing the number of nonzero digits [20]. An example is $105 = [10\bar{1}01001]_{CSD}$, which does not contain a CS. Replacing $[\bar{1}01]$ with $[0\bar{1}\bar{1}]$ results in a MSD representation, $[100\bar{1}\bar{1}001]$ with the same minimum number of nonzero digits but contains the subexpression $[100\bar{1}]$ and its negated form $[\bar{1}001]$ resulting in a reduction of the additions from 3 to 2. Similarly, the subexpression $[1000\bar{1}]$ and its negative form can also be found. The results obtained by Lefevre showed that both his original and improved algorithms outperform Bernstein's algorithm [19].

2.3.2 Simplification by Redundant Signed Digit Representation

A new method was introduced in Reference [8] to search for more CSs for elimination from all SD representations with up to a predefined maximum number of nonzero digits above the number of nonzero digits of CSD. However, it is observed that the search for CS in the SD representations does not help in further adder reduction after a certain number of nonzero digits above CSD. There are no advantages of considering SD representations with more than two extra nonzero digits above CSD for up to 16-bit constant multipliers. An example where it is advantageous to use an additional nonzero digit is $363 = [\bar{1}0100\bar{1}0\bar{1}0\bar{1}]_{CSD}$, which has five nonzero digits and no recurrent subexpression. It requires four LOs. With one additional nonzero digit, the SD representation $[10\bar{1}0\bar{1}10\bar{1}0\bar{1}]$ contains the subexpression $[10\bar{1}0\bar{1}]$ twice and the LO count is reduced to three, two for the subexpression and one for adding the subexpression to its shifted version.

Reference [21] introduces the $H(k)$ algorithm. This algorithm applies the Hartley algorithm [15] to all SD representations of a constant with up to k more nonzero digits than the CSD representation, and selects the one that requires the fewest adders. The Hartley algorithm was originally targeted for multiple constant multiplication, which will be described in Section 2.3.3. Besides the visible redundant digit patterns, as described before, there are redundant patterns that remain obscure in any representation of the constant. In [22], one such obscure pattern was noted and called clashing without further pursuit. An example of clashing is $39 = [10100\bar{1}]_{CSD}$. It can be decomposed into $[100100]_{CSD} + [00010\bar{1}]_{CSD}$ and constructed by $[1001] \ll 2 + [10\bar{1}]$. In this addition, the two 1 digits at the third bit position "clash" and form a 1 digit at the fourth bit position of 39. In Reference [23], more such obscure patterns are characterized and referred to as overlapping digit patterns (ODPs). The $H(k)$ algorithm was extended to handle ODPs and shown to further reduce the LO count for SCM.

2.3.3 Simplification by Adder Graph Approach

Even by searching for several SD representations or by including ODPs in the search, the CSs that can be found are limited by the representation. Such a

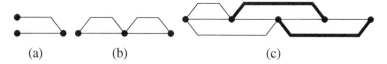

(a) (b) (c)

Figure 2.2 Classification of adder graph topologies, (a) additive, (b) multiplicative, and (c) leapfrog graphs

limitation can be overcome by a numerical search that is independent of the representation of constant. Under this premise, Dempster and Macleod [24] have applied the primitive operator graph representation introduced by Reference [25] for MCM, which will be described in Section 2.3.4, to formulate a minimized adder graph (MAG) approach. MAG finds the minimal set of adders to perform SCM with constants of magnitude up to 2^{12}. This gives an improvement of nearly 16% of LO count over the direct CSD implementation. Furthermore, 32 possible graph topologies for cost-1 to cost-4 fundamentals are shown. In order to find the minimum LO for a constant, all these graph topologies have to be considered. Reference [26] extends the optimal method for all constants up to a precision of 19 bits and reduces the number of graph topologies to only 19 by exploiting graph equivalence to directly reduce the computational effort needed to find a SCM solution with the minimum LO. After a classification of graph representations (based on Eq. (2.6) ($f_i = e_{ji} \times f_j + e_{ki} \times f_k$)) into additive (($f_j \vee f_k = 1$) ∩ ($f_j \neq f_k$), see Figure 2.2a), multiplicative ($f_j = f_k$, see Figure 2.2b) and leapfrog graphs, all 34 possible graph topologies for cost-5 fundamentals are presented. A leapfrog graph contains at least one edge that bypasses one level of distance from the input vertex, but does not originate in the input vertex. In other words, in a leapfrog graph exists a vertex f_i, which is a child of two different internal vertices, $f_j \neq 1$ and $f_k \neq 1$, where f_k is equal to one of the parents of f_j, which both are not the input vertex (($f_j = e_{mj} \times f_m + e_{nj} \times f_n$) ∩ ($f_k = \{f_m \vee f_n\} \neq 1$)). Figure 2.2c shows a leapfrog graph with the two leapfrog edges highlighted.

In Reference [22], previous algorithms [8,20,21,24,26–29] are reviewed, a detailed analysis of all possible graphs is given and an optimal and a heuristic algorithm are proposed. These algorithms performed better than those published before them. The graph topologies are reviewed and informative observations are given. They are used to reduce the number of topologies to be considered when searching for the solution with minimal number of adders. One of these observations is that the internal vertices (additions) can be reordered without changing the value of output vertex.

2.3.4   State of the Art in SCM

The graph-based algorithm of Reference [22] and the subexpression-based algorithm of Reference [23], to the best knowledge of the authors, are the best algorithms for single constant multiplication to date. For all integers up to a maximum word-size of 19 bits, the exact minimum number of additions for SCM is shown in References [24,26], and for higher word-size the lower bound of LOs for SCM is shown in Reference [30] to be

$$LO_{min} = \lceil \log_2 S(C) \rceil \tag{2.7}$$

This minimal bound can be used to compare the results of heuristic algorithms. For C with $S(C) = 2^i$, the minimal bound is only achievable if the adder structure shown in Figure 2.2b is used, which requires C to be a product of only weight-1 fundamentals. A trivial upper bound is $LO_{max} = S(C) - 1$, which is equal to the cost of a straightforward implementation using CSD/MSD representation of C. Reference [31] presents a asymptotic upper bound that gives the number of adders in a shift-and-add implementation with $O(n/(\log n)^{\alpha})$ where n is the bit width of the constant multiplier and $\alpha < 1$.

In Reference [32], a method for proving optimality for SCM solutions up to six adders was proposed. If the optimality cannot be proven, a search is performed that results in provably optimal SCM solutions. The results indicate that a SCM solution for any integer up to 32 bits can be realized using at most six adders. However, this conclusion is based on the experimentation of a large number of random constants.

2.4   Algorithms for Multiple Constant Multiplications

There are many different application scenarios where a single input is multiplied by several constants. Examples are constant matrix multiplication [11,33–35], discrete cosine transform (DCT) or fast fourier transform (FFT) [36], and polyphase FIR filters and filter banks [37]. Since the FIR filters are very often encountered in various DSP applications, we discuss the use of MCM for their implementation.

2.4.1   MCM for FIR Digital Filter and Basic Considerations

The output $y[n]$ of a Nth order FIR filter at instant n is given by

$$y[n] = \sum_{k=0}^{N-1} h_k x[n-k] \tag{2.8}$$

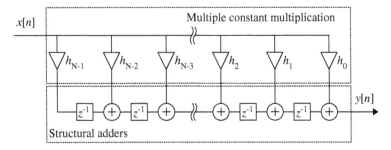

Figure 2.3 Transposed direct form FIR filter.

where N is the filter order, h_k is the kth coefficient and $x[n]$ is input sample available to the filter at the nth instant. To transform the computation of filter output $y[n]$ into the form of parallel multiplication and a sequence of additions on a delay line, Eq. (2.8) can be expressed in a recursive form as

$$y[n] = h_0 x[n] + D\big(h_1 x[n] + D\big(h_2 x[n] + \dots + D\big(h_{(N-1)} x[n]\big)\big)\big) \qquad (2.9)$$

where D is the unit delay operator.

The computation of Eq. (2.9) can be expressed by the schematic of a transposed direct form filter (Figure 2.3). Since the input sample $x[n]$ is multiplied by several constants simultaneously, the hardware minimization of the multiplier block can be directly modeled as an MCM problem [11,25]. The two-operand adder that adds the product of a filter coefficient and its corresponding input sample with the accumulated sum of such products pertaining to the preceding filter coefficients, is referred to as the structural adder (SA), and the complete tapped-delay adder unit is called the SA block.

In all AG algorithms and in some CSE algorithms, a solving set S consisting of all output fundamentals and if required, one or more nonoutput fundamentals (NOFs) is generated by an algorithm. The set of constant multiplicands G are first mapped to a target set of positive and odd fundamentals T. If the multiplier is followed by an adder (a structural adder in case of FIR filters), the sign-inversion pertaining to a negative constant is moved to the adder, such that the adder is changed to a subtractor and the constant is treated as positive. Since every even number has at least one constant "0" at the least significant bit (LSB), they can be divided by two until the result is odd, and shifted back by appropriate wiring. In Table 2.2, we have illustrated the different steps involved in the generation of target set T consisting of only odd and positive fundamentals for a 25 tap linear phase FIR filter taken from example 1 of Reference [38].

The resulting reduced, odd and positive set of fundamentals represents the minimal number of constant multiplications required to implement the filter.

Table 2.2 Illustration of generation of odd and positive fundamentals for a 25 tap linear phase FIR filter (example 1 of Reference [38]).

Steps	Description of functions
1.	Place all coefficients into the set G_1. In case of symmetrical filter, place only the first $\left\lceil \frac{N-1}{2} \right\rceil$ into the set to obtain: $G_1 = \{1, 3, -1, -8, -7, 10, 20, -1, -40, -34, 56, 184, 246\}$
2.	Take the absolute values of all elements in G_1 to obtain: $G_2 = \{1, 3, 1, 8, 7, 10, 20, 1, 40, 34, 56, 184, 246\}$
3.	Remove the repetitions and all zeros in G_2 to obtain: $G_3 = \{1, 3, 8, 7, 10, 20, 40, 34, 56, 184, 246\}$
4.	Divide all even elements in G_3 by 2 until the results are odd to obtain: $G_4 = \{1, 3, 1, 7, 5, 5, 5, 17, 7, 23, 123\}$
5.	Remove the repetitions in G_4 and sort in ascending order of implementation cost or magnitude to obtain: $T = \{1, 3, 5, 7, 17, 23, 123\}$

The amount of redundancy across the filter coefficients depends on many factors like the transfer function of the filter, design algorithm, word size, and rounding considerations in fixed-point approximation, and so on. It appears that the filter-length is not a good indicator of the complexity of the MCM block, but the size of the target fundamental set T and the complexity of its elements are more closely related to the implementation cost.

2.4.2 The Adder Graph Approach

2.4.2.1 The Early Developments

The first AG algorithm introduced by Bull and Horrocks [25] is called Bull–Horrocks (BH) algorithm. It introduced the graphical solution to the MCM problem. To improve the limited search capability of BH algorithm, an improved version called Bull–Horrocks-modified (BHM) algorithm was suggested in Reference [27]. The AG algorithm begins with an initial solved set S (sources) consisting of a single value "1" and a target set T containing all remaining odd fundamentals (sinks). In the next step, it selects some specific elements of T and moves them to S. A generic pseudocode for populating the solving set S is shown in Algorithm 2.1. The main difference between algorithms lies in how the heuristic FindNOF subroutine selects additional elements required to implement the shift-and-add network. The use of Eq. (2.6) for the adder graph construction makes these algorithms independent of number representation, as they manipulate directly with the numerical values of the fundamentals. The resulting graph shows the arithmetic relationship between the operators and the timing delay in terms of LO count on the path to output fundamentals.

Algorithm 2.1 Pseudocode for a generic adder graph algorithm

Inputs: Target set T, Solved set $S = \{1\}$

AdderGraphAlgorithm(S, T)

1: **while** true **do**

2:   $(S, T) = \text{Synthesize}(S, T)$ **if** $T = \emptyset$ **then**

3:     **return** (S)

4:   **else**

5:     $r = \text{FindNOF}(S, T)$

6:     $S \leftarrow S \cup \{r\}$

7:   **end If**

8: **end while**

Synthesize(R, T) {optimal part}

1: **if** $t_k \in T$ can be calculated by elements from R **then**

2:   $R \leftarrow R \cup \{t_k\}$

3:   $T \leftarrow T \setminus \{t_k\}$

4:   $(R, T) = \text{Synthesize}(R, T)$

5: **end if**

6: **return** (R, T)

FindNOF(R, T) {heuristic part}

1: find non-output fundamental r based on heuristic

2: **return** r

2.4.2.2 *n*-Dimensional Reduced Adder Graph (RAG-n) Algorithm

Dempster and Macleod have introduced a systematic approach called n-dimensional reduced adder graph (RAG-n) algorithm [27], which consists of an optimal stage and a heuristic stage. At first, all possible fundamentals are synthesized in the optimal stage, and whenever necessary, the heuristic stage is invoked thereafter. The optimal stage guarantees the minimal adder cost for fundamentals in the range from 1 to 4096. The heuristic stage uses a new measure called the "adder distance" to estimate the number of extra adders needed to generate the requested vertex from the existing vertices of the incomplete graph. It selects the vertex with the minimum adder distance and adds them progressively until all the fundamentals are generated. The adder distances are generated by an exhaustive search, using minimum adder graph (MAG) algorithm [24], over all possible graph topologies, and stored in a LUT. The utilization rate of the optimal part of RAG-n algorithm has been improved in Reference [39] to achieve lower adder cost and shorter runtime. It is shown in Reference [40] that the BHM algorithm can be used here as a subroutine to find the maximally sharable graph for a fundamental in an easier way. The major limitation of RAG-n algorithm arises due to the bit width limit of the adder distances stored in the LUT.

2.4.2.3 Maximum/Cumulative Benefit Heuristic Algorithm

The RAG-n algorithm was considered as a reference algorithm for a long time until Reference [41] introduced the maximum benefit heuristic (H_{maxb}) and the cumulative benefit heuristic (H_{cub}), which are free from the bit width limitation. Besides, to streamline the search process, Voronenko and Puschel have introduced a new notion called "$\mathcal{A}$-operation" as an operation on fundamentals, given by

$$\mathcal{A}_p(u, v) = \left| 2^{l_1}u + (-1)^s \times 2^{l_2}v \right| \times 2^{-r} \qquad (2.10)$$

where integers u and v are input fundamentals, l_1, l_2, and r are integers while $s \in 0, 1$. An $\mathcal{A}$-operation basically performs a single addition or subtraction, and an arbitrary number of shifts which do not truncate nonzero bits of the fundamental.

Based on the "$\mathcal{A}$-operation", the "$\mathcal{A}$-distance" is defined, which is analogous to the adder distance used in RAG-n. It gives either the exact or a heuristic estimate for the distance used in their algorithms, and produces significantly improved results than previously heuristics. The H_{cub} algorithm finds consistently better results than H_{maxb} and is the best reference algorithm to be competed against. Besides, it is shown that H_{cub} provides solutions very close to the minimal solution [42]. The HDL code as well as the soft code of this algorithm are available at Reference [43], which could be used to verify the results easily.

2.4.2.4 Minimization of Logic Depth

The AG algorithms synthesize the fundamentals sequentially and greedily reduce the LO count. As new fundamentals are generated from the existing ones, these algorithms tend to design solutions with high LD and longer critical path, while CSE algorithms either guarantee minimal or near minimal LD solutions but they have in general a higher LO count. High LD increases the likelihood to produce and propagate glitches [16,44]. The step-limiting BHM and step-limiting RAG-n are thus proposed to reduce the LD [16]. LD is reduced in the C1 algorithm of Reference [44] at the expense of a few additional adders without guaranteeing minimum LD. The latest release of H_{cub} [43] provides the possibility to restrict the maximal allowable LD, but to the best of our knowledge, this option in the updated version has not been described or reported in any archived publication so far. Another possibility to speed up the additions of multiple operands for each constant multiplication by using carry save adders (CSAs) instead of CPAs [45,46]. Each CSA has only one full adder delay irrespective of the word length of the CSA, which reduces the critical path delay through these adders. However, this speedup is offset by the need for a final CPA to merge the stored carry and sum vectors. Otherwise, two sets of registers are required to store the sum and carry vectors for the transposed direct form filter before they are accumulated to the sum and carry vectors of the next coefficient multiplier in the tap-delay line[4].

The minimum adder depth (MAD) algorithms [48,49], which include an exact and two heuristic methods, guarantee the minimum LD for all fundamentals. The MAD algorithms use LUTs to enumerate all possible combinations to form fundamentals with minimal LD. The results show a higher LO count and lower power dissipation than the unrestricted AG algorithms. The exact algorithm is, however, applicable only up to 40 fundamentals and the LUTs restrict the maximum size for the fundamentals. In Reference [50], the minimal logic depth (MinLD) algorithm is introduced to generate the fundamentals without increasing the LD. The MinLD algorithm does not require LUTs, since it uses CSD representation to check and ensure that the LD is not increased. The LD results are similar to those of the MAD algorithm but MinLD provides a further reduction of power dissipation.

2.4.2.5 Illustration and Comparison of AG Approaches

The fundamentals optimized by different AG algorithms BHM ([27]), C1 ([44]), H_{cub} ([41]), and MinLD ([50]) are shown in Figure 2.4, where the constants 556, 815, and 1132 are reduced to their odd fundamentals 139, 815, and 283. Since this small set of odd fundamentals does not contain any fundamentals of the form $(2^n \pm 1) \in N$, the solution cannot be obtained directly from the initial solved set and suitable NOFs need to be sought. The BHM algorithm is found to use the NOFs $\{3, 17, 51, 57\}$ (Figure 2.4a). Algorithm C1 puts all fundamentals $(2^k \pm 1) \in N$ into the solved set S, and eliminates unused ones before generating the final solution. It uses the NOFs $\{3, 13, 17, 257\}$ (Figure 2.4b),

---

4 More discussion on the use of CPA and CSA can be found in Reference [47].

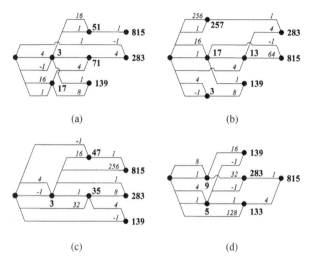

(a)   (b)

(c)   (d)

Figure 2.4 The fundamentals 139, 283, and 815 optimized by different adder graph techniques with highlighted subexpressions. (a) BHM [27] using seven LOs and having LDs of {2,3,3}, (b) C1 [44] using seven LOs and having LDs of {2,3,3}, (c) H_{cub} [41] using six LOs and having LDs of {3,3,3}, and (d) MinLD [50] using six LOs and having minimal LDs of {2,2,3}.

where there is one more NOF of the form $(2^k \pm 1) \in N$. The H_{cub} algorithm on the other hand requires only three NOFs $\{3, 35, 47\}$ (Figure 2.4c), at the cost of a higher LD. The MinLD algorithm also uses only three NOFs $\{5, 9, 133\}$ (Figure 2.4d), and constrains all the three output fundamentals to their minimal LD. Figure 2.4 also shows that the output fundamentals can be reused. This is more commonly encountered in larger fundamental sets.

2.4.3 Common Subexpression Elimination Algorithms

The common subexpression elimination (CSE) algorithms require the constants to be represented in a certain format where common patterns can be sought and eliminated. This is in stark contrast with the graph-based algorithms that operate independent of the number of representations. The challenge for a CSE approach is the search for the most beneficial patterns/subexpressions within the MCM block. The pseudocode for a generic algorithm CSE algorithm is shown in Algorithm 2.2.

Algorithm 2.2 Pseudocode for a generic CSE algorithm

Inputs: Target set T, Subexpression set $CS = \emptyset$

CSEAlgorithm(T, CS)

1:   $M \leftarrow T$ in a specific number representation $\{M$ could be a matrix$\}$

2:   **while** true **do**

3:      generate subexpression statistic

4:      **if** no subexpression exists more than once **then**

5:        **return** (M, CS)

6:      **else**

7:        heuristically select subexpression s_i based on statistic

8:        replace suitable occurrences of s_i in M with placeholder p_i

9:        update CS with relation $s_i \leftrightarrow p_i$

10:      (optional) insert p_i into M

11:      **end if**

12: **end while**

The fundamental idea of CSE can be illustrated by a simple example of $\{556, 815\}$. In the CSD representation of $\{556, 815\} = \{[010010\bar{1}0\bar{1}00], [10\bar{1}010\bar{1}00\bar{1}]\}_{\text{CSD}}$ the subexpression $s_1 = [10\bar{1}] \leftrightarrow p_i$ occurs three times. Hence, $556 = [010000p_1 0\bar{1}00]_{\text{CSE}}$ and $815 = [00p_1 000p_1 00\bar{1}]_{\text{CSE}}$ and the shift-and-add implementation can be expressed by $p_1 = (1 \ll 2) - 1$, $556 = (1 \ll 8) + (p_1 \ll 5) - (1 \ll 2)$, and $815 = (p_1 \ll 7) + (p_1 \ll 3) - 1$. The selection of the most beneficial subexpression is not simple, as selecting one subexpression for elimination can inhibit the reuse of several other subexpressions. For example, the contention arises when a single constant $[1\,0\,\bar{1}\,0\,0\,1]$ contains more than one subexpressions $[1\,0\,\bar{1}]$, $[\bar{1}\,0\,0\,1]$, and $[1\,0\,0\,0\,0\,1]$, but the selection of one subexpression prohibits other available subexpressions from being selected since one or more of their bits overlap with the selected subexpression. Higher order subexpressions or multiple representations of subexpressions add to the intricacy of contention.

Similar to the single constant case, the number representation influences the number of different types of subexpressions that can be found by a CSE algorithm. The canonical representations like CSD or binary representation result in a unique bit matrix for the search for CSs. This is good for the runtime reduction, but limits the possibilities of finding quality solutions with the minimum number of LOs. In the search for better solutions, CSE algorithms have moved from canonical CSD representation to the noncanonical or inconsistent MSD representation. Doing so requires more complicated pattern search and

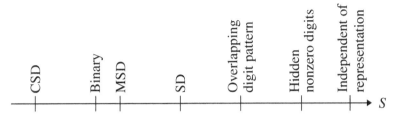

Figure 2.5 Representations in increasing order of complexity and search space.

matching. Besides the exploration of nonuniqueness in constant representations, Reference [51] generalizes the use of overlapping digit pattern (ODP) in SCM to MCM problems to allow more CSs to be found. Figure 2.5 shows different possibilities to represent numbers for optimizing constant multiplications ordered with increasing size of search space in the direction of S, whereas the unrestricted search space contains all possible "$\mathcal{A}$-operation" that form one or several fundamentals. The CSD and binary number representations are the most restricted form, since both representations are unique and the number of nonzero digits is fixed. The MSD representation is less restrictive. Both subpatterns $3 = [011]$ and $3 = [10\bar{1}]$ are allowable in MSD, which results in additional forms of representation for certain integers. Without bit width restriction, SD representation can have infinitely many forms for a number. When the sign of all nonzero digits are positive for a SD representation, it becomes identical to the binary representation. Therefore, the binary representation is a subset of SD representation, but only partially intersects with CSD/MSD space. The intersection contains all the binary representations that have the minimum number of nonzero digits. The ODP space includes the subspace of a number representation where all bit or digital patterns are visible plus the obscured patterns defined as ODPs. The concept of ODP was extended in Reference [50] to include the hidden nonzero digit patterns. The extension covers additional possibilities of digit patterns that cannot be directly detected from a number representation. These two search spaces can be seen as the expansion of search space from a fixed number representation toward an unrestricted search space that is independent of number representation. Along with the increased search space, the search complexity increases together with the runtime and the chance to find a better solution. Additional constraints like minimal logic depth [48–50] or the search for only selected patterns, for example, [52,53] can reduce the runtime significantly at the expense of compromising the optimality of the solutions.

2.4.3.1 Basic CSE Techniques in CSD Representation

The concept of subexpression sharing is quite old in software context for optimal usage of arithmetic unit in programmable machines [54]. The work of

Hartley is one of the first to use the CSE algorithms for constant multiplications in dedicated hardware realization of FIR filters [15,55]. The algorithm simply records the frequency of all subexpressions, eliminates the CS which occurs most and then recursively eliminates the next most frequently occurred CS. This process also eliminates higher weight (refers to the hamming weight) CSs, which are also called super-subexpressions. In Reference [56], CSE is used in an algebraic speed-up procedure, where the critical path is reduced to improve the throughput rate. A significant advancement on the transformation using CSE was achieved by Potkonjak *et al.* [11]. They have formulated the MCM problem in high level synthesis by considering the multiplications of one variable with several constants at a time and also reduced the number of shifts and additions based on an iterative pairwise matching procedure. Their iterative matching (ITM) algorithm finds the maximum number of coincidences between two fundamentals represented in a specific SD representation. It looks for the same nonzero digits within the fundamentals with the least number of shifts from the least significant bit position. A simple algorithm has been proposed in Reference [57] to iteratively extract CSs with two nonzero digits, which was modified and extended by Pasko *et al.* [58,59]. Pasko's algorithm differs from Hartley's algorithm, in that the higher weight CSs are eliminated first. The higher weight CSs are treated as additional constants such that the CS of lower weights can be eliminated subsequently. The CS selection is based on an exhaustive search for pattern frequencies, and a steepest descent approach is used to select the most frequent patterns. Both global and local binary trees are used to store the frequency statistics, where a local tree exists for each constant and a global tree is used for the entire set of constants. During pattern elimination, the trees are updated and used for the selection. This technique provides considerable reduction of LOs while maintaining the shortest LDs, although it does not guarantee the minimum LD. The algorithm of Reference [35] is an extension and enhancement of Reference [59], where previous results are generalized by dealing with the optimization problem of the multiplication of a vector with a constant matrix. The nonrecursive signed common subexpression elimination (NRSCSE) algorithm [60] does not search for the CS recursively but allows the CS to have more than two nonzero digits. It guarantees the minimum LD provided that only CSs of weight-two are used. The algorithm of Reference [61] starts with the CSD representation of the fundamentals but uses Eq. (2.6) to synthesize the fundamentals and ensures that the LD does not exceed a preset maximum.

2.4.3.2 Multidirectional Pattern Search and Elimination

Besides eliminating CSs that are contained in a single constant, Reference [15] proposes to search for CSs with their nonzero digits spanning across different coefficients. An array of constants can be stored in a bit matrix, where each row

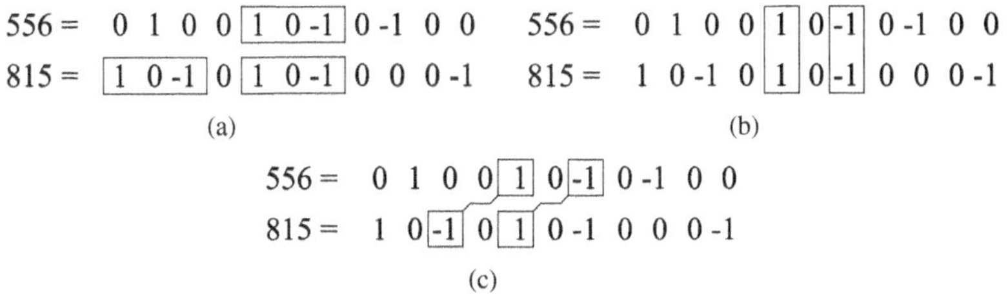

556 =   0 1 0 0 | 1 0 -1 | 0 -1 0 0   556 =   0 1 0 0 |1| 0 |-1| 0 -1 0 0
815 = | 1 0 -1 | 0 | 1 0 -1 | 0 0 0 -1   815 =   1 0 -1 0 |1| 0 |-1| 0 0 0 -1
         (a)   (b)

556 =   0 1 0 0 | 1 | 0 |-1| 0 -1 0 0
815 =   1 0 |-1| 0 | 1 | 0 -1 0 0 0 -1
                        (c)

Figure 2.6 The three different types of CS: (a) horizontal, (b) vertical, and (c) oblique.

of the matrix is a constant in some positional representation (binary, SD, etc.). A horizontal subexpression has all its bits confined within the same row of the bit matrix, a vertical subexpression that has all its bits confined in the same column of the bit matrix and an oblique subexpression has its bits spanned in different rows and columns of the bit matrix. Figure 2.6 shows three different types of CS that can be found in an array of two integers 556, 815.

Some algorithms that use vertical or oblique CS for CSE are References [34,35,62–64]. Vinod *et al.* [65] shows by example that the vertical CSE transcends the horizontal CS if adjacent constants have similar patterns in the MSB portion. However, the vertical CSE alone does not guarantee greater hardware savings over the conventional horizontal CSE methods in some cases, for example, linear-phase filters, where the vertical CSs destroy the symmetry of linear-phase filters [66]. Furthermore, the cost of delay elements required for vertical and oblique CSs is often neglected. The use of vertical or oblique CSs also prohibits the reduction to odd fundamentals, as the relative shift of CS bit patterns between constants must be preserved. A more generalized multidirectional search has been suggested in Reference [63] that combines (i) horizontal CSE, (ii) vertical CSE, (iii) horizontal super-subexpression elimination, and (iv) vertical super-subexpression elimination. Horizontal and vertical super-subexpressions eliminate the redundant horizontal and vertical CSs, respectively, with identical shifts between them. The design examples of channelizers of multistandards communication demonstrate the potential of this approach for LO reduction.

2.4.3.3 CSE in Generalized Radix-2 SD Representation

In Reference [67] the use of CSE in CSD, MSD, and binary representations of constants has been compared and it was shown that MSD is superior to CSD, and binary can be advantageous over CSD. These results were refined further in [68]. The CSE algorithms presented in References [52,69] operate solely on binary representation and are called binary subexpression elimination (BSE) algorithms. The use of binary offers some statistical advantage over CSD [68] at the expense of a somewhat higher LD. Detail comments on it have been discussed in a recent paper [53].

Using all possible MSD representations for the constants, the algorithm of Reference [7] synthesizes one fundamental at a time, based on the previously generated fundamentals. The multi-root binary partition graph (MBPG) algorithm [70] is the first algorithm to use an information theoretic approach to enhance the optimality of CSE. In this algorithm, instead of composing the output fundamentals from nonoutput fundamentals as in AG algorithms, the constants in SD representation are decomposed into the elementary digits $\{\bar{1}, 0, 1\}$. The entropy and the conditional entropy are calculated from the probability of occurrence of signed digit pairs, and the entropy information is then used to decide which nonzero digits should be assigned first to the children nodes when the parent nodes are decomposed.

2.4.3.4 The Contention Resolution Algorithm

The contention resolution algorithm (CRA) uses a data structure called "the admissibility graph" to allow the greedily selected poor CSs to be replaced when there is a contention with a good CS for synthesizing the optimal solution [64,71]. This allows the algorithm to move out of local minima. The latest version of CRA allows options to set the maximum hamming weight of the CSs and to include different configurations of horizontal, vertical, and oblique subexpressions.

2.4.3.5 Examples and Comparison of MCM Approaches

The CSD representations for the numbers 556, 815, and 1132 are shown in Figure 2.7a. The subexpressions that exist more than once are evaluated and are shown in Figure 2.7b, where the subexpression statistic is obtained by counting the maximum number of each type of subexpressions without contention. For example, the subexpression $[10\bar{1}]$ appears three times in 815, but at most two of these three subexpressions can be implemented due to the overlap. A supersubexpression $[10\bar{1}01]$ can also be found twice in the same constant, but as the two occurrences overlap in two nonzero digits, it cannot be considered as a CS.

556 = 0 1 0 0 1 0 -1 0 -1 0 0	2x	1 0 1
	3x	1 0 -1
815 = 1 0 -1 0 1 0 -1 0 0 0 -1	2x	1 0 0 1
	2x	1 0 0 0 1
1132 = 1 0 0 1 0 0 -1 0 -1 0 0	2x	1 0 0 0 0 -1
	3x	1 0 0 0 0 0 -1
(a)	(b)	

Figure 2.7 The constants 556, 815, and 1132 in CSD representation (a) and the subexpressions that occur more than once (b).

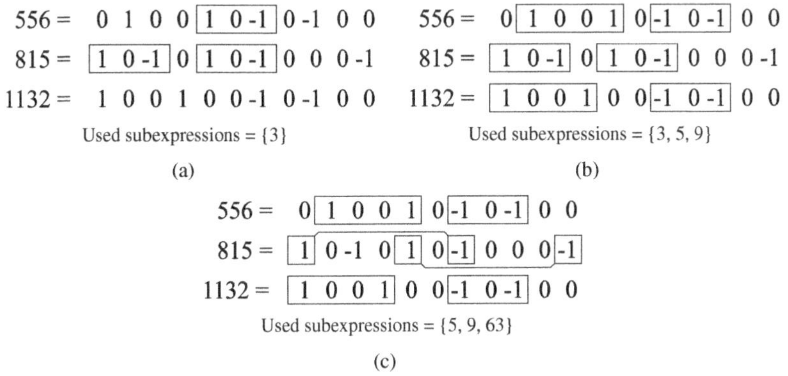

556 = 0 1 0 0 [1 0 -1] 0 -1 0 0   556 = 0 [1 0 0 1] 0 [-1 0 -1] 0 0

815 = [1 0 -1] 0 [1 0 -1] 0 0 0 -1   815 = [1 0 -1] 0 [1 0 -1] 0 0 0 -1

1132 = 1 0 0 1 0 0 -1 0 -1 0 0   1132 = [1 0 0 1] 0 0 [-1 0 -1] 0 0

Used subexpressions = {3}   Used subexpressions = {3, 5, 9}

(a)   (b)

556 = 0 [1 0 0 1] 0 [-1 0 -1] 0 0

815 = [1] 0 -1 0 [1] 0 [-1] 0 0 0 [-1]

1132 = [1 0 0 1] 0 0 [-1 0 -1] 0 0

Used subexpressions = {5, 9, 63}

(c)

Figure 2.8 The constants 556, 815 and 1132 optimized by different CSE algorithms with highlighted subexpressions. (a) Hartley's [15], Pasko's [59], and NRSCSE [60] only find the subexpression 3 while (b) CRAH-2 [71] and (c) MBPG [70] find two additional subexpressions, but use 3 only twice.

There is no super-subexpression that occurs more than once in this example. Super-subexpressions are more commonly found in large array of constants. The algorithms [15,59,60] that make use of such subexpression statistic produce the same result, as they all select the most frequent CS, $[10\bar{1}] = 3$ for elimination first, as shown in Figure 2.8a. After this subexpression is eliminated from all constants, no subexpression appears more than once. This is a good example of the contention problem mentioned previously. Still the LO count is reduced by two over the direct CSD implementation, which requires 10 adders. The newer, and contention aware, Reference [71] algorithm also eliminates the subexpression, $[10\bar{1}]$ in 556 and 815, but later revokes the elimination of $[10\bar{1}]$ in 556 as it finds the contending CSs $[101] = 5$ and $[1001] = 9$ in 556 and 1132, respectively. The final subexpression selection is shown in Figure 2.8b and it reduces the LO count by three over the direct CSD implementation. The Reference [70] algorithm selects the subexpression $[100000\bar{1}] = 63$ instead of $[10\bar{1}]$. The saving, in terms of LO, is identical and the result is shown in Figure 2.8c. The solutions in Figure 2.8 clearly show, albeit with an example of only three constants, that the solutions can deviate depending on how an algorithm evaluates the subexpressions, how the candidate CS are selected and their order of elimination.

Comparing the results of AG and CSE algorithms in Figure 2.4 and Figure 2.8, it can be seen that the best AG algorithms require one adder less than the best CSE algorithm. By comparing the results of this single example, no conclusion regarding the algorithms can be drawn. In the literature, a brief comparison of CSE and AG algorithms can be found in References [72,73], but the conclusions are not consistent. While advantages of CSE algorithms for complex problems is highlighted in Reference [72], the advantages for AG algorithms for larger word lengths are inferred in Reference [73]. An important conclusion of Reference [73] is that "even where one algorithm gives the lowest average adder cost, other algorithms may give a lower cost for particular filters."

2.4.4 Difference Algorithms

The algorithms in References [74–80] do not attempt to implement the fundamentals directly. Instead, they examine the differences between the fundamentals and order them such that the differences are minimized in order to reduce the implementation cost. The minimization of the adder cost to produce the differences can again be viewed as an MCM problem and the earlier techniques can be applied repeatedly. For the difference algorithm using CSE, the search space changes significantly from the original set of fundamentals, whereas for the graph-based difference algorithms, the search space does not change but the direction of search changes.

2.4.5 Reconfigurable and Time-Multiplexed Multiple Constant Multiplications

The notion of reconfigurable multiplier block (ReMB) for FPGA implementation is introduced in Reference [81] for the time-sharing of resources of an MCM computing block. It is shown that by suitable control of the multiplexers, the macro cells of an FPGA device can be reconfigured to generate more than one partial sum, resulting in a better area utilization of the device [82]. It is further demonstrated that the efficient implementation of an FIR filter using ReMB can reduce the area requirement on FPGA by nearly 20% over the implementation with conventional multipliers. The design methodologies in References [81,82] are developed for single input single output (SISO) ReMB, where all possible ways of constructing a constant by grouping of the existing partial sums are searched and listed. The partial sums that have common input value and shift amount are grouped into a basic cell of the FPGA. Finally, the outputs of the newly constructed basic cells are combined with the existing partial sums to construct the remaining constants. A systematic synthesis method [83] has also been proposed for both SISO and single input multiple output (SIMO) ReMB using a different approach.

A directed acyclic graph (DAG) fusion approach has been proposed in Reference [84] for multiplexed multiple constant multiplications. It does not depend on the specialized cells of FPGA, and has been optimized for ASIC mapping. The multiplexed multiple constant multiplication is modeled as a time-multiplexed multiplication of an input variable and one of several selective integer constants. Each integer constant is expressed as a fixed DAG from an existing algorithm [22]. The core idea is to find the best fusion of the DAGs of all constant multipliers. The best solution from all possible DAGs for the individual constant multipliers with the least-estimated area cost is selected as the final DAG for that constant. The design is completed by inserting all the required multiplexers and transforming all nodes in DAG into adders/subtractors. With no constraint on the basic cells of FPGA, the DAG fusion is capable of achieving significant saving in area cost. However, there are two shortcomings of DAG fusion. First of all, the DAG representations for constants are determined and fixed by another algorithm. These DAGs are generated sequentially and individually without a holistic consideration of subexpression sharing in the complete set of constants. Second, the search complexity to cover all possible fusion solutions is huge, so that it becomes intractable when the problem size is large. On the other hand, limiting the number of iterations in this search will compromise the quality of the solution since the probability of finding a nonoptimal fusion becomes higher in that case.

A time-multiplexed data-flow graph (TDFG) has been proposed in Reference [85] to model this design problem. The operators and edges in this TDFG have a direct correlation with the adders and shifters of the reconfigurable constant multipliers so that the minimization problem is transformed into a high-level synthesis problem of optimal scheduling and allocation of operators in the TDFG. The mobile operators are scheduled into optimal control steps to maximally time-share as many operators in the same control-step as possible. New aggregate distance and disparity measures based on the distribution of signed digits in the addends were defined to evaluate the opportunity cost of a scheduling decision. Experimental results on application benchmarks and random constant sets showed that the algorithm of Reference [85] is capable of producing more area-time efficient solution over the DAG fusion without exhaustive enumeration and search for independently optimized AG of each constant.

Recently, a reconfigurable FIR filter architecture based on BSE method [52] has been proposed in Reference [86]. The proposed FIR filter architecture is capable of operating for different word lengths and filter coefficients without any overhead in the hardware circuitry. It has been shown that the dynamically reconfigurable filters can be efficiently implemented by using CSE algorithms. Design examples showed that the architectures proposed in Reference [86] offer good area and power reductions and speed improvement compared to the existing reconfigurable FIR filter implementations.

Other time-multiplexed and reconfigurable implementations of MCM include the use of carry-save adders on FPGA devices in Reference [87], efficient FPGA realization of multiplications of a variable with a small possible set of constants in Reference [88] and programmable adder graph (PAG) approach for FPGA implementation of shift-and-add network in Reference [89]. The latter does not require prior knowledge of the multiplier value and can be used for adaptive filtering applications.

2.5 Optimization Schemes and Optimal Algorithms

The methods presented in the previous sections are mostly heuristic, where we cannot guarantee the optimality. In this section we will initially discuss how subexpression sharing can be solved optimally and extended for representation independent formulations.

2.5.1 Optimal Subexpression Sharing

The first optimal approach to subexpression sharing was proposed in Reference [90]. There, the best first-level two-term subexpression was modelled as an integer linear programming problem and solved. A subexpression can be described by two variables i and j, where variable i refers to the difference between the the location of a pair of nonzero digits in the signed-digit representation of the constant; while variable j indicates whether an addition or a subtraction is required to be performed for implementing the subexpression. For the subexpression 101, $i = 2$, and for $1000\bar{1}$, $i = 4$. We can assume the convention for the variable j, such that $j = 0$ indicates addition and $j = 1$ indicates subtraction to be performed for implementing the subexpression, that is, if the subexpression is of the form $10\ldots01$ then $j = 0$; or if it is of the form $10\ldots0\bar{1}$ then $j = 1$. The weight of a subexpression can therefore be written in a generalized form as $w_{i,j} = 2^i + (-1)^j$. Consider a constant A_m that can be written using first-level two-term subexpressions as

$$A_m = \sum_{k,l} a_{k,l,m}(-1)^k 2^l + \sum_{i,j,k,l} s_{i,j,k,l,m} w_{i,j}(-1)^k 2^l \qquad (2.11)$$

where, $a_{k,l,m}, s_{i,j,k,l,m} \in \{0, 1\}$ is 1, if the corresponding digit or subexpression is used in the representation, $k = 0$ or 1, respectively, when the digit (or subexpression) is added or subtracted, and l denotes the position of the digit (or LSB of the subexpression). Now, let us introduce the variable $x_{i,j} \in \{0, 1\}$, which is 1 if and only if any subexpression characterized by i, j is used. If a subexpression is used once, it may be used again without any additional cost. Based on this we can relate $s_{i,j,k,l,m}$ and $x_{i,j}$ as

$$x_{i,j} \geq s_{i,j,k,l,m}, \quad \forall i, j, k, l, m \qquad (2.12)$$

The objective function to be minimized is then the number of terms in equations like Eq. (2.11) plus the number of subexpression computations. This gives

$$\underbrace{\sum_{k,l,m} a_{k,l,m}}_{\text{Unpaired digits}} + \underbrace{\sum_{i,j,k,l,m} s_{i,j,k,l,m}}_{\text{Using subexpressions}} + \underbrace{\sum_{i,j} x_{i,j}}_{\text{Computing subexpressions}} \qquad (2.13)$$

Hence, it is possible to formulate an optimization problem as

Minimize Eq. (2.13), subject to Eq. (2.11) and Eq. (2.12), $\forall\, m$ (2.14)

Note that in the above problem no representation is assumed. By defining a representation for each constant, many of the variables can be removed from the problem by simply identifying that they will be zero. While not defining, a representation will, in general, produce slightly better results, it will also result in more variables and therefore a longer solution time. The optimization problem in Eq. (2.14) can be solved using standard integer linear programming approaches [91]. Alternatively, because of the binary variables and integer constraint coefficients, the problem in Eq. (2.14) is a so called pseudo-Boolean problem for which there exists solvers originating from the satisfiability problem [92].

By studying the above problem, it is easy to realize that no two $a_{k,l,m}$ variables with the same l and m can both be nonzero. This simply means that we do not want to choose a representation with two digits of opposite sign at the same position, which eventually will cancel. Introducing these constraints, called cuts from an optimization point of view, will neither affect the quality of the solution nor, for our case, the linear relaxation of the problem. However, they will typically help the solver eliminating paths in the search that will not give the optimal solution. A cut for the issue discussed above can simply look like

$$a_{0,l,m} + a_{1,l,m} \leq 1,\ \forall\, l, m \tag{2.15}$$

These cuts can be extended to include no adjacent nonzero digits as

$$a_{0,l,m} + a_{1,l,m} + a_{0,l+1,m} + a_{1,l+1,m} \leq 1,\ \forall\, l, m \tag{2.16}$$

and to include subexpressions. In their simplest form these cuts look like

$$\sum_k a_{k,l,m} + \sum_{i,j,k} s_{i,j,k,l,m} \leq 1,\ \forall\, l, m \tag{2.17}$$

Note that at most one digit or subexpression can be positioned at the lth bit-position for each constant, which can be utilized by adding cuts as in Eq. (2.17). This would forbid the solver to search for trivial intermediate solutions such as $5 = 101 + 10\bar{1} - 10\bar{1}$.

This approach has been generalized in Reference [93] to higher order subexpressions, by introducing subexpressions with more than two terms and the formulation to derive these subexpressions. To represent a three-term subexpression, we require two more indices, one for the additional shift and one to represent the sign of the third digit. Consider the subexpression $10100\bar{1}$ corresponding to $x_{i_1,j_1,i_2,j_2} = x_{5,0,3,1}$ with weight $w_{i_1,j_1,i_2,j_2} = 2^{i_1} + (-1)^{j_1} 2^{i_2} +$

$(-1)_2^j = 2^5 + (-1)^0 2^3 + (-1)^1 = 39$. This subexpression can be derived from any of the subexpressions 101 ($x_{2,0}$), $100\bar{1}$ ($x_{3,1}$), or $10000\bar{1}$ ($x_{5,1}$). This corresponds to the constraints

$$x_{2,0} + x_{3,1} + x_{5,1} \geq x_{5,0,3,1} \tag{2.18}$$

which effectively means that at least one of $x_{2,0}$, $x_{3,1}$, or $x_{5,1}$ must be one for $x_{5,0,3,1}$ to be one. It is straightforward to see that, in general, the possible two-term subexpressions are required to derive a three-term expressions. Similarly, it is possible to derive four-term expressions based on three-term subexpressions. It is worth noting that even though we may have an M-digit constant, it is not required to formulate a problem including M-term subexpressions, because each constant may be formed of several digits and subexpressions as in Eq. (2.11).

The derivation of higher order subexpressions discussed earlier requires the input subexpressions to be of exactly one term lower order. Hence, the method cannot be used to derive a four-term subexpression from two two-term subexpressions. A detailed formulation where the subexpressions are derived from pair of subexpressions is available in Reference [68]. In this approach, the additional variables are used for each possible pairing. These variables correspond to pairs of subexpressions that are both realized. For simplicity, from here onwards we will represent the subexpressions with their weights. Hence, the existence of the subexpression pair i and j, corresponding to the variables x_i and x_j, are represented by the variable $y_{i,j}$ (for normalization purposes we assume $i \leq j$). These variables are related as

$$x_i + x_j \geq 2y_{i,j} \tag{2.19}$$

which can be interpreted that both x_i and x_j should be one for $y_{i,j}$ to be one. Then, for each higher order subexpression, there exists a number of pairs which can be used to derive this subexpression. This is handled through constraints similar to Eq. (2.18), but with $y_{i,j}$ variables on the left-hand side. In Reference [68] the constants are not represented as a sum of the digits and subexpressions. Instead, constraints are introduced stating that the subexpressions corresponding to the constant values should be one.

2.5.2 Representation Independent Formulations

By using the weights instead of the digit-oriented subexpressions as identifiers, the representation dependence is reduced. Clearly, there are many more possible ways to derive a subexpression if the exact representation is not considered. This fact was also treated in Reference [68]. However, the method still requires an ordering of the constants. Due to the commutative property of addition and subtraction, if k can be derived from i and j, so can i be derived from j

and k, and j be derived from i and k. Even though it is possible to formulate constraints for all the three cases, solving the problem may result in a circular reasoning, where i, j, and k are derived from each other, for which there clearly are no practical solutions. Hence, it is required that an ordering is introduced such that only one of the three possible derivations are considered in the optimization problem. The method suggested in Reference [68] was to allow only the subexpressions with fewer nonzero digits to be used in the derivation. This guarantees a valid solution but at the same time it reduces the search space significantly.

This problem has been solved in Reference [94] where the constant multiplication problem is modeled as a hypergraph and shown that the solution to the problem is equivalent to finding a Steiner tree in the hypergraph. By keeping track of the depth of different subexpressions and introducing constraints similar to the MTZ constraints [95] used for various tree problems in plain graphs to avoid cycles, an exhaustive optimal solution could be found, subject to the long-run time associated with solving this large NP-hard problem. As a side-effect of this work the following important observations were made:

- The constant multiplication problem could be considered as a problem of finding a suitable set of extra subexpressions.
- Given the hypergraph and a valid set of extra subexpressions, finding a solution can be done with an execution time that is polynomial to the number of edges in the hypergraph.
- The minimum adder depth is completely defined by the set of extra subexpressions. Therefore, the only way to reduce the adder depth is to either change the extra subexpressions or add more extra subexpression.

Hence, a possible way of solving the constant multiplication problem would be to exhaustively search the possible extra subexpressions. An efficient way of doing this was proposed in Reference [96], which together with Reference [94] are the only two current approaches that can guarantee a result that is optimal in the number of adders.

While the algorithms discussed above are optimal in the number of adders, they may result in a large adder depth. To reduce the adder depth it might be required to use more number of subexpressions or change those selected. Hence, there is an obvious trade-off between the number of adders and the adder depth, although it is not clear where on this trade-off curve the "best" position is. In References [48,49], it is suggested that each subexpression should have as low adder depth as possible (as defined in Reference [30]) to obtain a low-power realization.

2.6 Applications

Several applications of constant multiplication methods have been discussed in the literature in the last three decades, most of which are for some computation relating to DSP algorithms applied in diverse areas of technology. We shall, however, focus our discussion on a few of the key contributions. These described techniques can however be used for matrix multiplications and inner-product; and can be applied easily to image/video processing and graphics applications.

2.6.1 Implementation of FIR Digital Filters and Filter Banks

The implementation of FIR filter is almost the defacto standard for the comparison of efficacy of MCM techniques. Most of the applications of constant multiplication methods on FIR filters relate to transposed direct form implementation. The technique of Reference [57] can be used for the direct form as well as the transposed direct form FIR filter, which exploits the common computations of all the multiplications. Simultaneous optimization of delay as well as area is discussed in Reference [16]. The use of MCM techniques for low-complexity design of broadband systems are found in an early paper [59] and in a more recent paper [63]. The use of MCM in polyphase FIR filters and filter banks is discussed in Reference [37]. The FIR filter design method for the shift-and-add realization of software defined radio (SDR) receiver with reduced system delay is presented in Reference [97]. Low-power realization of FIR filters based on differential-constant MCM is discussed in References [74,78,79,98], and power-efficient implementation of linear-phase FIR filter based on vertical CSE is suggested in Reference [62]. The area optimization of FIR filters using gate-level metrics is presented in References [99,100], which is claimed to be more relevant than the popularly used LO and LD metrics. Power consumption in high-order FIR filters is addressed in Reference [69]. A scheme for optimization of structural adders in FIR filters is proposed in Reference [101]. Reconfigurability along with low-complexity in the implementation of FIR filters are two mutually incompatible demands of multistandard wireless communication and SDR systems. To handle this issue, a low-complexity design of reconfigurable FIR filter is presented in a recent paper [86], where the authors have proposed the constant-shift method and programmable-shift method for reducing the complexity of multiplications. Implementation of filter banks can also be optimized by MCM techniques [102]. It is shown in Reference [63] that in the filter bank channelizer, which extracts multiple narrowband channels from the wideband signal, the number of LOs to implement the multiplications can be reduced considerably by forming three and four nonzero-bit super-subexpressions. For the filters used in the digital advanced mobile phone system (D-AMPS) and the personal digital cellular (PDC) channelizers, it is shown that the methods of

Reference [63] offer considerable reduction in the number of full adders when compared to conventional CSE methods. A reconfigurable frequency response masking filter of software defined radio (SDR) channelizer is proposed in Reference [103], where binary subexpression elimination has been used to achieve significant reduction of area and power consumption.

2.6.2 Implementation of Sinusoidal and Other Linear Transforms

The MCM optimizations of discrete orthogonal transforms such as FFT and DCT are also presented in Reference [11]. Various constant multiplication schemes have been used for implementation of linear transforms and matrix multiplication [59,104–106]. Low-power and high-speed implementation of DCT based on efficient sharing of common subexpressions in the CSD representation of elements of DCT matrix has been presented in Reference [107]. An advanced CSE method that uses the binary representation of the elements of DCT matrix has been presented in Reference [108]. The algorithm in Reference [108] chooses the maximum number of frequently occurring subexpressions to eliminate redundant computations in the DCT matrix and hence reduces the number of adders required to implement the multiplications. Design example of an 8×8 DCT using 16-bit input shows that the method in Reference [108] offers hardware reduction of 22% over the method in Reference [107]. Reference [109] proposes an efficient low power, high performance VLSI architecture for 2D-discrete wavelet transform (DWT) employing CSE technique. Besides optimizing the speed of transformation and reducing the area for hardware implementation, the architecture in Reference [109] also significantly reduces the power consumption in the DWT computation. The application of the MCM for the optimization of computation of Hadamard transform has been illustrated in Reference [11].

2.6.3 Other Applications

An independent system-level study of shift-and-add implementation of constant multiplications for power and hardware optimization was performed on several linear systems (controller in a small guidance system, elliptic wave filter, motor control systems, etc.) in Reference [110]. The numerically transformed systems were abstracted by a high-level DSP synthesis tool to perform flow-graph transformations involving retiming, distributivity, associativity, and common subexpression elimination. The operations were reduced by subexpression sharing in a shift-and-add decomposition procedure before hardware allocation and scheduling were performed to bind the operations onto hardware modules from a low-power library. Gate-level synthesis for minimum hardware implementation results indicated that the reduction in operations by subexpression elimination led to shorter schedule. The amount of capacitance switched and the energy at 5 V of the optimized system were reduced significantly to 78–37% of the original system without optimization of constant multiplications. The schedule slacks were exploited for voltage scaling to achieve further energy savings of 60–95%.

Another application of the MCM relates to the calculation of multiple polynomials over the same variable. This application has been illustrated in Reference [11] with the example of G2 blend surface calculation using quintic polynomials, which is an important and frequently used graphics task in mechanical CAD applications. CSE algorithm has been used for low-cost shift-and-add implementation of multiplication by constant matrices [34].

Boullis and Tisserand [35] have extended Lefevre method of pattern search to constant-vector multiplication and matrix-vector multiplications [20]. These techniques being very general in nature can be used for the computation of any discrete orthogonal transform, and FIR filter-based rational sampling-rate conversion [111]. Efficient use of MCM techniques for realization of Farrow structure for various applications are suggested in References [102,112,113]. Several interesting application scenarios have been identified in Reference [85], where the use of reconfigurable constant multiplier can lower the hardware cost significantly at the expense of little or no penalty to the overall system performance.

2.7 Pitfalls and Scope for Future Work

In this section, we discuss some of the shortcomings in the current research on constant multiplications, and possible scopes of improvement.

2.7.1 Selection of Figure of Merit

The combinational component of hardware increases with the LO count, while the computation-time of the multiplications increases linearly with the LD. The LO count is used as an indirect measure of area, but it is not a reliable indicator of area, because the adders/subtractors could be of widely different widths [99,100]. Assume the fundamental ripple-carry adders are used, the full-adder (FA) count [100] (given by $FA = \Sigma_{i=1}^{LO} FA_i$, where FA_i is the number of full adders used in the ith adder) is a better figure of merit and could be considered as an approximate measure of hardware cost. To account for the adder structure and storage/delay registers [70,99,100,114], the total gate count of the shift-add-network should instead be considered as a better measure of area-complexity but such fine-grained cost function is difficult to incorporate into high-level optimization algorithm.

Along with the increasing transistor density over the years, the minimization of power dissipation becomes more important than the area minimization. Based on these considerations, the switching power estimates like glitch path count [115], glitch path score [116] and power-index [117] have been developed. Most importantly, the dynamic power consumptions resulting from the switching activities involved in the shift-and-add process need to be reflected [118,119]. The power dissipation in a multiplication block on the other hand is closely related to the area of the circuits as well as the LD, since the switching activities increases with the number of adder steps and width of the adders. The total power dissipation, area consumption, and computation time are closely related, and cannot be optimized independently. Furthermore, the interconnect costs in MCM problems were not yet considered. A more holistic figure of merit based on all these metrics needs to be evaluated.

2.7.2 Benchmark Suites for Algorithm Evaluation

The design of benchmark suites for the competitive evaluation of algorithms is also an important issue. While a design can always establish itself for a given target application, the merit of the design could be more widely acknowledged if it is evaluated for a set of benchmarks. Only a few filters are commonly used for comparison of MCM algorithms, while in most of the cases a randomly generated sets of constants or an arbitrarily selected set of filters are used for comparison. While the evaluation of a design strategy for a given filter specification does not guarantee the merit of the approach, the random constants in most cases, deviate significantly from the actual distribution of the constant coefficients of practical relevance. To overcome the problem of poor repeatability, reliability, and acceptability of experimental results, FIRsuite: a Suite of Constant Coefficient FIR Filters [120] has been developed by M. Faust and C. H. Chang[5].

2.7.3 FPGA-Oriented Design of Algorithms and Architectures

Field programmable gate array (FPGA) has gained tremendous popularity as a hardware platform for the implementation of specialized applications with small or medium volumes of production, not only due to its reconfigurability and reusability but also for several other reasons like short time to market, attractive pricing, and widely available low-cost (or even free) software tools for design entry, synthesis, and implementation. The solutions generated by

---

5 This Web site lists coefficient values obtained from the literature and those provided by the authors annotated with information about the filter specification, the number of nonzero bits in binary and CSD representations as well as the results from some MCM optimization algorithms.

the surveyed constant and MCM optimization methodologies can be mapped directly onto FPGA, but FPGA devices have certain typical characteristics in terms of their logic and interconnect resources, and their way of implementation of combinational functions. They are built around sliced architecture, consisting of a large array of reconfigurable logic blocks (CLB) and a hierarchy of reconfigurable interconnects that allow the CLBs to be connected together in an optimal way to perform the necessary computations for specific applications. Usually, the logic resources of these devices are limited to LUTs, D flip-flops, and multiplexers embedded in the CLBs. Keeping these behaviors of FPGA devices in view, algorithms and architectures should be tailored for efficient utilization of FPGA resources [121]. Nevertheless, we find only a few work on FPGA implementation of shift-and-add implementation of constant multiplications [81–83,87–89,115,122]. The minimization of multiplier block logic depth and pipeline registers is shown to have the great effect in reducing FPGA area [123]. FPGA-specific optimization of MCM circuits is proposed in [124] to achieve higher performance with less resource usage. They have utilized fast carry-chains and 3-input adders along with the integration of LUT-based constant multipliers and embedded multipliers to achieve high performance in modern FPGAs. Looking at the popularity of FPGA devices for DSP and communication applications, research efforts need to be directed toward development of algorithms and optimized mapping of computation onto the FPGA fabric.

2.8   Conclusions

The optimization of shift-and-add network for constant multiplications is found to have great potential for reducing the area, delay, and power consumption of implementation of multiplications in several computation-intensive applications not only in dedicated hardware but also in programmable computing systems. Several techniques have been suggested in the last three decades for near optimal solutions of this problem, which could be placed under two broad classes on the basis of their optimization approach. One of them, called the CSE approach, searches for common pattern in specific representation of numbers with a constrained search space; while the other, called the adder graph approach, is independent of representation of numbers and opens up an enormous search space. The complexity of search increases with the size of the search space. The adder graph algorithms, in general, result in lower adder count but higher logic depth and higher power consumption compared with the CSE approach. In the existing literature, we can see that most of the algorithms aim at minimizing only one of the cost metrics. But, all the three metrics, namely, the area, delay, and power consumption are interdependent and all of them have their own importance in different applications and deployment environment. The future algorithms should, therefore, aim at more relevant and holistic figure of merit based on demand-driven trade-off across all these metrics. Moreover, the design solutions should either be optimized for specific applications or otherwise should be compared against suitable benchmarks for more general usage. Since the interconnect delay is going to be more prominent than the gate delay along with the progressive scaling of silicon devices, the constant multiplication algorithms need to give more emphasis on the minimization of interconnect length in the future. The role of CAD tools for synthesizing the designs cannot be undermined. Gate-level implementation cost of adders and subtractors influenced by the advanced low-level optimization capability of the CAD tools should be taken into consideration. FPGA is increasing in popularity and is used in combination with ASIC as well as general purpose microprocessors, therefore platforms and application specific solutions for realization of SCM and MCM on FPGA are needed.

References

1 J. G. Proakis and D. G. Manolakis, *Digital Signal Processing: Principles, Algorithms and Applications*, 4th ed., Pearson Prentice Hall, Upper Saddle River, NJ, 2007.

2 G. W. Reitweisner, Binary arithmetic, *Adv. Comput.*, vol. 1, 1960, pp. 232–308.

3 H. L. Garner, L. A. Franz, and R. Morris, Number systems and arithmetic, in *Advances in Computers*. Academic Press, New York, vol. 6, 1965, pp. 131–194.

4 K. Hwang, *Computer Arithmetic: Principles, Architecture and Design*. John Wiley & Sons, Inc., 1979.

5 A. Avizienis, Signed-digit number representation for fast parallel arithmetic, *IRE Trans. Electron. Comput.*, vol. 10, 1961, pp. 389–400.

6 S. Arno and F. S. Wheeler, Signed digit representations of minimal Hamming weight, *IEEE Trans. Comput.*, vol. 42, no. 8, 1993 pp. 1007–1010.

7 I.-C. Park and H.-J. Kang, Digital filter synthesis based on an algorithm to generate all minimal signed digit representations, *IEEE Trans. Comput. Aided Des. Integr. Circuits Syst.*, vol. 21, no. 12, 2002, pp. 1525–1529.

8 A. G. Dempster and M. D. Macleod, Generation of signed-digit representations for integer multiplication, *IEEE Signal Process. Lett.*, vol. 11, no. 8, 2004, pp. 663–665.

9 K. Nakayama, Permuted difference coefficient realization of FIR digital filters, *IEEE Trans. Acoust.*, vol. 30, no. 2, 1982, pp. 269–278.

10 L. Guerra, M. Potkonjak, and J. Rabaey, Concurrency characteristics in DSP programs, in *Proc. IEEE Int. Conf. Acoustics, Speech, and Signal Processing, (ICASSP-94)*, vol. II, Adelaide, SA, April 19–22, 1994, pp. 433–436.

11 M. Potkonjak, M. B. Srivastava, and A. P. Chandrakasan, Multiple constant multiplications: efficient and versatile framework and algorithms for exploring

common subexpression elimination, *IEEE Trans. Comput. Aided Des. Integr. Circuits Syst.*, vol. 15, no. 2, 1996, pp. 151–165.

12 P. R. Cappello and K. Steiglitz, Some complexity issues in digital signal processing, *IEEE Trans. Acoust.*, vol. 32, no. 5, 1984, pp. 1037–1041.

13 A. Matsuura and A. Nagoya, Formulation of the addition-shift-sequence problem and its complexity, in *Algorithms and Computation*, Lecture Notes in Computer Science, (eds H. Leong, H. Imai, and S. Jain), vol. 1350, Springer, Berlin, 1997, pp. 42–51.

14 K. D. Cooper, L. T. Simpson, and C. A. Vick, Operator strength reduction, *ACM Trans. Program. Lang. Syst.*, vol. 23, 2001, pp. 603–625.

15 R. I. Hartley, Subexpression sharing in filters using canonic signed digit multipliers, *IEEE Trans. Circuits Syst. II, Analog Digit. Signal Process.*, vol. 43, no. 10, 1996, pp. 677–688.

16 H.-J. Kang and I.-C. Park, FIR filter synthesis algorithms for minimizing the delay and the number of adders, *IEEE Trans. Circuits Syst. II, Analog Digit. Signal Process.*, vol. 48, no. 8, 2001, pp. 770–777.

17 Y. C. Lim, J. B. Evans, and B. Liu, Decomposition of binary integers into signed power-of-two terms, *IEEE Trans. Circuits Syst.*, vol. 38, no. 6, 1991, pp. 667–672.

18 M. Faust, O. Gustafsson, and C. H. Chang, Fast and VLSI efficient binary-to-CSD encoder using bypass signal, *Electron. Lett.*, vol. 47, no. 1, 2011, pp. 18–20.

19 R. Bernstein, Multiplication by integer constants, *Softw. Pract. Exp.*, vol. 16, no. 7, 1986, pp. 641–652.

20 V. Lefevre, Multiplication by an Integer Constant, Institut national de recherche en informatique et automatique (INRIA), Rapport de recherche RR-4192, May 2001. Available at http://www.vinc17.org/research/papers/rr_mulcst2.pdf.

21 A. G. Dempster and M. D. Macleod, Using all signed-digit representations to design single integer multipliers using subexpression elimination, in *Proc. IEEE Int. Symp. Circuits Syst. (ISCAS 2004)*, vol. 3, Vancouver, Canada, May 23–26, 2004, pp. 165–168.

22 O. Gustafsson, A. G. Dempster, K. Johansson, M. D. Macleod, and L. Wanhammar, Simplified design of constant coefficient multipliers, *Circuits, Syst. Signal Process.*, vol. 25, no. 2, 2006, pp. 225–251.

23 J. Thong and N. Nicolici, Time-efficient single constant multiplication based on overlapping digit patterns, *IEEE Trans. VLSI Syst.*, vol. 17, no. 9, 2009, pp. 1353–1357.

24 A. G. Dempster and M. D. Macleod, Constant integer multiplication using minimum adders, *IEE Proc. Circuits, Devices Syst.*, vol. 141, no. 5, 1994, pp. 407–413.

25 D. R. Bull and D. H. Horrocks, Primitive operator digital filters, *IEE Proc. Circuits, Devices Syst.*, vol. 138, no. 3, 1991, pp. 401–412.

26 O. Gustafsson, A. G. Dempster, and L. Wanhammar, Extended results for minimum-adder constant integer multipliers, in *Proc. IEEE Int. Symp. Circuits Syst. (ISCAS 2002)*, vol. 1, Scottsdale, AZ, May 26–29, 2002, pp. 73–76.

27 A. G. Dempster and M. D. Macleod, Use of minimum-adder multiplier blocks in FIR digital filters, *IEEE Trans. Circuits Syst. II, Analog Digit. Signal Process.*, vol. 42, no. 9, 1995, pp. 569–577.

28 A. G. Dempster and M. D. Macleod, General algorithms for reduced-adder integer multiplier design, *Electron. Lett.*, vol. 31, no. 21, 1995, pp. 1800–1802.

29 D. Li, Minimum number of adders for implementing a multiplier and its application to the design of multiplierless digital filters, *IEEE Trans. Circuits Syst. II, Analog Digit. Signal Process.*, vol. 42, no. 7, 1995, pp. 453–460.

30 O. Gustafsson, Lower bounds for constant multiplication problems, *IEEE Trans. Circuits Syst. II, Express Briefs*, vol. 54, no. 11, 2007, pp. 974–978.

31 R. Pinch, Asymptotic upper bound for multiplier design, *Electron. Lett.*, vol. 32, no. 5, 1996, pp. 420–421.

32 J. Thong and N. Nicolici, A novel optimal single constant multiplication algorithm, in *Proc. 47th ACM/IEEE Design Automation Conf. (DAC)*, Anaheim, CA, June 13–18, 2010, pp. 613–616.

33 A. Chatterjee, R. K. Roy, and M. A. d'Abreu, Greedy hardware optimization for linear digital circuits using number splitting and refactorization, *IEEE Trans. VLSI Syst.*, vol. 1, no. 4, 1993, pp. 423–431.

34 M. D. Macleod and A. G. Dempster, Common subexpression elimination algorithm for low-cost multiplierless implementation of matrix multipliers, *Electron. Lett.*, vol. 40, no. 11, 2004, pp. 651–652.

35 N. Boullis and A. Tisserand, Some optimizations of hardware multiplication by constant matrices, *IEEE Trans. Comput.*, vol. 54, no. 10, 2005, pp. 1271–1282.

36 H. Safiri, M. Ahmadi, G. A. Jullien, and W. C. Miller, A new algorithm for the elimination of common subexpressions in hardware implementation of digital filters by using genetic programming, *J. VLSI Signal Process.*, vol. 31, no. 2, 2002, pp. 91–100.

37 O. Gustafsson and A. G. Dempster, On the use of multiple constant multiplication in polyphase FIR filters and filter banks, in *Proc. Nordic Signal Processing Symp. (NORSIG 2004)*, Espoo, Finland, June 9–11, 2004, pp. 53–56.

38 H. Samueli, An improved search algorithm for the design of multiplierless FIR filters with powers-of-two coefficients, *IEEE Trans. Circuits Syst.*, vol. 36, no. 7, 1989, pp. 1044–1047.

39 F. Xu, C. H. Chang, and C. C. Jong, Modified reduced adder graph algorithm for multiplierless FIR filters, *Electron. Lett.*, vol. 41, no. 6, 2005, pp. 302–303.

40 J.-H. Han and I.-C. Park, FIR filter synthesis considering multiple adder graphs for a coefficient, *IEEE Trans. Comput. Aided Des. Integr. Circuits Syst.*, vol. 27, no. 5, 2008, pp. 958–962.

41 Y. Voronenko and M. Puschel, Multiplierless multiple constant multiplication, *ACM Trans. Algorithms*, vol. 3, no. 2, 2007, p. 11.

42 L. Aksoy, E. O. Gunes, and P. Flores, An exact breadth-first search algorithm for the multiple constant multiplications problem, in *Proc. IEEE NORCHIP Conf*, Tallinn, Estonia, November 16–17, 2008, pp. 41–46.

43 Spiral: Software/Hardware Generation for DSP Algorithms Multiplierless Constant Multiplication, Technical Report, 2011. Available at http://www.spiral.net/hardware/multless.html.

44 A. G. Dempster, S. S. Dimirsoy, and I. Kale, Designing multiplier blocks with low logic depth, in *Proc. IEEE Int. Symp. Circuits Syst. (ISCAS 2002)*, vol. 5, Scottsdale, Arizona, May 26–29, 2002, pp. 773–776.

45 O. Gustafsson, H. Ohlsson, and L. Wanhammar, Minimum-adder integer multipliers using carry-save adders, in *Proc. IEEE Int. Symp. Circuits Syst. (ISCAS 2001)*, vol. 2, Sydney, NSW, May 6–9, 2001, pp. 709–712.

46 O. Gustafsson, A. G. Dempster, and L. Wanhammar, Multiplier blocks using carry-save adders, in *Proc. IEEE Int. Symp. Circuits Syst, (ISCAS 2004)*, vol. 2, Vancouver, Canada, May 23–26, 2004, pp. 473–476.

47 O. Gustafsson, Comments on 'A 70 MHz multiplierless FIR Hilbert transformer in 0.35 μm standard CMOS library', *IEICE Trans. Fundamentals Electron. Commun. Comput. Sci.*, vol. E91-A, no. 3, 2008, pp. 899–900.

48 K. Johansson, Low power and low complexity shift-and-add based computations Ph.D. dissertation, Linkoping Studies in Science and Technology, Linkoping University, 2008.

49 K. Johansson, O. Gustafsson, L. S. DeBrunner, and L. Wanhammar, Minimum adder depth multiple constant multiplication for low power FIR filters, May 15–18, 2011.

50 M. Faust and C. H. Chang, Minimal logic depth adder tree optimization for multiple constant multiplication, in *Proc. IEEE Int. Symp. Circuits Syst. (ISCAS 2010)*, Paris, France, May 2010, pp. 457–460.

51 O. Gustafsson, K. Khursheed, M. Imran, and L. Wanhammar, Generalized overlapping digit patterns for multi-dimensional sub-expression sharing, in *Proc. Int. Conf. Green Circuits Syst. (ICGCS)*, Shanghai, China, June 21–23, 2010, pp. 65–68.

52 R. Mahesh and A. P. Vinod, A new common subexpression elimination algorithm for realizing low-complexity higher order digital filters, *IEEE Trans. Comput. Aided Des. Integr. Circuits Syst.*, vol. 27, no. 2, 2008, pp. 217–229.

53 C. H. Chang and M. Faust, On a new common subexpression elimination algorithm for realizing low-complexity higher order digital filters, *IEEE Trans. Comput. Aided Des. Integr. Circuits Syst.*, vol. 29, no. 5, 2010, pp. 844–848.

54 A. V. Aho, S. C. Johnson, and J. D. Ullman, Code generation for expressions with common subexpressions, *J. ACM*, vol. 24, no. 1, 1977, pp. 146–160.

55 R. Hartley, Optimization of canonic signed digit multipliers for filter design, in *Proc. IEEE Int. Symp. Circuits Syst. (ISCAS 1991)*, vol. 4, Singapore, June 11–14, 1991, pp. 1992–1995.

56 Z. Iqbal, M. Potkonjak, S. Dey, and A. Parker, Critical path minimization using retiming and algebraic speed-up, in *Proc. 30th ACM/IEEE Design Automation Conf. (DAC '93)*, Dallas, TX, june 14–18, 1993, pp. 573–577.

57 M. Mehendale, S. D. Sherlekar, and G. Venkatesh, Synthesis of multiplier-less FIR filters with minimum number of additions, in *Proc. IEEE/ACM Int. Conf. Comput. Aided Design, (ICCAD-2005)*, San Jose, CA, November 5–9, 1995, pp. 668–671.

58 R. Pasko, P. Schaumont, V. Derudder, and D. Durackova, Optimization method for broadband modem FIR filter design using common subexpression elimination, in *Proc. 10th Int. Symp. Syst. Synthesis*, Antwerp, Belgium, September 17–19, 1997, pp. 100–106.

59 R. Pasko, P. Schaumont, V. Derudder, S. Vernalde, and D. Durackova, A new algorithm for elimination of common subexpressions, *IEEE Trans. Comput. Aided Des. Integr. Circuits Syst.*, vol. 18, no. 1, 1999, pp. 58–68.

60 M. Martnez-Peiro, E. I. Boemo, and L. Wanhammar, Design of high-speed multiplierless filters using a nonrecursive signed common subexpression algorithm, *IEEE Trans. Circuits Syst. II, Analog Digit. Signal Process.*, vol. 49, no. 3, 2002, pp. 196–203.

61 C.-Y. Yao, H.-H. Chen, T.-F. Lin, C.-J. Chien, and C.-T. Hsu, A novel common-subexpression-elimination method for synthesizing fixed-point FIR filters, *IEEE Trans. Circuits Syst. I Regul. Pap.*, vol. 51, no. 11, 2004, pp. 2215–2221.

62 Y. Jang and S. Yang, Low-power CSD linear phase FIR filter structure using vertical common sub-expression, *Electron. Lett.*, vol. 38, no. 15, 2002, pp. 777–779.

63 A. P. Vinod and E. M. K. Lai, On the implementation of efficient channel filters for wideband receivers by optimizing common subexpression elimination methods, *IEEE Trans. Comput. Aided Des. Integr. Circuits Syst.*, vol. 24, no. 2, 2005, pp. 295–304.

64 F. Xu, C. H. Chang, and C. C. Jong, Contention resolution: A new approach to versatile subexpressions sharing in multiple constant multiplications, *IEEE Trans. Circuits Syst. I Regul. Pap.*, vol. 55, no. 2, 2008, pp. 559–571.

65 A. P. Vinod, E. M. K. Lai, A. B. Premkuntar, and C. T. Lau, FIR filter implementation by efficient sharing of horizontal and vertical common subexpressions, *Electron. Lett.*, vol. 39, no. 2, 2003, pp. 251–253.

66 A. P. Vinod and E. M.-K. Lai, Comparison of the horizontal and the vertical common subexpression elimination methods for realizing digital filters, in *Proc. IEEE Int. Symp. Circuits Syst. (ISCAS)*, Kobe, Japan, May 21–24, 2005, pp. 496–499.

67 P. Flores, J. Monteiro, and E. Costa, An exact algorithm for the maximal sharing of partial terms in multiple constant multiplications, in *Proc. IEEE/ACM Int. Conf. Comput. Aided Design, (ICCAD 2005)*, San Jose, CA, November 6–10, 2005, pp. 13–16.

68 L. Aksoy, E. da Costa, P. Flores, and J. Monteiro, Exact and approximate algorithms for the optimization of area and delay in multiple constant multiplications, *IEEE Trans. Comput. Aided Des. Integr. Circuits Syst.*, vol. 27, no. 6, 2008, pp. 1013–1026.

69 K. G. Smitha and A. P. Vinod, Low power realization and synthesis of higher-order FIR filters using an improved common subexpression elimination method, *IEICE Trans. Fundam. Electron. Commun. Comput. Sci.*, vol. E91-A, no. 11, 2008, pp. 3282–3292.

70 C. H. Chang, J. Chen, and A. P. Vinod, Information theoretic approach to complexity reduction of FIR filter design, *IEEE Trans. Circuits Syst. I, Regul. Pap.*, vol. 55, no. 8, 2008, pp. 2310–2321.

71 F. Xu, C. H. Chang, and C. C. Jong, Contention resolution algorithm for common subexpression elimination in digital filter design, *IEEE Trans. Circuits Syst. II, Express Briefs*, vol. 52, no. 10, 2005, pp. 695–700.

72 A. G. Dempster, M. D. Macleod, and O. Gustafsson, Comparison of graphical and subexpression methods for design of efficient multipliers, in *Proc. 38th Asilomar Conf. Signals Syst. Comp.*, vol. 1, Pacific Grove, CA, November 7–10, 2004, pp. 72–76.

73 M. D. Macleod and A. G. Dempster, Multiplierless FIR filter design algorithms, *IEEE Signal Process. Lett.*, vol. 12, no. 3, 2005, pp. 186–189.

74 H. Choo, K. Muhammad, and K. Roy, Complexity reduction of digital filters using shift inclusive differential coefficients, *IEEE Trans. Signal Process.*, vol. 52, no. 6, 2004, pp. 1760–1772.

75 O. Gustafsson, H. Ohlsson, and L. Wanhammar, Improved multiple constant multiplication using a minimum spanning tree, in *Proc. 38th Asilomar Conf. Signals Syst. Comp.*, vol. 1, Pacific Grove, CA, November 7–10, 2004, pp. 63–66.

76 O. Gustafsson, A difference based adder graph heuristic for multiple constant multiplication problems, in *Proc. IEEE Int. Symp. Circuits Syst. (ISCAS 2007)*, New Orleans, LA, May 27–30, 2007, pp. 1097–1100.

77 K. Muhammad and K. Roy, A graph theoretic approach for synthesizing very low-complexity high-speed digital filters, *IEEE Trans. Comput. Aided Des. Integr. Circuits Syst.*, vol. 21, no. 2, 2002, pp. 204–216.

78 N. Sankarayya, K. Roy, and D. Bhattacharya, Algorithms for low power and high speed FIR filter realization using differential coefficients, *IEEE Trans. Circuits Syst. II, Analog Digit. Signal Process.*, vol. 44, no. 6, 1997, pp. 488–497.

79 A. P. Vinod, A. Singla, and C. H. Chang, Low-power differential coefficients-based FIR filters using hardware-optimised multipliers, *IET Circuits, Devices Syst.*, vol. 1, no. 1, 2007, pp. 13–20.

80 Y. Wang and K. Roy, CSDC: a new complexity reduction technique for multiplierless implementation of digital FIR filters, *IEEE Trans. Circuits Syst. I Regul. Pap.*, vol. 52, no. 9, 2005, pp. 1845–1853.

81 S. S. Demirsoy, A. G. Dempster, and I. Kale, Design guidelines for reconfigurable multiplier blocks, in *Proc. IEEE Int. Symp. on Circuits Syst. (ISCAS 2003)*, vol. 4, Bangkok, Thailand, May 25–28, 2003, pp. 293–296.

82 S. S. Demirsoy, I. Kale, and A. G. Dempster, Efficient implementation of digital filters using novel reconfigurable multiplier blocks, in *Proc. 38th Asilomar Conf. Signals Syst. Comp.*, vol. 1, Pacific Grove, CA, November 7–10, 2004, pp. 461–464.

83 S. Demirsoy, I. Kale, and A. G. Dempster, Reconfigurable multiplier blocks: structures, algorithm and applications, *Circuits, Syst. Signal Process.*, vol. 26, no. 6, 2007, pp. 793–827.

84 P. Tummeltshammer, J. C. Hoe, and M. Puschel, Time-multiplexed multiple-constant multiplication, *IEEE Trans. Comput. Aided Des. Integr. Circuits Syst.*, vol. 26, no. 9, 2007, pp. 1551–1563.

85 J. Chen and C. H. Chang, High-level synthesis algorithm for the design of reconfigurable constant multiplier, *IEEE Trans. Comput.-Aided Des. Integr. Circuits Syst.*, vol. 28, no. 12, 2009, pp. 1844–1856.

86 R. Mahesh and A. P. Vinod, New reconfigurable architectures for implementing FIR filters with low complexity, *IEEE Trans. Comput. Aided Des. Integr. Circuits Syst.*, vol. 29, no. 2, 2010, pp. 275–288.

87 R. Gutierrez, J. Valls, and A. Perez-Pascual, FPGA-implementation of time-multiplexed multiple constant multiplication based on carry-save arithmetic, in *Proc. Int. Conf. Field Programmable Logic Appl. (FPL 2009)*, Prague, Czech Republic, August 31, 2009, pp. 609–612.

88 N. Sidahao, G. Constantinides, and P. Cheung, Power and area optimization for multiple restricted multiplication, in *Proc. IEEE Int. Conf. Field Programmable Logic Appl.* Tampere, Finland, August 24–26, 2005, pp. 112–117.

89 C. Howard and L. DeBrunner, High speed DSP block for FPGA devices using a programmable adder graph, in *Proc. 13th IEEE Digital Signal Processing Workshop and 5th IEEE Signal Processing Education Workshop (DSP/SPE 2009)*, Marco Island, FL, January 4–7, 2009, pp. 490–494.

90 A. Yurdakul and G. Dundar, Multiplierless realization of linear DSP transforms by using common two-term expressions, *J. VLSI Signal Process.*, vol. 22, no. 3, 1999, pp. 163–172.

91 G. L. Nemhauser and L. A. Wolsey, *Integer and Combinatorial Optimization*, Wiley-Interscience, New York, NY, 1999.

92 E. Boros and P. L. Hammer, Pseudo-boolean optimization, *Discrete Appl. Math.*, vol. 123, no. 1–3, 2002, pp. 155–225.

93 O. Gustafsson and L. Wanhammar, Ilp modelling of the common subexpression sharing problem, in *Proc. IEEE Int. Conf. Elec. Circuits Syst. (ICECS 2002)*, vol. 3, Dubrovnik, Croatia, December 10, 2002, pp. 1171–1174.

94 O. Gustafsson, Towards optimal multiple constant multiplication: a hypergraph approach, in *Proc. 42nd Asilomar Conf. Signals Syst. Comp.*, Pacific Grove, CA, October 26–29, 2008, pp. 1805–1809.

95 M. Desrochers and G. Laporte, Improvements and extensions to the miller-tucker-zemlin subtour elimination constraints, *Operations Res. Lett.*, vol. 10, no. 1, 1991, pp. 27–36.

96 L. Aksoy, E. O. Gunes, and P. Flores, Search algorithms for the multiple constant multiplications problem: exact and approximate, *Microprocess. Microsyst.*, vol. 34, no. 5, 2010, pp. 151–162.

97 K. S. Yeung and S. C. Chan, The design and multiplier-less realization of software radio receivers with reduced system delay, *IEEE Trans. Circuits Syst. I Regul. Pap.*, vol. 51, no. 12, 2004, pp. 2444–2459.

98 A. P. Vinod, A. Singla, and C. H. Chang, Improved differential coefficients-based low power FIR filters. Part I. fundamentals, in *Proc. IEEE Int. Symp. Circuits Syst. (ISCAS)*, Island of Kos, Greece, May 21–24, 2006, pp. 617–620.

99 K. Johansson, O. Gustafsson, and L. Wanhammar, A detailed complexity model for multiple constant multiplication and an algorithm to minimize the complexity, in *Proc. 2005 European Conf. Circuit Theory Design*, vol. 3, Cork, Ireland, August 28, 2005, pp. 465–468.

100 L. Aksoy, E. Costa, P. Flores, and J. Monteiro, Optimization of area in digital FIR filters using gate-level metrics, in *Proc. 44th ACM/IEEE Design Automation Conf. (DAC '07)*, San Diego, CA, June 4–8, 2007, pp. 420–423.

101 M. Faust and C. H. Chang, Optimization of structural adders in fixed coefficient transposed direct form FIR filters, in *Proc. IEEE Int. Symp. Circuits Syst. (ISCAS 2009)*, Taipei, Taiwan, May 24–27, 2009, pp. 2185–2188.

102 A. G. Dempster and N. P. Murphy, Efficient interpolators and filter banks using multiplier blocks, *IEEE Trans. Signal Processing*, vol. 48, no. 1, 2000, pp. 257–261.

103 R. Mahesh and A. P. Vinod, Reconfigurable frequency response masking filters for software radio channelization, *IEEE Trans. Circuits Syst. II, Express Briefs*, vol. 55, no. 3, 2008, pp. 274–278.

104 M. D. Macleod, Multiplierless implementation of rotators and FFTs, *EURASIP J. Appl. Signal Process.*, vol. 2005, no. 17, 2005, pp. 2903–2910.

105 W. Han, A. T. Erdogan, T. Arslan, and M. Hasan, High-performance low-power FFT cores, *ETRI J.*, vol. 30, no. 3, 2008, pp. 451–460.

106 U. Meyer-Base, H. Natarajan, and A. G. Dempster, Fast discrete Fourier transform computations using the reduced adder graph technique, *EURASIP J. Adv. Signal Process.*, vol. 2007, no. 67360, 2007, p. 8.

107 A. P. Vinod and E. M. K. Lai, Hardware efficient DCT implementation for portable multimedia terminals using subexpression sharing, in *Proc. IEEE Region 10 Conf.*, vol. 1, Chiang Mai, Thailand, November 21–24, 2004, pp. 227–230.

108 S. Vijay and A. P. Vinod, A new algorithm to implement low complexity DCT for portable multimedia devices, in *Proc. Int. Conf. Signal Process. Commun. Syst.*, Gold Coast, Australia, December 2007, pp. 171–176.

109 P. P. Dang and P. M. Chau, A high performance, low power VLSI design of discrete wavelet transform for lossless compression in JPEG 2000 standard, in *Proc. Int. Conf. Consum. Electron. (ICCE 2001)*, Los Angeles, CA, June 19–21, 2001, pp. 126–127.

110 H. Nguyen and A. Chattejee, Number-splitting with shift-and-add decomposition for power and hardware optimization in linear DSP synthesis, *IEEE Trans. Very Large Scale Integration (VLSI) Syst.*, vol. 8, no. 4, 2000, pp. 419–424.

111 O. Gustafsson and H. Johansson, Efficient implementation of FIR filter based rational sampling rate converters using constant matrix multiplication, in *Proc. 40th Asilomar Conf. Signals, Syst. Comput.*, Pacific Grove, CA, October 29, 2006.

112 C. K. S. Pun, Y. C. Wu, S. C. Chan, and K. L. Ho, On the design and efficient implementation of the Farrow structure, *IEEE Signal Process. Lett.*, vol. 10, no. 7, 2003, pp. 1680–1685.

113 H. Johansson, O. Gustafsson, J. Johansson, and L. Wanhammar, Adjustable fractional-delay FIR filters using the Farrow structure and multirate techniques, in *Proc. IEEE Asia Pacific Conf. Circuits Systems (APCCAS)*, Singapore, December 4–7, 2006.

114 A. P. Vinod and E. M. K. Lai, An efficient coefficient-partitioning algorithm for realizing low-complexity digital filters, *IEEE Trans. Comput. Aided Des. Integr. Circuits Syst.*, vol. 24, no. 12, 2005, pp. 1936–1946.

115 S. S. Demirsoy, A. Dempster, and I. Kale, Transition analysis on FPGA for multiplier-block based FIR filter structures, in *Proc. IEEE Int. Conf. Elec. Circuits Syst. (ICECS 2000)*, vol. 2, Jounieh, Lebanon, December 17–20, 2000, pp. 862–865.

116 S. S. Demirsoy, A. G. Dempster, and I. Kale, Power analysis of multiplier blocks, in *Proc. IEEE Int. Symp. Circuits Syst. (ISCAS 2002)*, vol. 1, Scottsdale, Arizona, May 26–29, 2002, pp. 297–300.

117 J. Chen, C. H. Chang, and H. Qian, New power index model for switching power analysis from adder graph of FIR filter, in *Proc. IEEE Int. Symp. Circuits Syst. (ISCAS 2009)*, Taipei, Taiwan, May 24–27, 2009, pp. 2197–2200.

118 K. Johansson, O. Gustafsson, and L. Wanhammar, Switching activity estimation for shift-and-add based constant multipliers, in *Proc. IEEE Int. Symp. Circuits Syst. (ISCAS 2008)*, Seattle, WA, May 18–21, 2008, pp. 676–679.

119 K. Johansson, O. Gustafsson, and L. S. DeBrunner, Estimation of the switching activity in shift-and-add based computations, in *Proc. IEEE Int. Symp. Circuits Syst. (ISCAS 2009.)*, Taipei, Taiwan, May 24–27, 2009, pp. 3054–3057.

120 FIRsuite: Suite, of Constant Coefficient FIR Filters, Technical Report, 2016. Available at http://www.firsuite.net.

121 S. Mirzaei, R. Kastner, and A. Hosangadi, Layout aware optimization of high speed fixed coefficient FIR filters for FPGAs, *Int. J. Reconfigurable Comput.*, vol. 2010, no. 697625, 2010, p. 17.

122 U. Meyer-Base, J. Chen, C. H. Chang, and A. G. Dempster, A comparison of pipelined RAG-n and DA FPGA-based multiplierless filters, in *Proc. IEEE Asia Pacific Conf. on Circuits and Systems (APCCAS)*, Singapore, December 2006, pp. 1555–1558.

123 K. Macpherson and R. Stewart, Area efficient FIR filters for high speed FPGA implementation, *IEEE Proc.-Vision, Image Signal Process.*, vol. 153, no. 6, 2006, pp. 711–720.

124 Martin Kumm, *Multiple Constant Multiplication Optimizations for Field Programmable Gate Arrays*, Springer Fachmedien Wiesbaden, 2016.

3

DA-Based Circuits for Inner-Product Computation

Mahesh Mehendale[1], Mohit Sharma[2], and Pramod Kumar Meher[3]

[1] *Texas Instruments Inc.*

[2] *Broadcom Limited*

[3] *Independent Hardware Consultant*

3.1   Introduction

Distributed arithmetic (DA) is a technique for multiplier less implementation of inner product of two vectors, particularly when one of the vectors is a constant and known *a prior*. Inner product, also known as dot product or simply the weighted sum $\left(\sum A[i] \times X[i]\right)$, is a core computation at the heart of many digital signal processing algorithms. In DSP application like finite impulse response (FIR) filter, where the filter coefficients are constants and all data representation are done in fixed-point format, DA-based implementation has been shown to provide area, power, performance efficiency for both ASIC as well as FPGA-based implementations. DA-based implementation was first proposed in early 1970s [1–3]. Since then it has been widely used to implement multiple DSP functions including FIR filters [4–6], DCT [7], FFT [5,8], DWT [9], Walsh-Hadamard [10] transforms, image and video processing functions such as color space conversion [11], interpolation, scaling, and so on. Since DA replaces multipliers with table look-ups of precalculated partial products, it is the architecture of choice for mapping weighted-sum computations on FPGAs which use LUTs as the basic building blocks.

The rest of the chapter is organized as follows. In Section 2, we introduce the mathematical concepts which lead to the distributed-arithmetic architecture for inner-product computation, and show how the baseline DA architecture

---

[1] Mohit Sharma was with Texas Instruments Inc. when this chapter was written.

Arithmetic Circuits for DSP Applications, First Edition. Edited by Pramod Kumar Meher and Thanos Stouraitis.
© 2017 by The Institute of Electrical and Electronics Engineers, Inc. Published 2017 by John Wiley & Sons, Inc.

compares with the conventional MAC (Multiply-and-Accumulate) implementation. We then present techniques for optimization of area, performance and power consumption of the DA implementation in Section 3, 4, and 5, respectively. We finally conclude the chapter with Section 6, summarizing the key ideas presented in this chapter.

3.2 Mathematical Foundation and Concepts

Consider the following sum of products:

$$Y = \sum_{n=1}^{N} A[n] \times X[n] \tag{3.1}$$

where $A[n]$ for $n = 1, 2, ..., N$, are the fixed coefficients and $X[n]$ for $n = 1, 2, ..., N$, constitute the variable vector. If $X[n]$ is a 2's-complement binary number scaled such that $|x[n]| < 1$ and K is the word size of each $X[n]$, then it can be represented as

$$X[n] = -b_{n0} + \sum_{k=1}^{K-1} b_{nk} \times 2^{-k} \tag{3.2}$$

where the b_{nk} are the bits 0 or 1, b_{n0} is the sign bit and $b_{n,K-1}$ is the LSB. Using Eqs (3.1) in (3.2) we get

$$Y = \sum_{n=1}^{N} A[n] \times \left(-b_{n0} + \sum_{k=1}^{K-1} b_{nk} \times 2^{-k} \right)$$

Interchanging the order of the summations we get

$$Y = \sum_{k=1}^{K-1} \sum_{n=1}^{N} \left(A[n] \cdot b_{nk} \right) \times 2^{-k} - \sum_{n=1}^{N} A[n] \times b_{n0} \tag{3.3}$$

Consider the bracketed term $\left(A[n] \times b_{nk} \right)$ in the above expression. Since each b_{nk} may take on values of 0 and 1 only, the term can have 2^N possible values corresponding to N possible values of n for $n = 1, 2, ..., N$. Instead of computing these values online, they can be precomputed and stored in a look-up-table memory. The input data can then be used to directly access the look-up table and the result can be accumulated with appropriate shift. The output Y can thus be obtained after K such cycles with $(K - 1)$ additions and no multiplications.

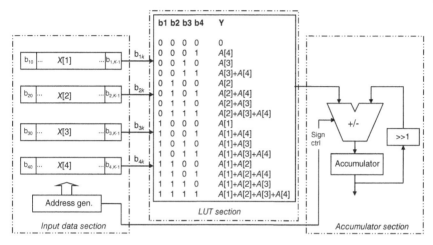

Figure 3.1 DA-based implementation of a four-term inner-product computation.

Figure 3.1 shows DA-based implementation of a four term weighted-sum computation ($\sum_{n=1}^{4} A[n] \times X[n]$). As can be seen from the figure, there are three components of the DA-based implementation, the input data section, the LUT section, and the accumulator section. The input data section holds the data ($X[1]$–$X[4]$ in this case) and makes available kth bits of each of the inputs, starting with LSB ($k = K - 1$) to MSB ($k = 0$). The output bits from the input data section serve as the address for the look-up table, thus for each of the K cycles, the corresponding data ($\sum_{n=1}^{N} A[n] \times b_{kn}$) are read out. Figure 3.1 also shows the contents of the look-up table for each of the addresses. The accumulator section adds/subtracts the output from the LUT section to/from the present value in the accumulator after right shifting it by 1. The add/subtract sign control is driven by the input data section. In case where the input data are stored in 2's complement representation, the LUT contents are added for all the bits, except for MSB ($k = 0$) for which they are subtracted. Thus, at the end of K cycles, the weighted-sum result is available in the accumulator.

For N term inner-product computation with data precision of K bits, the input data section requires the storage of ($N \times K$) bits. The input data section also needs a mechanism to read out the data, one bit from each of the inputs every cycle. This can be accomplished using shift registers or by storing the input data in a specialized memory that supports data to be written row at a time, but read out column at a time. The DA-based implementation shown in Figure 3.1 accesses the input data in LSB first order. This results in the output being available in LSB first bit-serial order. This enables the next computation stage to start bit-serially processing of the DA output in a pipelined fashion, improving the system throughput. The LSB first access also helps in cases where the input data are available in LSB first bit-serial order. We will discuss this further in the context of linear-phase FIR filters in Section 3.4. It is possible to access input data in MSB first order as well, in which case the accumulator value needs to be left shifted each cycle before adding to the LUT output. With MSB first access, the output is available only at the end of K cycles. While MSB first access is hence not preferred, we will discuss in Section 5, how it can be leveraged to achieve power versus quality tradeoffs.

The number of entries of the LUT is given by 2^N and width of the LUT is given by $(B + \log_2 N)$, where B is the bit-precision of the coefficients/weights in fixed-point format. There are multiple approaches for implementing the LUT and the choice is driven by the area, performance, power goals, target implementation technology (ASIC vs FPGA), and the application requirement in terms of whether the coefficients/weights are constant or varying. If the coefficients are fixed, for example, discrete cosine transform (DCT) in JPEG or the filter coefficients for finite impulse response (FIR) filters, the LUT can be implemented as a hardwired logic using logic synthesis to drive appropriate area versus delay tradeoff. In case the coefficients are fixed, but application requires the flexibility to specify them/change them late in the design cycle or post silicon thro' metal only revision, the LUT can be implemented as a ROM. For applications where the coefficients need to be specified at run time, SRAM-based implementation is preferred. The SRAM-based implementation is also helpful in cases where the same DA module is to be reused with different sets of coefficients, as in case of a multistandard video decoder, where the coefficients for the spatial to frequency domain transformation used vary with standards (e.g. it's DCT for MPEG2 and arithmetic transform for H.264). In applications where the coefficients change at run-time (e.g. adaptive filters), the LUT could preferably be implemented using adders. The adder-based implementation also enables area-efficient implementation in cases where the number of coefficients is high (e.g., 1024 tap FIR filter).

The adder/subtractor in the accumulator section supports input data precision of $(B + \log_2 N)$ bits, and the accumulator register has $(K + B + \log_2 N)$ bits assuming the accumulation is to be done without loss of any precision. Since the accumulator performs $K - 1$ successive additions where one of the inputs to the adder is left shifted relative to the other input (achieved by accumulator right shifted by 1 bit position), these additions can be performed using carry-save addition technique. Thus, except for the last cycle when the carry and sum vectors need to be added, for all the first $(K - 2)$ cycles, the delay of the addition is that of a 1 bit full adder. Such an implementation is thus efficient in terms of both power as well as performance (in terms of critical path delay of the accumulator unit), and also makes the delay of the accumulator unit independent (except for the last cycle) of the bit precision of the LUT entries.

Figure 3.2 shows a conventional implementation of weighted-sum computation using a multiply and accumulate (MAC) unit. During each cycle a

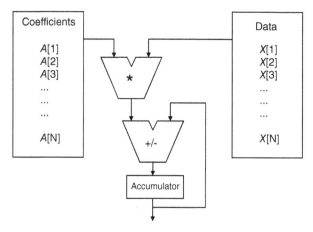

Figure 3.2 MAC (multiply-accumulate)-based implementation of inner-product computation.

{coefficient $(A[i])$, data $(X[i])$} pair is multiplied and added to the accumulator, thus an N term inner product can be computed over N cycles with N multiplication and $N-1$ additions. The following table summarizes the characteristics of the two (i.e., conventional and DA-based) implementations.

3.3 Techniques for Area Optimization of DA-Based Implementations

In this section, we present techniques for optimizing the area of the DA-based implementations. As can be seen from the Table 3.1, the LUT size of the baseline DA implementation grows exponentially with the length of inner product. Various techniques have been developed to reduce the LUT size, and make the relationship between the LUT size and the inner product length linear instead of an exponential one. Most of these techniques, however, result in performance and/or power increase, so the overall DA-based architecture for a given weighted-sum computation needs to be driven by the desired area–power-performance tradeoffs.

3.3.1 Offset Binary Coding

The size of the LUT can be reduced to half by using offset binary coding [5] that assumes each bit of the input to have a value of either 1 or −1 (as against 1 or 0 in case of the conventional binary coding).

Table 3.1 MAC-based versus DA-based weighted-sum computation.

No. of coefficients: N Data precision: K bits Coefficient precision: B bits	Conventional (MAC-based) implementation	DA-based implementation with single LUT
Number of multiplications	$N\{K$ bits $\times B$ bits$\}$	0
Number of additions	N - 1 $\{(K + B)$ bits + $(K + B + \log_2 N)$ bits$\}$	K - 1 $\{(B + \log_2 N)$ bits + $(B + \log_2 N)$ bits$\}$
LUT size	0	$2N$ rows, $(B + \log_2 N)$ columns

Consider the following sum of products, as discussed in Section 2:

$$Y = \sum_{n=1}^{N} A[n] \times X[n] \qquad (3.4)$$

where $A[n]$ for n = 1, 2, …, N, are the fixed coefficients and the $X[n]$ for n = 1, 2, …, N, constitute the variable vector. If $X[n]$ is a 2's-complement binary number scaled such that $|x[n]| < 1$ and K is word size of each $X[n]$, then it can be represented as

$$X[n] = -b_{n0} + \sum_{k=1}^{K-1} b_{nk} \times 2^{-k} \qquad (3.5)$$

where the b_{nk} are the bits 0 or 1, b_{n0} is the sign bit and $b_{n,K-1}$ is the LSB. $X[n]$ can be written as follows:

$$X[n] = \frac{1}{2}\left(X[n] - (-X[n])\right) \qquad (3.6)$$

From Eqs (3.5) and (3.6), we get

$$X[n] = \frac{1}{2}\left(-\left(b_{n0} - \bar{b}_{n0}\right) + \sum_{k=1}^{K-1} \left(b_{nk} - \bar{b}_{nk}\right) \times 2^{-k} - 2^{-(K-1)}\right) \qquad (3.7)$$

where $\bar{b}_{nk}$ denotes the complement of bit b_{nk}.

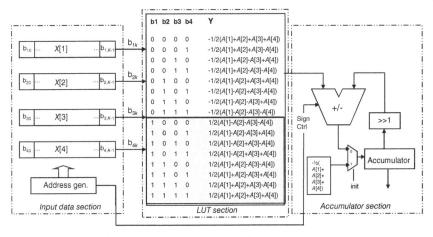

Figure 3.3 DA-based implementation using OBC (offset binary coding).

Let $c_{nk} = \left(b_{nk} - \bar{b}_{nk}\right)$ for $k \neq 0$ and $C_{n0} = -\left(b_{n0} - \bar{b}_{n0}\right)$, then from Eqs (3.7) and (3.4), we get

$$Y = \sum_{k=0}^{K-1} \left(\sum_{n=1}^{N} \frac{A[n]}{2} \times c_{nk}\right) \times 2^{-k} - \left(\sum_{n=1}^{N} \frac{A[n]}{2}\right) \times 2^{-(K-1)} \qquad (3.8)$$

It can be noted that $(b_{nk} = 0)$ translates to $(C_{nk} = -1)$ and $(b_{nk} = 1)$ translates to $(C_{nk} = 1)$. The first term in Eq. (3.8) can thus be computed using the LUT as shown in Figure 3.3. The second term is a constant that can be added to the accumulator at the end of the accumulation of LUT outputs, to get the final result.

It can be noted that the contents of the upper half of the LUT are same as the lower half, but with the sign reversed. This symmetry can be exploited to store only the upper half of the LUT, and using XORs to reverse the sign when required. The resultant DA structure is shown in Figure 3.4.

As can be seen from Figure 3.4, the LUT size is reduced to half, but there is an overhead of XOR gates. The power is impacted by the "N" XORs performed every cycle, and also due to one more addition required $\{-1/2 \times (A[1] + A[2] + A[3] + A[4])\}$ over the baseline DA implementation.

It can be noted that $(-1/2 \times A[1])$ term is common to all the entries of the LUT in Figure 3.3. Taking this term out, the contents of the upper half of LUT are same as the contents of the lower half, but with sign reserved. The LUT size can thus be further reduced by half, as shown in Figure 3.5. While the LUT size is reduced to the 1/4th of the baseline implementation, there is further increase in power due to additional XORs and an addition or subtraction of $\{-1/2 \times (A[1])\}$ performed every cycle. The critical path delay between the

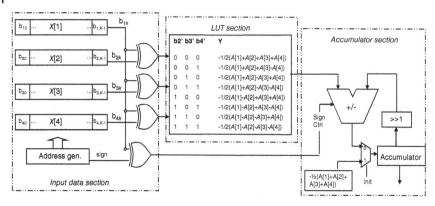

Figure 3.4 LUT size reduction using OBC (offset binary coding).

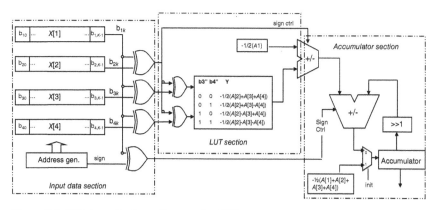

Figure 3.5 LUT size reduction with recursive OBC.

input data and the final stage adder is also higher in this implementation due to the XOR gates and the adder. This performance impact can be reduced with pipelining that, however, adds to the area and latency.

It can be noted that the terms $\{-1/2 \times A[2]\}$ is common to all the terms of the LUT in Figure 3.5. Also, with this common term removed, the contents of the upper half of the LUT are same as the lower half, so the LUT size can be further reduced with addition of XORs at the input and an adder at the output. This technique can thus be recursively applied [13]. If applied all the way, this results in an LUT-less implementation. Such an implementation may not always be the most area efficient, as the overhead of XOR gates and the additional adders may offset the reduction in the LUT size. The most area optimal point can vary depending on the target implementation style, such as ASIC and FPGA, and also how the LUT is implemented.

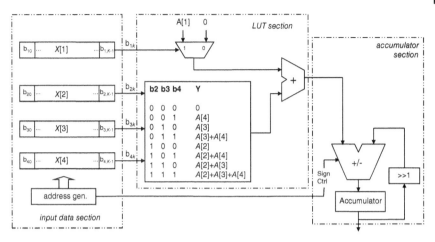

Figure 3.6 LUT size reduction using an adder.

3.3.2   Adder-Based DA

It can be noted that the contents of the lower half of the LUT of the baseline DA implementation shown in Figure 3.1, can be derived from the upper half by adding $A[1]$. The LUT size can thus be reduced by half, by storing only the upper half of the LUT, and adding $A[1]$ to the LUT output whenever the bit $\{b_{1k}\}$ is 1. The resultant implementation is shown in Figure 3.6.

It can be noted that the contents of the lower half of the LUT in Figure 3.6, can be derived from the contents of the upper half by adding $\{A[2]\}$. The LUT size can thus be reduced by storing only the upper half of the LUT and adding $\{A[2]\}$ to the LUT output whenever bit $\{b_{2k}\}$ is 1. Such an optimization can be recursively applied to completely eliminate the LUT. The resultant structure as shown in Figure 3.7 is called the "adder-based DA" implementation [14].

3.3.3   Coefficient Partitioning

For an N term inner-product computation, instead of using all N bits to address one LUT (results in LUT requiring 2^N rows), the number of address bits can be split into two partitions each accessing a separate LUT. Each of the two LUTs requires $2^{(N/2)}$ rows, thus the total LUT size reduces by the factor of $2^{(N/2-1)}$. Figure 3.8 shows DA-based implementation of a 4-term inner-product computation, using two LUTs. As can be seen, the total LUT size is 8 rows (4 rows per LUT) as against 16 rows of the LUT in the baseline DA implementation (Figure 3.1). It can also be noted that the number of columns of the two LUTs is $(B + \log_2(N/2))$ which 1 less than the LUT of the baseline DA implementation, which has $(B + \log_2 N)$ columns.

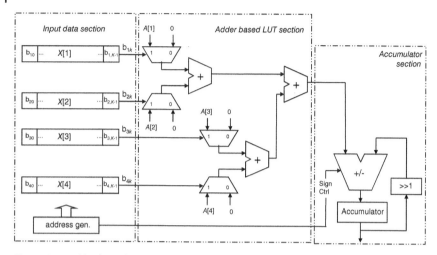

Figure 3.7 Adder-based DA implementation of a four-term inner-product computation.

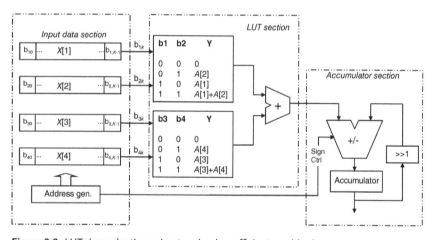

Figure 3.8 LUT size reduction using two-bank coefficient partitioning.

While the LUT size reduces with 2-bank coefficient partitioning, it requires an extra addition to be performed every cycle, thus impacting power dissipation as well as the critical path delay.

Computing an inner-product as the sum of two half-length inner products, results in significant reduction in the LUT size. But for higher values of N, the resultant LUT area could be very high. For example, for $N = 32$ the LUT in the baseline implementation requires 4,29,49,67,296 (2^{32}) rows. With 2-bank splitting the implementation requires two LUTs each with 65,536 (2^{16}) rows, which is still prohibitively large. Thus, for higher values of N, the coefficients can

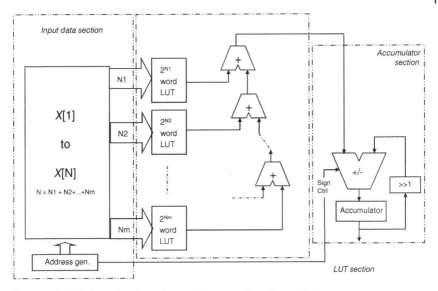

Figure 3.9 LUT size reduction using multibank coefficient partitioning.

be split into multiple banks. Figure 3.9 shows such a multibank implementation [5,15], where the coefficients are split into M groups.

For $N = 32$, and $M = 8$ (equal size partitions), the resultant multibank implementation requires eight LUTs each with 16 rows, which is significantly smaller compared to the two-bank implementation. This 8-bank implementation, however, requires seven extra additions to be performed every cycle. Thus, the area reduction comes with a significant impact to power and performance.

It can be noted that if the coefficients are split to the extreme of $M = N$ (i.e., 1 bit per LUT), it results in the adder-based DA structure shown in Figure 3.7.

3.3.4 Exploiting Coefficient Symmetry

Finite impulse response (FIR) filters are one of most commonly used DSP functions. FIR filtering is achieved by convolving the input data samples with desired unit impulse response of the filter. The output $Y[n]$ of an N-tap FIR filter is given by the weighted sum of the latest N input data samples as follows:

$$Y[n] = \sum_{i=0}^{N-1} A[i] \times X[n-i]$$

The FIR filter, being a weighted-sum computation, maps well onto a DA-based implementation. The output requires N latest data samples, $N - 1$ of the N inputs used to compute current output, are reused for computing the next

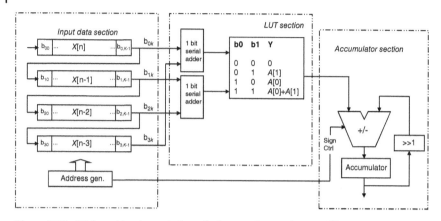

Figure 3.10 DA-based implementation of a four-tap linear-phase FIR filter.

output. This can be supported in the input data section, by organizing all the inputs into one long shift register chain. For input data precision of K bits, at the end of K shifts, not only the output $Y[n]$ is available, but also the data from $X[i]$ location is also shifted to $X[i - 1]$ location.

Among the commonly used FIR filters are linear-phase FIR filters, which have the characteristics of coefficient symmetry, that is, $A[i] = A[N - 1 - i]$, for an N-tap filters with coefficients $A[0]$ to $A[N - 1]$. This coefficient symmetry can be exploited to compute the output as follows: $Y[n] = \sum_{i=0}^{N/2-1} A[i] \times (X[n - i] + X[n - (N - 1 - i)])$ for N even. The corresponding DA-based implementation requires LUT with $2^{N/2}$ rows supporting $N/2$ coefficients $A[0]$ to $A[N/2 - 1]$. This represents reduction in the LUT size by a factor of $2^{N/2}$. The addition of input samples can be done bit serially, resulting in a structure shown in Figure 3.10.

It can be noted that LSB first bit-serial access makes it easier to implement the bit-serial addition of the input data. It can also be noted that since two K bit inputs are added, the DA computation involves $(K + 1)$ LUT accesses, and hence takes an extra cycle over the baseline DA implementation.

The coefficient symmetry property is found in many other transforms as well. For example, the interpolation performed in video processing to get pixel values at half-pel positions or the filtering performed to both upscale as well as downscale the images involve weighted-sum computations with coefficient symmetry.

3.3.5 LUT Implementation Optimization

The choice of implementation of the LUT is driven by the target technology, for example, ASIC or FPGAs, and also the flexibility requirements. If the coeffi-

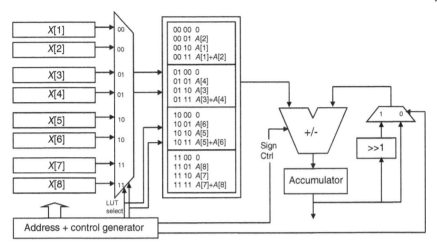

Figure 3.11 Single LUT-based implementation of a eight-term inner product with four coefficient partitions.

cients are fixed, for example, for DCT computation, the LUT can be synthesized into hardwired gates. Such an implementation is optimal in terms of area, but is least flexible. For FPGAs where LUTs form the building blocks, the DA LUT is best realized using these LUTs. If the coefficient programmability is required, the LUT can be implemented using either an SRAM or a nonvolatile memory such as FLASH or FRAM. In most CMOS technologies there is a lower limit on the size of embedded SRAM. If the required LUT size is smaller than this minimum size, the LUT is implemented using storage elements such as flip-flops or latches. Such an implementation, however, is not as area efficient (in terms of area per bit) as SRAM. This needs to be considered while deciding on the coefficient partitioning for LUT size reduction.

The area of a multipartitioned DA implementation can be optimized, if the LUTs are accessed sequentially, thereby eliminating the need for the extra adders. For an implementation with M partitions, since each of the M LUTs are accessed each cycle, $M - 1$ adders are required to sum up their outputs to feed the accumulator. With sequential access, these $M - 1$ adders can be completely eliminated. Also, since the LUTs are accessed sequentially, all the LUTs can be combined into one big LUT which can be implemented using a single ported SRAM thus further improving area efficiency. Figure 3.11 shows such an implementation for an eight-term weighted-sum computation.

It can be noted that the aggressive area reduction with this scheme results in significant increase in the number of clock cycles. For M partitions, the number of clock cycles to compute the output goes up by a factor of M. If such a performance degradation is not acceptable, an intermediate approach can be adopted where, for example, instead of one LUT for M partitions, two LUTs for

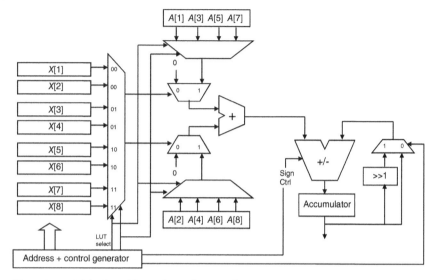

Figure 3.12 Adder-based DA with LUT functionality implemented using a single adder.

$M/2$ partitions can be used. This requires an extra adder, but the performance improves by a factor of two.

3.3.6 Adder-Based DA Implementation with Single Adder

The area optimization technique discussed in Section 3.3.5, can be applied to adder-based DA structure as well, such that only one adder is required independent of the number of coefficients. Instead of selecting all input data bits corresponding to coefficients of different partitions, the process can be done sequentially, where during each clock cycle only two input data bits are used. Figure 3.12 shows such an implementation for an eight-term inner product computation.

Along the lines of what we discussed in Section 3.3.5, this scheme also results in significant performance degradation over the adder-based DA scheme presented in Section 3.2. For N coefficients, the number of cycles goes up by a factor of $N/2$. This performance impact can be reduced to a factor of $N/4$, by using two adders instead of one to implement the LUT functionality.

3.3.7 LUT Optimization for Fixed Coefficients

As discussed in Section 3.3.5, in case of weighted-sum computations where the coefficients are fixed and known *a priori*, the LUT can be implemented by hardwired logic with minimum area consumption. The LUT can be treated as a truth-table which a logic synthesis tool can optimize. As discussed earlier, the number of columns of the LUT is a function of the bit-precision of the coefficients and the number of coefficients. It has the upper bound of $(B + \log_2 N)$. With fixed coefficients, the dynamic range of the entries in the truth table can be computed, and the number of columns can be set accordingly. For example, consider the coefficient set $\{-4, 53, 18, -3\}$ used for the bi-cubic interpolation to compute Luma quarter-pel values during motion compensation in video processing as per the VC1 standard. These coefficients require seven bits ($B = 7$) to represent in 2's complement representation. With $N = 4$, the upper bound on the bits required to represent the resultant LUT entries is nine ($7 + \log_2 4$). However, with these four coefficients, the dynamic range of the LUT entries is -7 to 71, which requires only eight bits. The reduction in the number of output bits of the LUT also helps reducing the precision of the adders required in the accumulator section, there by further reducing the area. Even for fixed coefficients, if the number of coefficients is large, the area of the hardwired LUT implementation can still be high, so to get to the area efficient implementation the LUT size reduction techniques need to be applied and all those smaller size LUTs should be optimized by suitable hardwired implementation.

Consider the LUT size reduction technique shown in Figure 3.6, which uses an adder to reduce the LUT size by half. While the figure shows that coefficient A1 being added whenever the bit of $X[1]$ is "1," there is flexibility in terms of the coefficient-data pair to be used for this addition. For the four tap bi-cubic interpolation discussed above, if the coefficient -4 is used for the adder, the number of rows of the LUT reduces to eight, but the number of columns remains the same as eight, to support the dynamic range of $\{-3 \text{ to } 71\}$. Therefore, the adder is required to support a 3 bit + 8 bit addition. If coefficient 51 is picked, it results in LUT six columns to support the dynamic range of $\{-7 \text{ to } 18\}$, and requires the adder to support 6 bit + 6 bit addition. Thus, the flexibility of coefficient selection for the adder can be used to further optimize the area.

Similar flexibility can be exploited with the coefficient partitioning technique. For example, for the 2-bank coefficient partitioning scheme, the specific coefficients to be grouped in the two banks can be selected so as to reduce the number of columns of the two LUTs and also to reduce the area of the LUTs when synthesized into a hardwired logic. Even with the number of columns remaining the same, different grouping results in different bit patterns for the two LUTs, and as a result the area of the resultant logic can vary. Figure 3.13 shows an alternate grouping of coefficients for two bank implementation of a four-term inner product computation.

The results presented in Reference [16] show that with an optimal coefficient splitting, the LUT logic area can be reduced by more than 20% over the simple coefficient splitting.

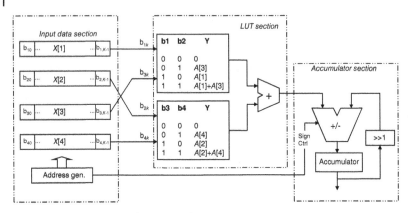

Figure 3.13 Alternate coefficient split for two bank DA structure.

3.3.8 Inner-Product with Data and Coefficients Represented as Complex Numbers

Several functions in communication systems such as channel equalization, modulation, and demodulation deal with data streams and coefficients represented as complex numbers. Moreover, multiply-accumulate (MAC) is a commonly used computation in these applications. The output of a weighted-sum computation of complex numbers is also a complex number. The computation can be expressed as

$$Y_R + j \times Y_J = \sum_{n=1}^{N} \left(A_R[n] + j A_J[n] \right) \times \left(X_R[n] + j X_J[n] \right)$$

where A_R, X_R, Y_R are the real components and A_J, X_J, Y_J are the imaginary components of the coefficient, input data and the output, respectively. The outputs Y_R and Y_J can be computed as follows:

$$Y_R = \sum_{n=1}^{N} A_R[n] \times X_R[n] - \sum_{n=1}^{N} A_J[n] \times X_J[n]$$

$$Y_J = \sum_{n=1}^{N} A_R[n] \times X_J[n] + \sum_{n=1}^{N} A_J[n] \times X_R[n]$$

Each of the four weighted-sums in the above equations can be mapped onto the baseline DA structure shown in Figure 3.1. Since the same LUTs can be used in both (Y_R and Y_J) the computations, the total LUT area can be reduced by half, by computing the two outputs sequentially, thereby sharing the same LUTs. The resultant structure is shown in Figure 3.14.

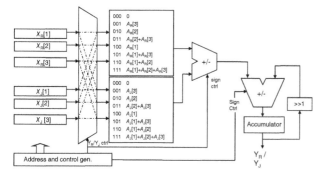

Figure 3.14 DA structure for inner-product computation of complex numbers.

3.4 Techniques for Performance Optimization of DA-Based Implementations

In this section we present techniques for improving the performance of the baseline DA structure (Figure 3.1) for inner-product computation.

3.4.1 Two-Bits-at-a-Time (2-BAAT) Access

Instead of accessing the input data one bit at a time, which requires K cycles (K being the input bit-precision), two bits can be accessed each cycle, so that the computation could be completed in $(K/2 + 1)$ cycles $(K + 1)/2$ cycles when K is odd). When K is even, the input data is sign extended by one bit and prepended by a zero, when K is odd, the input data is prepended by zero. If two bits of input data are accessed in each clock cycle (2-BAAT scheme [5]), the number of address bits for the LUT gets doubled. For a single LUT structure, this results in increase of LUT size by a factor of 2^N. The resultant structure is shown in Figure 3.15. Alternatively, if bits of two different weights access two different LUTs the total LUT size will increase by a factor of two only. Such a structure for 2-BAAT scheme is shown in Figure 3.16.

The number of bits accessed in each clock cycle can be increased to three or even higher to get further speed-up, where the LUT size would increase linearly with the number of bits, by using a structure with multiple LUTs.

3.4.2 Coefficient Distribution over Data

The adder-based DA implementation shown in Figure 3.7, requires K clock cycles (K being the bit-precision of inputs), and N adders (N being the number of coefficients) adding B bit coefficients (B being the bit-precision of coefficients). Instead of distributing the data over the coefficient as done in the conventional

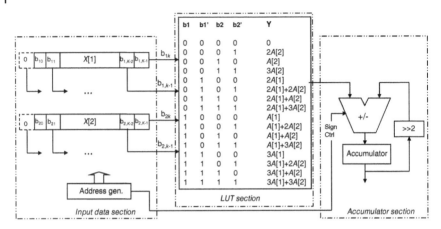

Figure 3.15 2 BAAT DA structure with single LUT.

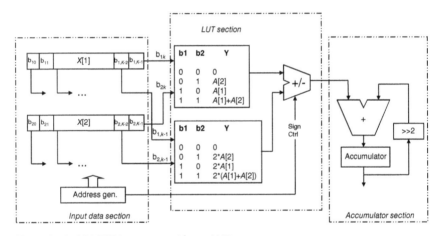

Figure 3.16 2 BAAT DA structure with two LUTs.

DA implementation, the adder-based DA structure, enables distributing coefficients over data [17]. In other words, instead of accessing data one bit at a time, coefficients can be accessed one bit at a time, and the data value added if the bit of the corresponding coefficient is "1." Such an implementation can be accomplished by simply swapping data and coefficients in the adder-based DA structure. The resultant structure is shown in Figure 3.17.

This implementation takes B cycles to compute, and require N adders to take in K bits of data. In case of implementations where the bit-precision of coefficients is smaller than the bit precision of input data ($B < K$), the adder-based DA which accesses coefficients bit-serially instead of data, results in

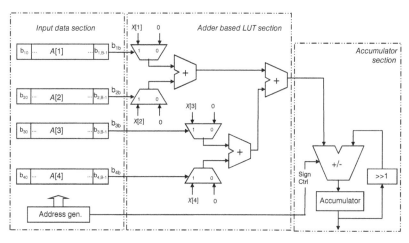

Figure 3.17 Adder-based DA with coefficients distributed over data.

$(K - B)$ fewer clock cycles. However, since the multiplexers and adders now need to handle data of higher precision than the coefficients, the performance improvement using this technique comes with an associated area increase.

3.4.3 RNS-Based Implementation

Residue number system (RNS)-based implementation of DSP algorithms have been presented in the literature [18–20] as a technique for high-speed realization. In a residue number system (RNS), an integer is represented as a set of residues with respect to a set of integers called the moduli. Let $\{m_1, m_2, m_3, \dots, m_n\}$ be a set of relatively prime integers called the moduli set. An integer X can be represented as $\{X_1, X_2, X_3, \dots, X_n\}$, where $X_i = X$ modulo m_i, for $i = 1, 2, \dots, n$.

We use the notation X_i to represent $jX_{j_{mi}}$ the residue of X w.r.t. m_i. Given the moduli set, the dynamic range (M) is given by the LCM of all the moduli. If the elements are pair-wise relatively prime, the dynamic range is equal to the product of all the moduli. The bit-precision of a given moduli set is given by $\log_2 M$. So the determination of moduli set is based on the bit-precision needed for the computation. For example, for 19-bit precision the moduli set $\{5, 7, 9, 11, 13, 16\}$ can be used [21].

Let X, Y, and Z have the residue representations $X = \{X_1, X_2, X_3, \dots, X_n\}$, $Y = \{Y_1, Y_2, Y_3, \dots, Y_n\}$, and $Z = \{Z_1, Z_2, Z_3, \dots, Z_n\}$, respectively and $Z = X <op> Y$, where $<op>$ is any of addition, multiplication, or subtraction. Then we have in RNS, $Z_i = jX_i < op > Y_i j_{mi}, for\ i = 1, 2, \dots, n$.

Since, X_i's and Y_i's require lesser precision than X and Y, the computation of Z_i's can be performed faster than the computation of Z. Moreover, since the computations are independent of each other, they can be executed in parallel resulting in significant performance improvement. For example, consider the moduli set $\{5, 7, 11\}$.

Let $X = 47 = \{j47_5, j47_7, j47_{11}\} = \{2, 5, 3\}$, $Y = 31 = \{j31_5, j31_7, j31_{11}\}$ $= \{1, 3, 9\}$, and $Z = X + Y = 47 + 31 = 78 = \{j78_5, j78_7, j78_{11}\} = \{3, 1, 1\}$. Then Z_i's can be computed, independent of each other, using RNS as $Z = \{j2 + 1_5, j5 + 3_7, j3 + 9_{11}\} = \{3, 1, 1\}$.

While in RNS, basic operations like multiplication, addition, and subtraction can be performed with high speed, operations like division and magnitude comparison require several basic operations and hence are slower and more complex to implement. Since most DSP functions (such as FIR and IIR filtering, FFT, correlation, and DCT) do not need these operations to be performed, this limitation of RNS does not apply.

Figure 3.18 shows the RNS-based implementation of an N-term weighted-sum computation. The implementation is generic and assumes K moduli (M_1 to M_K) selected to meet the desired precision requirements. The coefficient memory stores the precomputed residues of the coefficients $A[1]$ to $A[N]$ for each of the moduli M_1 to M_K. The data memory stores the residues of the data values for each of the moduli M_1 to M_K. The data needs to be converted from binary to RNS before loading in the data memory. In case of an FIR filter implementation with N taps, each of the input data read is used in the computation of next N outputs, thus per each output computation only one data needs binary-to-RNS conversion.

The weighted-sum computation is performed as a series of modulo MAC operations. During each cycle the coefficient address generator and the data address generator provide the appropriate coefficient-data pairs to the K modulo MAC units that perform the modulo computations in parallel.

Figure 3.19 shows an implementation of a modulo MAC unit. It consists of a modulo multiplier and a modulo adder connected via an accumulator so as to perform repeated modulo MAC operations and make the result available to the RNS-to-binary converter every N cycles. The modulo multiplication and modulo addition is typically implemented as a look-up-table (LUT) for small moduli. Figure 3.19 also shows the look-up-table-based implementation of a "modulo 3" multiplier.

The RNS implementation shown in Figure 3.18 requires N clock cycles to compute the output in RNS domain of an N-term inner product computation. Each of the modulo MAC computation can be implemented using distributed arithmetic technique as well, where the residues of the input data are accessed bit-serially to generate address bits of an LUT which contains residues of the precomputed partial product terms. The data is accessed MSB-first order, and the LUT output is modulo added twice with the accumulator content. The resultant implementation is shown in Figure 3.20.

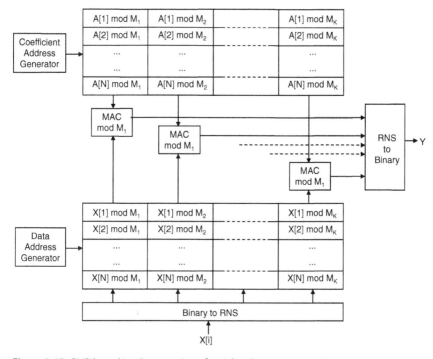

Figure 3.18 RNS-based implementation of weighted-sum computation.

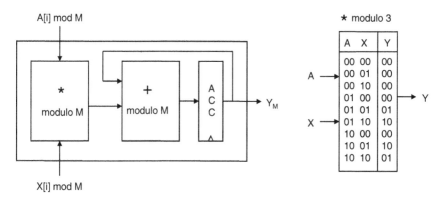

Figure 3.19 Module MAC unit and LUT-based implementation of modulo 3 multiplication.

While the RNS-based implementation in Figure 3.18 takes N clock cycles, the number of clock cycles required for the DA-based scheme shown in Figure 3.20, is given by the bit-precision of input data residues. Since the bit-resolution to store residues for small moduli is small (3 bits to store residues w.r.t. 5 and 7,

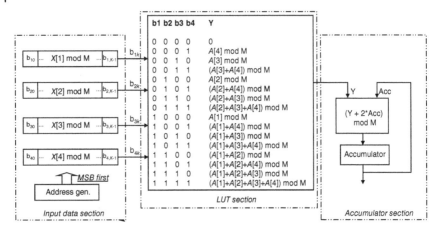

Figure 3.20 DA-based architecture for modulo MAC computation in RNS-based scheme.

4 bits to store residues w.r.t. 11 and 13, etc.), the DA-based approach improves the performance further. For example, for an eight tap filter, while "modulo 11 MAC" takes eight cycles with baseline RNS scheme, using DA, the cycles can be reduced to four (number of bits required to store residues w.r.t. 11).

It can also be noted that the structure in Figure 3.20 works even if the input data are not binary-to-RNS converted, that is, $X[i]$ can directly be used in place of $X[i]$ mod M values. Such a scheme eliminates the need for binary-to-RNS conversion, but results in higher number of cycles (input data bit precision as against bit-precision for mod M values).

3.5 Techniques for Low Power and Reconfigurable Realization of DA-Based Implementations

In this section we discuss techniques to reduce power dissipation of DA-based implementation. The sources of power dissipation in CMOS circuits can be classified into dynamic switching power, short circuit power, and leakage power. The first two types of power dissipation happen when a signal in a logic circuit toggles (i.e., changes the state from 1 to 0 or vice versa). The leakage power is dissipated when the circuit is in powered-on condition. Typically in logic circuits, the dynamic switching power is the major part of the total power dissipated. The dynamic power consumption in a CMOS circuit is given by $P_{\text{dynamic}} = \alpha.C.V^2.f$, where V is the supply voltage, C is the switched capacitance, α is the switching activity factor, and f is the operating frequency. In the following section we discuss the techniques for the reduction of switching activity for reducing the total capacitance switched during the inner-product computation.

3.5.1 Adder-Based DA with Fixed Coefficients

Adder-based DA scheme presented in Section 3.2 has been shown to be area efficient, especially for large N (number of coefficients). However, it performs N additions per iteration. To compute the weighted-sum, it requires $K \times N$ number of additions (K is the bit-precision of input data). Since the number of additions required is N times the number of additions required with the baseline single LUT structure, the adder-based DA scheme results in significantly higher power. For the implementation for fixed coefficients, specific bit pattern of coefficients can be exploited to reduce the power dissipation. This is accomplished by first switching to the adder-based DA scheme presented in Section 3.4.2, where coefficients are distributed over data. In this scheme, the total number of additions performed is given by the number of 1 s in the binary representation of the coefficients. The number of nonzero (1 or −1) bits can be reduced using suitable representation of numbers such as canonical sign digit (CSD) or nega-binary number representation. The number of additions can be further reduced by exploiting the redundancy in computation across different iterations. For example, consider two eight bit coefficients $A[1] = 01\underline{1}01\underline{1}01$ and $A[2] = 00\underline{1}01011$. This results in addition of $X[1] + X[2]$ for LSB (i.e., bit 0), and also for bits 3 and 5. If the result of $X[1] + X[2]$ is stored after the first iteration, it can be reused for the two iterations corresponding to bit 3 and bit 5, thereby reducing the number of additions performed by two. This approach can be extended to extract common sub-expressions which add different combinations of inputs and also add three or even higher number of inputs per term. As an example, consider the following coefficient matrix for four term weighted-sum computation, with coefficient bit-precision of eight.

$$
\begin{array}{rcccccccc}
 & 7 & 6 & 5 & 4 & 3 & 2 & 1 & 0 \\
A[1] & 0 & 0 & 1 & 1 & 1 & 0 & 1 & 1 \\
A[2] & 0 & 0 & 1 & 0 & 1 & 0 & 1 & 1 \\
A[3] & -1 & 0 & 1 & 1 & 0 & 0 & 1 & 1 \\
A[4] & -1 & 1 & 0 & 0 & 1 & 0 & 1 & 0 \\
\end{array}
$$

The adder-based DA implementation of this weighted-sum computation requires 18 add/subtract operations. The coefficient matrix has multiple common subexpressions including $(X[1] + X[2])$ which is common across bits 0,1, 3 and 5, $(X[3] + X[4])$ which is common across bits 1 and 7, $(X[1] + X[2] + X[3])$ which is common across bits 0, 1, and 5, $(X[1] + X[3])$ which is common across bits 0, 1, 4, and 5, $(X[1] + X[2] + X[4])$ which is common across bits 1 and 3. By iter-

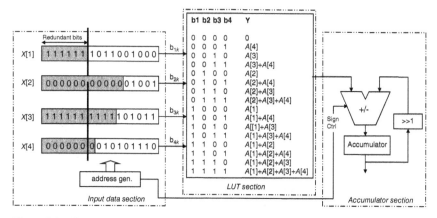

Figure 3.21 DA computation over minimum required input bit precision.

atively detecting and eliminating these common sub-expressions the number of additions can be reduced to 13. The common sub-expression elimination approaches have been discussed extensively in the literature [22,23], and the same can be applied in this context.

This common sub-expression elimination-based technique for reducing power consumption of adder-based DA implementation, can be further extended in the context of vector matrix multiplications used in orthogonal transforms, for example DCT, DST, DHT.

3.5.2 Eliminating Redundant LUT Accesses and Additions

The bit precision of the input variables is considered to find the largest possible values (i.e., the entire dynamic range). In most signal processing applications, the signals are generally correlated and only few samples actually require the largest of the bit precision for representation. For example, average bit precision required by a typical speech file is around 6.9 bits [24], while the filtering circuits are designed to support maximum precision of 16 bits. This property can be exploited to save energy without loss in accuracy. The proposed scheme is shown in Figure 3.21.

When a group of inputs being processed, the minimum required bit-precision can be computed by looking at the leading 0's or 1's (sign extension bits for negative numbers). In the figure above, while the input precision is 16 bits, the minimum precision to represent inputs $X[1]$ to $X[4]$ is 10 bits. Thus, using LSB first approach, the DA computation can be completed in 10 cycles, saving six LUT accesses and six accumulations.

This technique not only saves the power but also reduces the number of cycles required to compute the output. The reduction in number of cycles can be

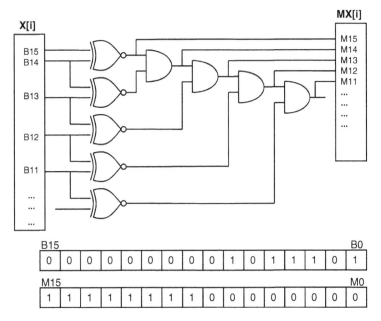

Figure 3.22 Extracting sign-extension bits.

exploited to further reduce power by running the circuit at lower frequency and lower supply voltage. Since the minimum bit precision and hence the number of clock cycles to compute is data dependent, it can vary from one output to the next. Hence, voltage–frequency scaling is required to be performed dynamically and adaptively. One approach to achieve this is by using two LUT-accumulate units operating at different voltages. Such a scheme is presented in Reference [24] and shows significant power reduction.

Figure 3.22 shows the circuit [24] to extract sign extension bits into a mask register, for each of the inputs. The masks can be bit wise ANDed across all the inputs, to identify minimum bit-precision for a given set of inputs. For input bit precision of K bits, and N inputs, the mask computation requires $N \times (K - 1)$ XORs and $N \times (K - 1)$ AND gates. Moreover, the mask computation logic has the critical path of 1 XOR and $(K - 1)$ AND gates. One way to reduce this overhead is to compute the mask for $2K/3$ (or even $K/2$) most significant bits. In FIR filters, where each input is used in the computation of N outputs, the mask can be computed once and stored in a mask register. With a mask register for each of the inputs, the mask computation logic and the power dissipation in mask computation can be reduced by a factor of N.

The area overhead of sign-extension computation logic and mask registers, can be significantly reduced by computing the mask bit-serially, as shown in Figure 3.23 for N = 4.

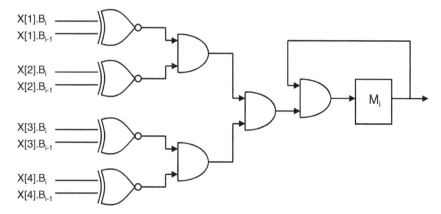

Figure 3.23 Bit-serial extraction of sign-extension bits.

In this scheme the inputs are accessed in MSB first order. If the current bit is detected as sign-extension bit across all the inputs, the LUT access and accumulation is not performed. This approach saves dynamic switching related power, but since the number of cycles is not reduced, additional power reduction with voltage/frequency scaling cannot be achieved. The scheme can be modified to do LSB first accesses, and compute sign-extension bits for the top $K/2$ bits. Thus, after the first $K/2$ cycles, the mask for the top $K/2$ is available, can be used to determine the minimum number of additional bits to be accessed to complete the computation. Since this approach saves the cycles as well, it enables further power reduction using frequency and voltage scaling.

3.5.3 Using Zero-Detection to Reduce LUT Accesses and Additions

Since the LUT entry for address "0" (all inputs bits 0) is 0, a zero-detect logic on the LUT address bus can be used to disable LUT access as well as additions as shown in Figure 3.24.

It can be noted that this scheme can be used in conjunction with the scheme discussed in Section 3.3.2. The amount of power reduction using this technique is dependent on the number of times the input bits at a given location are all zero. The probability of all zero vectors can be increased by recoding the inputs so as to increase the number of 0s in their representation. CSD (canonical sign digit) representation uses the digit set $\{0,1,-1\}$ to represent each bit, and results in a representation which maximizes the number of 0s. For example, eight bit representation of 30 (00011110) has four zeros, and the corresponding CSD representation (0010000-1) has six zeros. The CSD representation however does not map well on the DA structure, as it requires representing one out of three values $\{0,1,-1\}$ at each of the bit location, instead of one out of two values

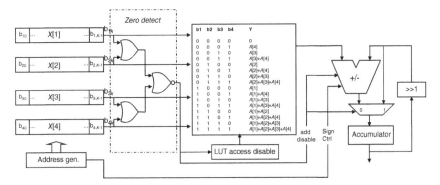

Figure 3.24 Using zero-detect logic to reduce LUT reads and additions.

in case of 2's complement binary representation. Input recoding to -2 radix representation has been explored in Reference [25] as a technique to increase the number of all zero vectors over 2's complement representation.

3.5.4 Nega-Binary Coding for Reducing Input Toggles and LUT Look-Ups

In addition to the LUT accesses and accumulation, the switching activity in the input data section also contributes to the power dissipation. The input variables are typically stored in shift registers and right shifted each cycle to get the address for the LUT. Thus, per each output computation, there are $N \times (K - 1)$ shifts performed in the input data section. If the adjacent bits are different, the shift operation results in toggles. The power dissipated in case of toggles is significantly higher than when the same bit value is shifted.

For applications where the distribution of data values is known, a suitable data coding scheme can be derived which results in less toggles in the shift registers. The main constraint is to have a scheme that results in toggle reduction with minimal hardware overhead (implying power dissipation overhead as well). The second important constraint is that the coding scheme should be programmable so that the same hardware can be used for different distribution profiles of data values, and also can be made adaptive to match changing characteristics of the input data. The nega-binary scheme [26] discussed here satisfies these two constraints and can be directly incorporated into the DA structure.

Nega-binary numbers belong to a generic case of a 2's complement representation. Consider a K bit 2's complement number. Only the most significant bit (MSB) has a weight of -1 while all others have a weight of $+1$. A K-bit nega-binary number is a weighted sum of $\pm 2^i$. As a special case, consider a weighted $(-2)i$ series, where nb_i denotes the ith bit in the nega-binary representation of a number X. The value of X is then given by: $X = \sum_{i=0}^{K-1} nb_i \times (-2)^i$. In this case, powers of 2 alternate in signs. Let us consider a simple example to see how this nega-binary scheme can result in reduced number of toggles. Consider an eight-bit 2's complement number 10101010_B. The corresponding nega-binary representation is 11111111_{NB}. While the first 2's complement representation results in maximum possible toggles, the nega-binary representation has minimum toggles. If instead, if we consider the number was 01010101_B, the corresponding nega-binary representation will be the same, and hence result in the same number of toggles as the 2's complement representation. However, if a different nega-binary scheme given by $X = \sum_{i=0}^{K-1} -nb_i \times (-2)^i$ is selected, it results in a nega-binary representation 11111110_{NB} with just one toggle. Thus, it can be noted that different nega-binary schemes have different "regions" in their entire range resulting in fewer toggles. Therefore, depending on the data distribution one can choose a scheme that minimizes toggling without altering the basic DA-based structure.

While the 2's complement representation has the range of $\left[-2^{(K-1)}, 2^{(K-1)} - 1\right]$, the nega-binary scheme $X = \sum_{i=0}^{K-1} nb_i \times (-2)^i$ has the range of $\left[-\left(4^{\lfloor K/2 \rfloor} - 1\right) \times 2/3, \left(4^{\lceil K/2 \rceil} - 1\right)/3\right]$. It can be noted that the nega-binary scheme results in a different range of numbers than the 2's complement representation for the same number of bits.

Thus, there can be a number that has a K bit 2's complement representation but does not have a K bit nega-binary representation. For example, number 100 which has a eight-bit 2's complement representation, but cannot be represented in eight-bit nega-binary scheme.

While the term "nega-binary" typically refers to binary representation with radix -2. The definition of the term can be extended to encompass all possible representations obtained by using $\pm 2^i$ as weights for the ith bit. Hence, for a K-bit precision there exist $2K$ different nega-binary schemes. The range of numbers covered by each of the $2K$ nega-binary schemes is different. The bit precision hence needs to be increased to $K + 1$, so as to have a range which accommodates all K bit numbers in 2's complement representation. It can be shown that out of $2(K+1)$ nega-binary schemes, there exist $(2K + 1)$ schemes which can represent the entire range of the K bit 2's complement representation. For example, from a total of 32 different 5-bit nega-binary schemes, 17 schemes cover the entire range of 4-bit 2's complement representation.

For a given profile of input data distribution a nega-binary scheme can be selected, out of the ones which overlap with the 2's complement representation, such that it minimizes the total weighted toggles (i.e., the product of the number of toggles in a data value and the corresponding probability of its occurrence). Figure 3.25 shows the distribution profile of a typical audio data extracted from an audio file.

The nonuniform nature of the distribution is quite apparent in Figure 3.25. A nega-binary scheme that has the minimum number of toggles in data values with very high probability of occurrence will substantially reduce power

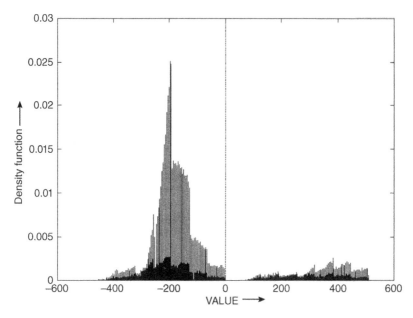

Figure 3.25 Typical data distribution across 25,000 samples extracted from an audio file.

consumption. Further, each of the (2^{K+1}) $(K + 1)$ bit nega-binary schemes have different "regions" of minimum toggle over the range, which implies that there exists a nega-binary representation that minimizes total weighted toggles corresponding to a data distribution peaking at a different "region" in the range.

Figure 3.26 illustrates the difference in number of toggles for a 6-bit, 2's complement representation and two different 7-bit, nega-binary representations (NB_A and NB_B) for each data value. Figure 3.27 shows two profiles for 6-bit Gaussian distributed data (GD_X and GD_Y).

The nega-binary scheme NB_A can be used effectively for a distribution like GD_X resulting in 34.2% toggle reduction. Similarly, nega-binary scheme NB_B can be used for a distribution like GD_Y resulting in 34.6% toggle reduction [26]. The nega-binary representations NB_A and NB_B depict two out of a total of 65 possibilities. Each of these peaks (i.e., the corresponding nega-binary scheme) has fewer toggles compared to the 2's complement case differently, and hence, for a given distribution, a nega-binary scheme can be selected to reduce power dissipation.

The programmable nega-binary recoding of inputs can be incorporated in the DA structure, by using a bit-serial 2's complement to nega-binary converter as shown in Figure 3.28. In case of FIR filters, only one input is read-in per output, the other $N - 1$ inputs are reused. Thus, it suffices to have this bit-serial

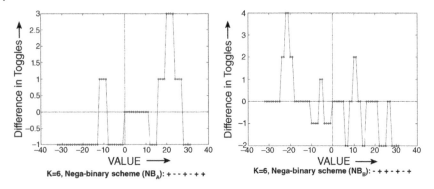

Figure 3.26 Toggle difference between nega-binary and 2's complement representations of numbers (−32 to 31).

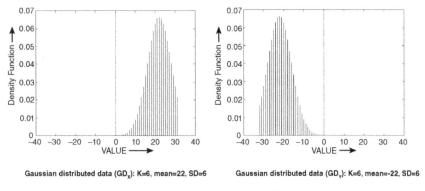

Figure 3.27 Two six-bit Gaussian data distributions with different mean values.

converter for the latest input, resulting in minimum area and power overhead of input recoding.

Since the input recoding with an appropriate nega-binary scheme aims at minimizing toggles during the shifts of the input data, such a coding also helps increase the number of identical successive LUT addresses. This not only reduces the power consumption due to LUT reads, but also results in identical LUT output. Since one input of the adder does not toggle in such cases, it leads to saving of power consumption in the adder as well. The results presented in Reference [26] show that the nega-binary coding can help reduce LUT toggles by 12–25%.

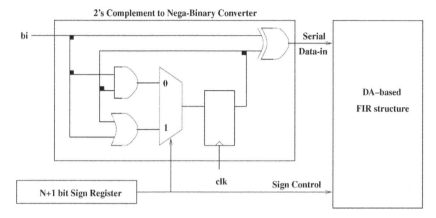

Figure 3.28 Programmable bit-serial 2's complement to nega-binary conversion.

3.5.5   Accuracy versus Power Tradeoff

One approach for power reduction is to employ "approximate signal processing" techniques [27] that drive appropriate quality versus arithmetic complexity tradeoffs dynamically. In the context of FIR filters, this can be accomplished either by reducing the number of taps or by reducing the bit-precision of inputs or coefficients or both. The multibank DA structure shown in Figure 3.9, can be used to support variation in the number of taps. For the dynamic reduction in the input bit precision, DA structure with MSB first access is ideally suited, as it supports stochastically monotonic successive approximation properties. Thus, starting with MSB first access, the DA computation can be terminated at the desired bit precision depending on the error which can be tolerated. A detailed scheme based on this technique is presented in Reference [27].

It can be noted that the approximate signal processing technique can be applied in input-recoding context as well. Instead of using $K + 1$ bit nega-binary representation to represent all values of K bit 2's complement number, K bit nega-binary scheme can be adopted, and the values falling outside the range saturated to the most positive or the most negative number as appropriate. This approach not only makes the power reduction possible by input toggle reduction and zero detect schemes but also ensures that the number of additions remain the same (K instead of $K + 1$). Given that in a typical signal processing application, there is a smaller percentage of data values using the entire bit precision, the error introduced with input data saturation can be small. If the application can tolerate the error, it can further help reduce the power consumption.

3.5.6 Reconfigurable DA-Based Implementations

Reconfigurable finite impulse response (FIR) filter whose filter coefficients change dynamically during runtime plays an important role in the software defined radio (SDR) systems [33,34], multichannel filters [35], and digital up/down converters [36]. In order to activate reconfgurability in the DA-based FIR filter, the lookup tables (LUTs) are required to be implemented in RAM, and the RAM-based LUT is found to be costly for ASIC implementation. A register-based shared-LUT design has been proposed in Reference [37] for cost-effective implementation of reconfigurable DA-based FIR filter. Instead of using separate registers to store the possible results of partial inner products for DA processing of different bit positions, registers are shared by the DA units for bit slices of different weightage.

Adaptive filters are widely used in several digital signal processing (DSP) applications. The tapped-delay line finite impulse response (FIR) filter whose weights are updated by the famous Widrow–Hoff least mean square (LMS) algorithm is the most popularly used adaptive filter not only due to its simplicity but also due to its satisfactory convergence performance [38]. The direct form configuration on the forward path of the FIR filter results in a long critical-path due to an inner-product computation to obtain a filter output. Therefore, when the input signal has a high sampling rate, it is necessary to reduce the critical-path of the structure so that the critical path could not exceed the sampling period. Hardware-efficient DA-based design of LMS adaptive filter has been suggested by Allred *et al.* [39] using two separate LUTs for filtering and weight-update. Guo *et al.* [40,41] have improved that further by using only one LUT for filtering as well as weight-updating. Parallel lookup table update and concurrent implementation of filtering and weight-update operations are proposed in Reference [42], to achieve high throughput rate. The adder-based shift-accumulation for DA-based inner-product computation has been replaced by conditional signed carry-save accumulation in order to reduce the sampling period and area-complexity. It is found that the design in Reference [42] results in significantly less power and less area-delay product (ADP) compared to the other DA-based designs.

3.6 Conclusion

In this chapter, we have presented distributed arithmetic (DA)-based circuits. DA-based multiplier-less implementation provides a compelling option for low-power implementations and the LUT-based nature lends itself for efficient FPGA implementation. It can be noted that area, power, performance advantage of DA over conventional MAC-based implementation is heavily dependent on the characteristics of the weighted-sum computation, in terms of number of terms, bit-precision of the input data and the coefficients and the desired coefficient programmability. In this chapter, we have also discussed techniques for area, power, performance optimization/tradeoffs w.r.t. the basic DA implementation. The selection of the optimization techniques is driven by the relative area, power, performance requirements of the target application, and also the implementation technology, for example, ASIC or FPGA. Apart from inner-product computation, DA has also been discussed in the literature in the context of adaptive filters [28] and also leveraged for implementing fault-tolerance [29]. There are also circuit level techniques and optimizations applicable in DA context, specifically in terms of efficient implementation of LUT and the transpose memory for the input data (write row-wise, read-column wise memory). More than 30 years since its first application, DA continues to be a compelling architecture for efficient implementation of inner-product computation.

References

1 S. Zohar, New hardware realization of nonrecursive digital filters, *IEEE Trans. Comput.*, vol. C-22, no. 4, 1973, pp. 328–338.

2 A Croisier, D. J. Esteban, M. E. Levilion, and V. Rizo, Digital filter for PCM encoded signals, U.S. Patent 3 777 130, Dec. 4, 1973.

3 A Peled and B. Liu, A new hardware realization of digital filters, IEEE Trans. Acoust. Speech, Signal Process., vol. 22, no. 6, 1974, pp. 456–462.

4 S. Burrus, Digital filter structures described by distributed arithmetic, *IEEE Trans. Circuits Syst.*, vol. 24, no. 12, 1977, pp. 674–680.

5 S. A. White, Applications of distributed arithmetic to digital signal processing: a tutorial review, *IEEE ASSP Mag.*, vol. 6 no. 3, 1989, pp. 4–19.

6 S. Zohar, A VLSI implementation of a correlator/digital filter based on distributed arithmetic, *IEEE Trans. Acoust.*, Speech, Signal Process., vol. 37, no. 1, 1989, pp. 156–160.

7 Sungwook Yu and Earl E. Swartzlander JJr. DCT implementation with distributed arithmetic. *IEEE Trans. Comput.*, vol. 50, no. 9, 2001, pp. 985–991.

8 M. Rawski, M. Wojtynski, T. Wojciechowski, and P. Majkowski, Distributed arithmetic based implementation of fourier transform targeted at FPGA architectures, in *14th Int. Conf. Mixed Design Integr. Circuits Syst. (MIXDES '07)*, 2007, pp. 152–156.

9 Ali M. Al-Haj, Fast discrete wavelet transformation using FPGAs and distributed arithmetic. *Int. J. Appl. Sci. Eng.*, vol. 1, no. 2, 2003, pp. 160–171.

10 Amira, A. Bouridane, P. Milligan, and M. Roula, Novel FPGA implementations of Walsh-Hadamard transforms for signal processing, in *IEEE Proc.–Vis. Image Signal Process.* 2001, pp. 377–383.

11 F. Bensaali, A. Amira, and A. Bouridane, An efficient architecture for color space conversion using distributed arithmetic, *in Proc. 2004 Int. Symp. Circuits Syst. (ISCAS '04)*, vol. II, 2004, pp. 265–268.

12 Pramod Kumar Meher, Shrutisagar Chandrasekaran and Abbes Amira, FPGA realization of FIR filters by efficient and flexible systolization using distributed arithmetic. *IEEE Trans. Signal Process.*, vol. 56 no. 7 2008, pp. 3009–3017.

13 Jung-Pil Choi, Swung-Cheol Shin, and Jin-Gyun Chung, Efficient ROM Size reduction for distributed arithmetic. *IEEE Int. Symp. Circuits Syst.*, 2000, vol. II, pp. 61–64.

14 Heejong Yoo and David V. Anderson, Hardware-efficient distributed arithmetic architecture for high-order digital filters, in *Proc. IEEE Int. Conf. Acoustics, Speech, Signal Process. (ICASSP '05)*, 2005, pp. V-125–V-128.

15 Mahesh Mehendale, S. D. Sherlekar, and G. Venkatesh, Area-delay tradeoff in distributed arithmetic based implementation of FIR filters, in *Int. Conf. VLSI Design (VLSI Design'97)*, 1997, pp. 124–129.

16 Amit Sinha, and Mahesh Mehendale, Improving area efficiency of FIR filters implemented using distributed arithmetic, in *Int. Conf. VLSI Design (VLSI Design'98)*, 1998, pp. 104–109.

17 Soumik Ghosh, Soujanya Venigalla, and Magdy Bayoumi, Design and implementation of a 2D-DCT architecture using coefficient distributed arithmetic, in *Proc. IEEE Comput. Soc. Annu. Symp. VLSI New Front. VLSI Design*, 2005.

18 W. K. Jenkins and B. Leon, The use of residue number system in the design of finite impulse response filters, *IEEE Trans. Circuits Syst.*, vol. 24 no. 4 1977, pp. 191–201.

19 W. K. Jenkins, A highly efficient residue-combinatorial architecture for digital filters. *Proc. IEEE*, vol. 66 no. 6 1978, pp. 700–702.

20 N. S. Szabo and R. I. Tanaka, Residue Arithmetic and its Applications to Computer Technology, Mc-Graw Hill, 1967.

21 M. A. Soderstrand and K. Al-Marayati, VLSI implementation of very-high-order FIR filters, in *IEEE Int. Symp. Circuits Syst.*, 1995, pp. 1436–1439.

22 Mahesh Mehendale, S. D. Sherlekar, and G. Venkatesh, Synthesis of multiplier-less FIR filters with minimum number of additions, in *IEEE Int. Conf. Comput. Aided Design (ICCAD'95)*, 1995, pp. 668–671.

23 M. Potkonjak, Mani Srivastava, and Anantha Chandrakasan, Multiple constant multiplications: efficient and versatile framework and algorithms for exploring common subexpression elimination, *IEEE Trans. Comput. Aided Design*, vol. 15 no. 2 1996, pp. 151–165.

24 Amit Sinha and Anantha Chandrakasan, Energy Efficient Filtering using Adaptive Precision and Variable Voltage, 12th Annual IEEE ASIC conference, 1999.

25 J. Sacha and M.J. Irwin, Input recording for reducing power in distributed arithmetic in Proc. *IEEE Workshop Signal Process. Syst. (SIPS 98)*, 1998, pp. 599–608.

26 Mahesh Mehendale, Amit Sinha, and S. D. Sherlekar, Low power realization of fir filters implemented using distributed arithmetic, Asia and South Pacific Design Automation Conf. *(ASP-DAC'98)*, 1998, pp. 151–156.

27 R. Amirtharajah and A. Chandrakasan, A micropower programmable DSP using approximate signal processing based on distributed arithmetic, *IEEE J. Solid State Circuits*, vol. 39 no. 2 2004, pp. 337–347.

28 N. Tan, S. Eriksson, and L. Wanhammar, A power-saving technique for bit-serial DSP ASICs *(ISCAS)*, vol. IV, 1994, pp. 51–54.

29 K. Gaedke, J. Franzen and P. Pirsch, A fault-tolerant DCT-architecture based on distributed arithmetic. *IEEE Int. Symp. Circuits Syst., (ISCAS)*, 1993, pp. 1583–1586.

30 Patrick Longa and Ali Miri, Area-efficient FIR filter design on FPGAs using distributed arithmetic, *IEEE Int. Symp. Signal Process. Inf. Technol.*, 2006, pp. 248–252.

31 Sangyun Hwang, Gunhee Han, Sungho Kang, and Jaeseok Kim, New distributed arithmetic algorithm for low-power FIR filter implementation, *IEEE Signal Process. Lett.*, vol. 11 no. 5 2004, pp. 463–466.

32 K. Nourji and N. Demassieux, Optimization of real-time VLSI architectures for distributed arithmetic based algorithms: application to HDTV filters, in *Proc. IEEE Int. Symp. Circuits Syst. (ISCAS 1994)*, vol. 4, June 1994, pp. 223–226.

33 T. Hentschel, M. Henker, and G. Fettweis, The digital front-end of software radio terminals. *IEEE Personal Commun. Mag.*, vol. 6, no. 4, 1999, pp. 40–46.

34 K.-H. Chen and T.-D. Chiueh, A low-power digit-based reconfigurable FIR filter. *IEEE Trans. Circuits Syst.* II, vol. 53, no. 8, 2006, pp. 617–621.

35 L. Ming and Y. Chao, The multiplexed structure of multi-channel FIR filter and its resources evaluation, in 2012 *Int. Conf. Comp. Distributed Control Intelligent Environmental Monitoring (CDCIEM)*, Mar. 2012.

36 I. Hatai, I. Chakrabarti, and S. Banerjee, Reconfigurable architecture of a RRC FIR interpolator for multi-standard digital up converter, in *2013 IEEE 27th Int. Parallel Distributed Process. Symp. Workshops Ph.D. Forum (IPDPSW)*, May 2013.

37 S. Y. Parand P. K. Meher, Efficient FPGA and ASIC realizations of DA-based reconfigurable FIR digital filter. *IEEE Trans. Circuits Syst.* II, vol. 61, no. 7, 2014, pp. 511–515.

38 S. Haykin and B. Widrow, Least-Mean-Square Adaptive Filters, John Wiley & Sons, Inc, Hoboken, NJ, USA, 2003.

39 D. J. Allred, H. Yoo, V. Krishnan, W. Huang, and D. V. Anderson, LMS adaptive filters using distributed arithmetic for high throughput. *IEEE Trans. Circuits Syst.* I, Reg. Papers, vol. 52, no. 7, 2005, pp. 1327–1337.

41 R. Guo and L. S. DeBrunner, Two high-performance adaptive filter implementation schemes using distributed arithmetic. *IEEE Trans. Circuits Syst. II, Exp. Briefs*, vol. 58, no. 9, 2011, pp. 600–604.

41 R. Guo and L. S. DeBrunner, A novel adaptive filter implementation scheme using distributed arithmetic, in *Proc. Asilomar Conf. Signals, Syst., Comput.*, Nov. 2011, pp. 160–164.

42 S. Y. Park and P. K. Meher, Low-Power, High-Throughput, and Low-Area Adaptive FIR Filter Based on Distributed Arithmetic. *IEEE Trans. Circuits Syst. II, Exp. Briefs*, vol. 60, no. 6, 2013, pp. 346–350.

4

Table-Based Circuits for DSP Applications

Pramod Kumar Meher[1] and Shen-Fu Hsiao[2]

[1] *Independent Hardware Consultant*

[2] *Department of Computer Science and Engineering, National Sun Yat-sen University, Kaohsiung, Taiwan*

4.1   Introduction

A lookup table (LUT) can be formally defined as an array of storage cells that is initialized by a set of precomputed values. LUTs can be used for performing arithmetic operations such as multiplications with a constant, divisions by a constant, exponentiations of constants, inversion, evaluation of square and square root of integers, and so on. It can also be used for the evaluation of nonlinear functions such as sinusoidal functions, sigmoid functions, logarithms and antilogarithms, and so on. LUT-based computation is popularly used in a wide range of applications, not only in various application-specific platforms but also in programmable systems, where memory units are used either as a complete arithmetic circuit or as a part of that for the evaluation of nonlinear functions. The most fundamental concept underneath the table-based computing systems is that if the desired results of computations could be stored in a memory unit, then instead of performing actual arithmetic operations the results could be retrieved directly from appropriate memory locations. It is very much close to the way in which human brain performs the computations. Human brain functions in a way distinctively different from the conventional computer systems, at least for the execution of arithmetic operations. It is for sure that, there are no arithmetic circuits within the brain. Otherwise no one needs to memorise the tables of addition and the tables of multiplication and squaring. While doing manual calculations in our everyday life/work, we use a set of multiplication tables and addition tables that we have learnt from our

Arithmetic Circuits for DSP Applications, First Edition. Edited by Pramod Kumar Meher and Thanos Stouraitis.
© 2017 by The Institute of Electrical and Electronics Engineers, Inc. Published 2017 by John Wiley & Sons, Inc.

childhood. How simple and convenient indeed it might be, if we could remember the results of all arithmetic operations of any two numbers! Never the less, we know the methods of doing calculations involving large numbers with the help of small tables of results that we remember.

In LUT-based computing systems, the use of combinational arithmetic circuits is minimized, and in lieu of that, desired results of arithmetic operations are read out directly from a memory unit like ROM, RAM, or a register file. The LUT-based computing structures are inherently regular, since the memory cells are placed in regular layout within a chip. Besides, very often they involve a delay of only one memory access time to perform a desired operation, irrespective of the complexity of the operation. The LUT-based designs have greater potential for high-throughput and reduced-latency implementation, since the memory access-time very often could be shorter than the usual time for arithmetic operations or evaluation of functions by actual arithmetic circuits. Moreover, they involve very less dynamic power consumption due to less switching activities for memory read operations compared to the arithmetic circuits for executing those operations.

Along with the progressive scaling of silicon devices, over the years, semiconductor memory has become cheaper, faster, and more power-efficient. It has also been found that the transistor packing density of SRAM is not only high, but also increasing much faster than the transistor density of logic devices[1]. According to the requirement of different application environments, memory technology has been advanced in a wide and diverse manner. Radiation-hardened memories for space applications, wide temperature memories for automotive, high reliability memories for biomedical instrumentation, low-power memories for consumer products, and high-speed memories for multimedia applications are under continuous development process to take care of the special needs [1,2]. The development in memory technology provides an additional impetus to go in favor of designing efficient LUT-based computing systems not only for dedicated processors but also for the functional units for programmable processors. The integration of logic and memory components in the same chip paves the way for more effective utilization of LUT-based computing devices for improving the performance in terms of various incompatible and competitive design metrics for resource constrained application environments.

The LUT-based structures are well-suited for hardware realization of computation-intensive digital signal processing (DSP) applications, which involve multiplications with a fixed set of coefficients or inner-product computation with a fixed vector. Multiplication by constants is very often encountered in popular DSP applications that involve digital filtering or orthogonal transforms

---

1 International Technology Roadmap for Semiconductors Available at http://public.itrs.net/

such as discrete Fourier transform (DFT), discrete cosine transform (DCT), discrete sine transform (DST), and discrete Hartley transform (DHT) etc. [3–10]. LUTs have been utilized for evaluation of logarithms, antilogarithms, and trigonometric functions in programmable processors. It is a popular approach for the estimation of sigmoid functions that is one of the bottlenecks in the implementation of artificial neural networks [11–17]. LUT multipliers are used for multiplication with a fixed set of coefficients, while inner products with fixed vectors are evaluated by LUT-based structures using the principle of distributed arithmetic (DA) [18–20].

The objective of this chapter is to provide a brief overview of the key concepts, important developments, implementations, and applications of some popularly used LUT-based computing systems. Although the computations by DA approach is one of the most important applications of LUT-based computing, in this chapter we are not discussing that since that is discussed in Chapter 3. The remainder of this chapter is organized as follows. LUT design for evaluation of Boolean function is discussed in Section 4.2. The principle of LUT-based constant multipliers and their implementation are discussed in Section 4.3. The use of LUTs in evaluation of nonlinear functions is discussed in Section 4.4. Application of LUT-based designs for the implementation of circular convolution, calculation of reciprocals, and the evaluation of sigmoid function are discussed in Section 4.5. Finally, a short summary of this chapter is presented in Section 4.6.

4.2 LUT Design for Implementation of Boolean Function

We can find that an LUT of 1-bit width having the required number of entries can be used for the implementation of any Boolean function. It is the simplest and the most popular application of LUTs these days due to its widespread use in FPGA devices. Most of the combinational logic in FPGA devices are implemented by LUTs instead of implementing them by logic gates so that different logic functions can be implemented by the sáme device by changing the LUT entries. The LUT-based implementation, therefore, provides the desired reconfigurability to these devices.

For example, let us consider the computation of the sum and carry of a 1-bit full adder to add three bits a, b, and c, given by

$$\text{sum} = a \oplus b \oplus c \tag{4.1}$$
$$\text{carry} = (a \cdot b) + (c \cdot (a \oplus b)) \tag{4.2}$$

A combinational logic circuit of full adder to compute the sum and carry according to Eq. (4.1) is shown in Figure 4.1a.

A combined truth table of sum and carry given by Eq. (4.1) is shown in Table 4.1. If we store the bits in column 4 and 5 of Table 4.1 in an LUT of 2-bit width

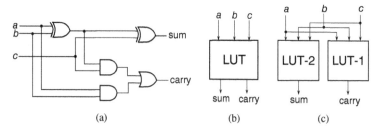

Figure 4.1 (a) A combinational circuits of 1-bit full adder. (b) LUT-based 1-bit full adder. (c) 1-bit full adder using LUTs of 1-bit width.

Table 4.1 Truth tables for sum and carry of a 1-bit full adder.

a	b	c	Carry	Sum
0	0	0	0	0
0	0	1	0	1
0	1	0	0	1
0	1	1	1	0
1	0	0	0	1
1	0	1	1	0
1	1	0	1	0
1	1	1	1	1

and use the bit pattern ($a\ b\ c$) as address of the LUT, we can read out the sum and carry bits. Instead of using a logic circuit for 1-bit full adder consisting of logic gates like that of Figure 4.1a, we can use an LUT of 2-bit width as shown in Figure 4.1b. LUTs can similarly be used for the implementation of any Boolean function and which in turn can be used to realize any combinational unit for arithmetic operation or control logic. The truth table of Table 4.1 for a three-variable logic function requires eight entries. In general, the number of LUT entries to realize a Boolean function of n Boolean variables is 2^n.

Any desired Boolean function is realized by an LUT of 1-bit width in FPGA devices as shown in Figure 4.1c for the carry and sum bits. Such 1-bit LUTs constitute the main computing resource of FPGA devices to implement combinational functions [21,22]. LUT-based approach for realization of Boolean functions is the core design strategy to achieve reconfigurability in FPGA devices since the functional behavior of the LUT can be easily modified by changing the LUT entries. The LUT content in an FPGA device can be modified when the FPGA is reprogrammed, which provides logic level reconfigurability to FPGA devices.

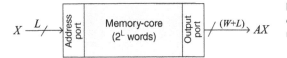

Figure 4.2 Functional model of basic lookup table multiplier.

4.3 Lookup Table Design for Constant Multiplication

The principle of LUT-based constant multiplication is shown in Figure 4.2. Let A be a known constant and X be an input word to be multiplied with A. Assuming X to be a positive binary number of word length L, there can be 2^L possible values of X, and accordingly, there can be 2^L possible values of product $C = A \cdot X$. Therefore, conventionally, multiplication of an L-bit input word X requires an LUT of 2^L words consisting of precomputed product values corresponding to all possible values of X. The product word $A \cdot X_i$ is stored at the location whose address is the same as binary representation of X_i for $0 \leq X_i \leq 2^L - 1$, such that if L-bit binary value of X_i is used as address word, then the corresponding product value $A \cdot X_i$ is available as LUT output.

The main difficulty of LUT-based approach is that the LUT size becomes too large when precision of input X is high, since it requires 2^L LUT entries when the input word length is L. Two simple optimization techniques can be used to reduce the LUT size for constant multiplication. Those are: (i) antisymmetric product coding (APC) scheme and (ii) odd multiple-storage (OMS) scheme. In the OMS scheme only odd multiples of the constant are stored, while the even multiples are derived by left-shift operations of one of those odd multiples [23]. Using the APC approach, the LUT size is reduced to half, where the product words are recoded as antisymmetric pairs [24]. The LUT-based implementation of multiplier using a combination of APC and OMS techniques is discussed in the following.

4.3.1 Lookup Table Optimizations for Constant Multiplication

The LUT optimizations for multiplication of a number X with a given constant A is possible for both sign-magnitude and 2's complement representations of X and A. Besides, both X and A could be fractions or integers in fixed-point format. But, for simplicity of discussion, we assume both X and A to be positive integers[2].

---

2 It could however, be easily extended for signed values of A and X in sign-magnitude form or two's complement form.

Table 4.2 The APC-based LUT allocation for different values of X for $L = 5$

Input, X	Product values	APC representation	Input, X	Product values	APC representation	Storage address $\widetilde{X} = \tilde{x}_3\tilde{x}_2\tilde{x}_1\tilde{x}_0$	Stored words 2's complement	Sign-magnitude
0 0 0 0 1	A	$(16A - 15A)$	1 1 1 1 1	$31A$	$(16A + 15A)$	1 1 1 1	$15A$	$15|A|$
0 0 0 1 0	$2A$	$(16A - 14A)$	1 1 1 1 0	$30A$	$(16A + 14A)$	1 1 1 0	$14A$	$14|A|$
0 0 0 1 1	$3A$	$(16A - 13A)$	1 1 1 0 1	$29A$	$(16A + 13A)$	1 1 0 1	$13A$	$13|A|$
0 0 1 0 0	$4A$	$(16A - 12A)$	1 1 1 0 0	$28A$	$(16A + 12A)$	1 1 0 0	$12A$	$12|A|$
0 0 1 0 1	$5A$	$(16A - 11A)$	1 1 0 1 1	$27A$	$(16A + 11A)$	1 0 1 1	$11A$	$11|A|$
0 0 1 1 0	$6A$	$(16A - 10A)$	1 1 0 1 0	$26A$	$(16A + 10A)$	1 0 1 0	$10A$	$10|A|$
0 0 1 1 1	$7A$	$(16A - 9A)$	1 1 0 0 1	$25A$	$(16A + 9A)$	1 0 0 1	$9A$	$9|A|$
0 1 0 0 0	$8A$	$(16A - 8A)$	1 1 0 0 0	$24A$	$(16A + 8A)$	1 0 0 0	$8A$	$8|A|$
0 1 0 0 1	$9A$	$(16A - 7A)$	1 0 1 1 1	$23A$	$(16A + 7A)$	0 1 1 1	$7A$	$7|A|$
0 1 0 1 0	$10A$	$(16A - 6A)$	1 0 1 1 0	$22A$	$(16A + 6A)$	0 1 1 0	$6A$	$6|A|$
0 1 0 1 1	$11A$	$(16A - 5A)$	1 0 1 0 1	$21A$	$(16A + 5A)$	0 1 0 1	$5A$	$5|A|$
0 1 1 0 0	$12A$	$(16A - 4A)$	1 0 1 0 0	$20A$	$(16A + 4A)$	0 1 0 0	$4A$	$4|A|$
0 1 1 0 1	$13A$	$(16A - 3A)$	1 0 0 1 1	$19A$	$(16A + 3A)$	0 0 1 1	$3A$	$3|A|$
0 1 1 1 0	$14A$	$(16A - 2A)$	1 0 0 1 0	$18A$	$(16A + 2A)$	0 0 1 0	$2A$	$2|A|$
0 1 1 1 1	$15A$	$(16A - A)$	1 0 0 0 1	$17A$	$(16A + A)$	0 0 0 1	A	$|A|$
1 0 0 0 0	16	$(16A - 0)$	1 0 0 0 0	$16A$	$(16A + 0)$	0 0 0 0	$16A$	$16|A|$

4.3.1.1 Antisymmetric Product Coding for LUT Optimization

The product words for different values of X for $L = 5$ are shown in Table 4.2. It may be observed in this table that the input word X on the first column of each row is 2's complement of that on the fourth column of the same row. Besides, the sum of product values corresponding to these two input values on the same row is $32A$. Let the product values on the second and the fifth columns of a row be u and v, respectively. Since one can write $u = \left[\frac{u+v}{2} - \frac{v-u}{2}\right]$ and $v = \left[\frac{u+v}{2} + \frac{v-u}{2}\right]$, for $(u + v) = 32A$, we can have

$$u = 16A - \left[\frac{v-u}{2}\right] \text{ and } v = 16A + \left[\frac{v-u}{2}\right] \tag{4.3}$$

The product values on the second and fourth columns of Table 4.3, therefore, have a *negative mirror symmetry*. This behavior of the product words can be used to reduce the LUT size, where instead of storing u and v, only $\left[\frac{(v-u)}{2}\right]$ is stored for a pair of input on a given row. The 4-bit LUT addresses and corresponding coded words are listed on seventh and eighth columns of the table, respectively. Since the representation of product is derived from the antisymmetric behavior of the products, we can name it as *antisymmetric product code*. The 4-bit address $X' = (x_3' x_2' x_1' x_0')$ of the APC word is given by

$$X' = \begin{cases} X_L, & \text{if } x_4 = 1. \\ X_L', & \text{if } x_4 = 0 \end{cases} \tag{4.4}$$

where $X_L = (x_3 x_2 x_1 x_0)$ consists of four less significant bits of X, and X_L' is 2's complement of X_L. The desired product could be obtained by adding or subtracting the stored value $(v - u)/2$ to or from the fixed value $16A$, when x_4 is 1 or 0, respectively, that is,

$$\text{Product word} = 16A + (\text{sign-value}) \times (\text{APC word}) \tag{4.5}$$

where the sign-value $= 1$ for $x_4 = 1$ and sign-value $= -1$ for $x_4 = 0$. The product value for $X = (10000)$ corresponds to APC value "zero", which could be derived by resetting the LUT output, instead of storing that in the LUT.

4.3.1.2 Odd Multiple-Storage for LUT Optimization

It is easy to find that for the multiplication of any binary word X of size L, with a fixed coefficient A, instead of storing all the 2^L possible values of $C = A \cdot X$, only $(2^L/2)$ words corresponding to the odd multiples of A may be stored in the LUT, while all the even multiples of A could be derived by left-shift operations of one of those odd multiples [23]. Based on the above assumptions, the LUT for

Table 4.3 The APC words for different input values for $L = 5$.

Input, X	Product values	Input, X	Product values	Address $x_3' x_2' x_1' x_0'$	APC words
0 0 0 0 1	A	1 1 1 1 1	$31A$	1 1 1 1	$15A$
0 0 0 1 0	$2A$	1 1 1 1 0	$30A$	1 1 1 0	$14A$
0 0 0 1 1	$3A$	1 1 1 0 1	$29A$	1 1 0 1	$13A$
0 0 1 0 0	$4A$	1 1 1 0 0	$28A$	1 1 0 0	$12A$
0 0 1 0 1	$5A$	1 1 0 1 1	$27A$	1 0 1 1	$11A$
0 0 1 1 0	$6A$	1 1 0 1 0	$26A$	1 0 1 0	$10A$
0 0 1 1 1	$7A$	1 1 0 0 1	$25A$	1 0 0 1	$9A$
0 1 0 0 0	$8A$	1 1 0 0 0	$24A$	1 0 0 0	$8A$
0 1 0 0 1	$9A$	1 0 1 1 1	$23A$	0 1 1 1	$7A$
0 1 0 1 0	$10A$	1 0 1 1 0	$22A$	0 1 1 0	$6A$
0 1 0 1 1	$11A$	1 0 1 0 1	$21A$	0 1 0 1	$5A$
0 1 1 0 0	$12A$	1 0 1 0 0	$20A$	0 1 0 0	$4A$
0 1 1 0 1	$13A$	1 0 0 1 1	$19A$	0 0 1 1	$3A$
0 1 1 1 0	$14A$	1 0 0 1 0	$18A$	0 0 1 0	$2A$
0 1 1 1 1	$15A$	1 0 0 0 1	$17A$	0 0 0 1	A
1 0 0 0 0	$16A$	1 0 0 0 0	$16A$	0 0 0 0	0

For $X = (0\,0\,0\,0\,0)$, the encoded word to be stored is $16A$.

the multiplication of an L-bit input with W-bit coefficient could be designed by the following strategy:

- A memory unit of $[(2^L/2) + 1]$ words of $(W + L)$-bit width can be used to store the product values, where the first $(2^L/2)$ words are odd multiples of A, and the last word is zero.
- A barrel shifter for producing a maximum of $(L - 1)$ left-shifts is used to derive all the even multiples of A.
- The L-bit input word is mapped to $(L - 1)$-bit address of the LUT by an address encoder; and control-bits for the barrel shifter are derived by a control circuit.

In Table 4.4, it is shown that the eight odd multiples, $A \times (2i + 1)$ are stored as P_i for $i = 0, 1, 2, \ldots, 7$ at eight LUT locations. The even multiples, $2A, 4A$, and $8A$ are derived by left-shift operations of A. Similarly, $6A$ and $12A$ are derived by left-shifting $3A$, while $10A$ and $14A$ are derived by left-shifting $5A$ and $7A$, respectively. A barrel shifter for producing a maximum of three left-shifts could be used to derive all the even multiples of A. As required by Eq. (4.5), the word to be stored for $X = (00000)$ is not 0 but $16A$, which we

Table 4.4 OMS-based design of LUT of APC words for $L = 5$.

Input X' $x_3' x_2' x_1' x_0'$	Product value	# of shifts	Shifted input, X''	Stored APC word	Address $d_3 d_2 d_1 d_0$
0 0 0 1	A	0	0 0 0 1	$P0 = A$	0 0 0 0
0 0 1 0	$2 \times A$	1			
0 1 0 0	$4 \times A$	2			
1 0 0 0	$8 \times A$	3			
0 0 1 1	$3A$	0	0 0 1 1	$P1 = 3A$	0 0 0 1
0 1 1 0	$2 \times 3A$	1			
1 1 0 0	$4 \times 3A$	2			
0 1 0 1	$5A$	0	0 1 0 1	$P2 = 5A$	0 0 1 0
1 0 1 0	$2 \times 5A$	1			
0 1 1 1	$7A$	0	0 1 1 1	$P3 = 7A$	0 0 1 1
1 1 1 0	$2 \times 7A$	1			
1 0 0 1	$9A$	0	1 0 0 1	$P4 = 9A$	0 1 0 0
1 0 1 1	$11A$	0	1 0 1 1	$P5 = 11A$	0 1 0 1
1 1 0 1	$13A$	0	1 1 0 1	$P6 = 13A$	0 1 1 0
1 1 1 1	$15A$	0	1 1 1 1	$P7 = 15A$	0 1 1 1

can obtain from A by four left-shifts using a barrel shifter. However, if $16A$ is not derived from A, only maximum of three left-shifts are required to obtain all other even multiples of A. Maximum of three bit-shifts can be implemented by a two-stage logarithmic barrel shifter, but implementation of four shifts requires a three-stage barrel shifter. Therefore, it would be more efficient strategy to store $2A$ for input $X = (00000)$, so that the product $16A$ can be derived by three arithmetic left-shifts.

The product values and encoded words for input words $X = (00000)$ and (10000) are shown separately in Table 4.5. For $X = (00000)$, the desired encoded word $16A$ is derived by 3-bit left-shifts of $2A$ (stored at address (1000)). For $X = (10000)$, the APC word "0" is derived by resetting the LUT output using an active high RESET signal given by

$$\text{RESET} = (\overline{x_0 + x_1 + x_2 + x_3}) \cdot x_4 \qquad (4.6)$$

It may be seen from Tables 4.4 and 4.5 that the 5-bit input word X can be mapped into 4-bit LUT address $(d_3 d_2 d_1 d_0)$ by a simple set of mapping relations

$$d_i = x_{i+1}'', \text{ for } i = 0, 1, 2, \text{ and } d_3 = \overline{x_0''} \qquad (4.7)$$

Table 4.5 Products and encoded words for $X = (00000)$ and (10000).

Input X $x_4x_3x_2x_1x_0$	Product values	Encoded word	Stored values	# of shifts	Address $d_3d_2d_1d_0$
1 0 0 0 0	16A	0	– – –	– –	– – –
0 0 0 0 0	0	16A	2A	3	1 0 0 0

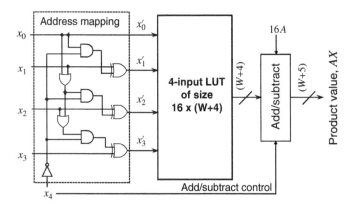

Figure 4.3 LUT-based multiplier for $L = 5$ using APC technique.

where $X'' = (x_3''x_2''x_1''x_0'')$ is generated by shifting out all the leading zeros of X' by arithmetic right-shift followed by address mapping, that is,

$$X'' = \begin{cases} Y_L, & \text{if } x_4 = 1. \\ Y_L', & \text{if } x_4 = 0 \end{cases} \qquad (4.8)$$

where Y_L and Y_L' are derived by circularly shifting out all the leading zeros of X_L and X_L', respectively. Note that, in case of $X = (00000)$ and $X = (10000)$, the circular shifting does not make any difference to X_L and X_L'.

4.3.2 Implementation of LUT-Multiplier using APC for $L = 5$

The structure and function of LUT-based multiplier for $L = 5$ using APC technique is shown in Figure 4.3. It consists of a 4-input LUT of 16 words to store the APC values of product words as given on the sixth column of Table 4.3,

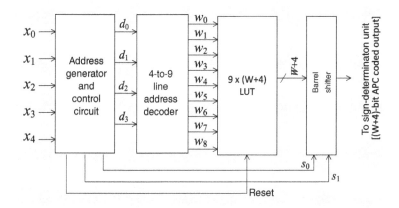

Figure 4.4 APC-OMS combined LUT design for the multiplication. Input bit-width, L=5.

except on the last row, where $2A$ is stored for input $X = (00000)$ instead of storing a "0" for input $X = (10000)$. Besides, it consists of an address mapping circuit, and an add/subtract circuit. The address mapping circuit generates the desired address $(x_3' x_2' x_1' x_0')$ according to Eq. (4.4). A straightforward implementation of address mapping can be done by multiplexing X_L and X_L' using x_4 as the control bit. The address mapping circuit, however, can be optimized to be realized by three XOR gates, three AND gates, two OR gates, and a NOT gate as shown in Figure 4.3. Note that the RESET can be generated by a control circuit (not shown in this figure) according to Eq. (4.6). The output of LUT is added with or subtracted from $16A$, for $x_4 = 1$ or 0, respectively, according to Eq. (4.5) by the add/subtract cell. Hence, x_4 is used as the control for the add/subtract cell.

4.3.3 Implementation of Optimized LUT using OMS Technique

The APC–OMS combined design of LUT for $L = 5$, and for any coefficient width W is shown in Figure 4.4. It consists of an LUT of nine words of $(W + 4)$-bit width, a 4-to-9 line address decoder, a barrel shifter, an address generation circuit, and a control circuit for generating the RESET signal and control word $(s_1 s_0)$ for the barrel shifter.

The precomputed values of $A \times (2i + 1)$ are stored as P_i for $i = 0, 1, 2, \dots, 7$ at the eight consecutive locations of the memory array as specified in Table 4.4, while $2A$ is stored for input $X = (00000)$ at LUT address "1000" as specified in Table 4.5. The decoder takes the 4-bit address from the address generator, and

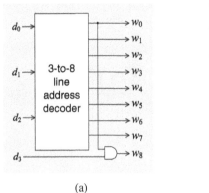
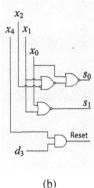

(a)   (b)

Figure 4.5 (a) 4-to-9 line address decoder. (b) Control circuit for generation of s_0, s_1 and RESET.

generates nine word-select signals, $\{w_i,$ for $0 \le i \le 8\}$, to select the referenced word from the LUT. The 4-to-9 line decoder is a simple modification of 3-to-8 line decoder as shown in Figure 4.5a. The control bits s_0 and s_1 to be used by the barrel shifter to produce the desired number of shifts of the LUT output are generated by the control circuit, according to the following relations:

$$s_0 = x_0 + \overline{(x_1 + \overline{x_2})} \tag{4.9a}$$
$$s_1 = \overline{(x_0 + x_1)} \tag{4.9b}$$

Note that $(s_1 s_0)$ is a 2-bit binary equivalent of the required number of shifts specified in Tables 4.4 and 4.5. The RESET signal given by Eq. (4.6), can alternatively be generated as $(d_3$ AND $x_4)$. The control circuit to generate the control word and RESET is shown in Figure 4.5b. The address generator circuit receives the five-bit input operand X, and maps that onto the four-bit address word $(d_3 d_2 d_1 d_0)$, according to Eqs (4.7) and (4.8). A simplified address generator is presented later in this section.

4.3.4 Optimized LUT Design for Signed and Unsigned Operands

The APC–OMS combined optimization of LUT can also be performed for signed values of A and X. When both the operands are in sign-magnitude form, the multiples of magnitude of the fixed coefficient are to be stored in the LUT, and the sign of the product could be obtained by XOR operation of sign-bits of both the multiplicands. When both the operands are in 2's complement representation, a 2's complement operation of the output of LUT

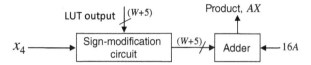

Figure 4.6 Modification of add/subtract cell of Figure 2 for 2's complement representation of product words.

is required to be performed for $x_4 = 1$. There is no need to add the fixed value $16A$ in this case, because the product values are naturally in antisymmetric form. The add/subtract circuit is not required in Figure 4.3. Instead of that a circuit is required to perform 2's complement operation of LUT output. For the multiplication of unsigned input X with singed as well as unsigned coefficient A, the products could be stored in 2's complement representation, and the add/subtract circuit of Figure 4.3 could be modified as shown in Figure 4.6. A straightforward implementation of sign-modification circuit involves multiplexing of LUT output and its 2's complement. To reduce the area–time complexity over such straightforward implementation, we discuss here a simple design for sign-modification of LUT output.

Note that except the last word, all other words in the LUT are odd multiples of A. The fixed coefficient could be even or odd, but if we assume A to be an odd number, then all the stored product words (except the last one) would be odd. If the stored value P is an odd number, it can be expressed as

$$P = P_{D-1}\, P_{D-2} \,\cdots\, P_1\, 1 \tag{4.10}$$

and its 2's complement is given by

$$P' = P'_{D-1}\, P'_{D-2} \,\cdots\, P'_1\, 1 \tag{4.11}$$

where P'_i is 1's complement of P_i for $1 \le i \le D - 1$, and $D = W + L - 1$ is the width of the stored words. If we store the 2's complement of all the product values, and change the sign of LUT output for $x_4 = 1$, the sign of last LUT word need not be changed. Based on Eq. (4.11), we can therefore, have a simple sign-modification circuit (shown in Figure 4.7a) when A is an odd integer. But the fixed coefficient A could be even, as well. When A is a nonzero even integer, we can express it as $A' \times 2^l$, where $1 \le l \le D - 1$ is an integer and A' is an odd integer. Instead of storing multiples of A, we can store multiples of A' in the LUT, and the LUT output can be left shifted by l bits by a hardwired shifter. Similarly, using (4.7) and (4.8), we can have an address generation circuit as shown in Figure 4.7b, since all the shifted address Y_L (except the last one) is an odd integer.

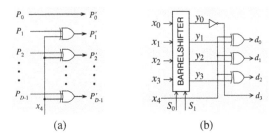

(a)   (b)

Figure 4.7 (a) Optimized circuit for sign-modification of odd LUT output. (b) Address generation circuit.

4.3.5 Input Operand Decomposition for Large Input Width

The memory size of LUT multiplier is reduced to nearly one-fourth by the APC and OMS-based optimization technique, but still it is not efficient for operands of very large widths. Therefore, the input word needs to be decomposed into certain number of segments or subwords, and the partial products pertaining to different subwords could be shift-added to obtain the desired product as discussed in the following text.

Let the input operand X be decomposed into T subwords, $\{ X_1 \ X_2 \ldots X_T \}$, where each $X_i = \{ x_{(i-1)S} \ x_{(i-1)S+1} \cdots x_{iS-1} \}$, is an S-bit subword for $1 \leq i \leq T - 1$, and $X_T = \{ x_{(T-1)S} \ x_{(T-1)S+1} \cdots x_{(T-1)S+S'-1} \}$ is the last subword of S'-bit, where $S' \leq S$, and x_i is the $(i + 1)$th bit of X. The product word $C = A \cdot X$ can be written as sum of partial products as

$$C = \sum_{i=1}^{T} 2^{S(i-1)} \cdot C_i \qquad (4.12)$$

$$C_i = A \cdot X_i \quad \text{for } 1 \leq i \leq T. \qquad (4.13)$$

A generalized structure for parallel implementation of LUT-based multiplication for input size $L = 5 \times (T - 1) + S'$ is shown in Figure 4.8, where $S' < 5$. The multiplicand X is decomposed into $(T - 1)$ more significant subwords X_1, $X_2, \ldots, X_{T-1}$ and the less significant S'-bit subword X_T. The partial products $C_i = A \cdot X_i$, for $1 \leq i \leq (T - 1)$ are obtained from $(T - 1)$ LUT multipliers optimized by APC and OMS. The Tth LUT multiplier is a conventional LUT, which is not optimized by APC and OMS, since it is required to store the sum of offset values $V = 16A \times 2^{S'} \left[\frac{2^{5(T-1)}-1}{2^5-1} \right]$ pertaining to all the $(T - 1)$ optimized LUTs used for the partial product generation[3]. The Tth LUT, therefore, stores the values $(A \cdot X_T + V)$. The sign of all the optimized LUT outputs are modified

---

3 Sum of $(T1)$ terms in GP with the first term $(2^{S'} \times 16A)$ and ratio 2^5.

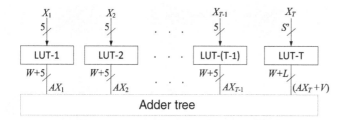

Figure 4.8 Proposed LUT-based multiplier for $L = 5(T - 1) + S'$.

according to the value of the most significant bit of corresponding subword X_i for $1 \leq i \leq T - 1$; and all the LUT outputs are added together by an adder tree as shown in Figure 4.8.

4.4 Evaluation of Elementary Arithmetic Functions

Several DSP and multimedia applications require efficient computation of elementary arithmetic functions, for example, reciprocal $(1/x)$, square root $(\sqrt{x})$, logarithm $(\log_a x)$, exponential (a^x), and trigonometric functions such as $\sin x$, $\cos x$, $\tan x$, and $\tan^{-1} x$ of any given fixed-point variable x. Dedicated hardware circuits are very often used to accelerate the evaluation of such functions. For example, special function units in modern graphics processing units (GPU) are used to compute the values of elementary functions such as reciprocal, square root, logarithm, and exponential [25–27]. Similarly, direct digital frequency synthesizer (DDS) in digital communication systems very often needs the evaluation of trigonometric functions [28,29]. LUT-based approach using lookup tables with some supporting arithmetic circuits like adders and/or multipliers are popularly used for evaluation of these arithmetic functions [30,31].

In general, LUT-based function evaluation could be divided into two categories: (i) piecewise polynomial approximation (PPA) method [32–37] and (ii) table-addition (TA) method [38–43]. The former method partitions the input interval into subintervals (or segments) and approximates the values of the function in each of those segments by a low-order (usually of degree-one or degree-two) polynomial whose coefficients or the boundary values are stored in an LUT. The latter method adds up values from several LUTs to find the desired result. PPA methods can be further divided into two subcategories, for example, uniform segmentation and nonuniform segmentation, depending on whether the sizes of the partitioned subintervals are equal or not. TA methods can also be divided into three subcategories, for example, bipartite (BP), multipartite (MP), and add-table-add (ATA). BP (MP) consists of indexing two (or

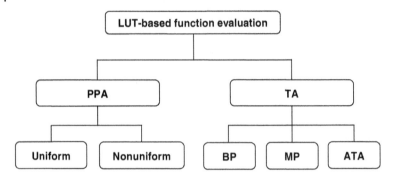

Figure 4.9 LUT-based function evaluation methods.

more than two) tables followed by an adder. ATA designs involve adders before and after the LUTs. Figure 4.9 shows the classification of LUT-based function evaluation methods.

4.4.1 Piecewise Polynomial Approximation (PPA) Approach for Function Evaluation

Without loss of generality we can consider the input variable x to be a fraction (*i. e.*, $0 \le x < 1$) in n-bit fixed-point representation. We can split the input variable x into two parts, as $x = x_m + x_l$, where x_m consists of m most significant bits (MSB) and x_l consists of $(n - m)$ least significant bits (LSB). Note that in fixed-point representation $x_l < 2^{-m}$, such that $x_m \le x < x_m + 2^{-m}$. If the value of the function $f(x)$ in any segment $[x_m, x_m + 2^{-m})$ of the input interval is approximated by a polynomial of degree 2, it can be expressed as follow:

$$f(x) \approx P(x) = p_0(x_m) + p_1(x_m) \cdot x_l + p_2(x_m) \cdot x_l^2 \qquad (4.14)$$

For approximation by a polynomial of degree 1, we can consider $p_2 = 0$ in Eq. (4.14). The coefficients p_0, p_1, and p_2 for different segments in the whole input interval could be stored in three different tables. x_m could be used to select the coefficients of the polynomial from the coefficient tables for the segment $[x_m, x_m + 2^{-m})$. Note that if the input interval is partitioned into s number of uniform segments then the number of entries in the coefficient tables should also be s.

The LUT-based circuit for function evaluation using piecewise polynomial approximation using approximation polynomial of degree 2 is shown in Figure 4.10. The coefficients p_0, p_1, and p_2 for all the input segments are stored in lookup tables T0, T1, and T2. The values of p_0, p_1, and p_2 are read from their respective LUTs by feeding x_m as address to the LUT. The squaring circuit computes x_l^2, and $p_1(x_m) \cdot x_l$ and $p_2(x_m) \cdot x_l^2$ are computed by the multipliers M1 and M2, respectively. The value of the function $f(x)$ at a given value of x is

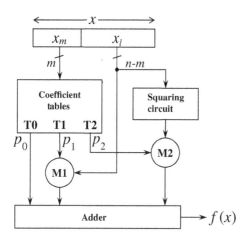

Figure 4.10 LUT-based circuit for function evaluation using piecewise polynomial approximation using approximation polynomial of degree 2.

finally obtained as the output of the 3-operand adder according to Eq. (4.14). If approximation polynomial is of degree 1, then $p_2 = 0$, so coefficient table T2, squaring circuit, and multiplier M2 in Figure 4.10 would not be required and a 2-operand adder will be required instead of 3-operand adders of Figure 4.10.

4.4.1.1 Accuracy of PPA Method

It is necessary to analyze all the sources of error in PPA in order to determine the bit width required for each hardware component in Figure 4.10 based on total error budget for the precision requirement. For example, in the degree-two approximation architecture of Figure 4.10, there are four types of errors: (i) approximation error ε_{apx}, (ii) quantization error ε_q, (iii) truncation error ε_{tr}, and (iv) rounding error ε_{rnd}. The approximation error ε_{apx} is the difference between the exact value of the function and the approximation polynomial in a segment. It is related to the number of partitioned segments and the degree of the approximation polynomial. For a given choice of degree, larger allowable approximation error means that the input interval can be divided into larger segments, leading to smaller number of entries in the LUT.

The quantization error ε_q is introduced due to the quantization of stored coefficients p_0, p_1, and p_2 in the LUTs T0, T1, and T2. It is related to the bit width of each entry in the LUTs. Truncation error ε_{tr} results from the truncation at the outputs of the multipliers M1 and M2, and at the squaring circuit. It determines the hardware cost of the internal partial product compression in the multipliers and squarer. Rounding error ε_{rnd} denotes the final rounding at the output of the multioperand adder. For faithfully rounded designs, the total error can be expressed as follows:

$$\varepsilon_{total} = \varepsilon_{apx} + \varepsilon_q + \varepsilon_{tr} + \varepsilon_{rnd} < 2^{-n} = 1\text{ulp}. \tag{4.15}$$

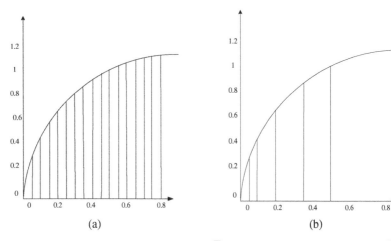

Figure 4.11 (a) Uniform segmentations of $\sqrt{x}$. (b) Nonuniform segmentations of $\sqrt{x}$.

where n is the number of fractional bits in the target accuracy and ulp represents the unit of last position (ulp), that is, 1 ulp $= 2^{-n}$.

In order to achieve the target precision, we can assign a maximum allowable error value for each individual error source, and then design the hardware components based on the allocated error budget. For example, if we choose the following error budget assignment:

$$|\varepsilon_{\text{apx}}| < \frac{1}{4}\text{ulp}, |\varepsilon_{\text{q}}| < \frac{1}{8}\text{ulp}, |\varepsilon_{\text{tr}}| < \frac{1}{8}\text{ulp}, |\varepsilon_{\text{rnd}}| < \frac{1}{2}\text{ulp}. \tag{4.16}$$

so that the total error $\varepsilon_{\text{total}} < 1$ ulp. Based on the above error assignment, we can determine the number of entries and the bit width of each entry in LUT; the number of partial product bits to be compressed in the multipliers and squarer; and the number of input bits and output bits in the final multioperand adder. A rule of thumb for the number of LUT entries is around $2^{n/3}$ for degree-two polynomial approximation and around $2^{n/2}$ for degree-one polynomial approximation [44].

4.4.1.2 Input Interval Partitioning for PPA

Conventionally the input interval is partitioned into segments of equal lengths in PPA method, which is referred to as uniform segmentation. Uniform segmentation is not efficient in terms of the number of LUT entries and LUT area particularly for evaluating functions with high nonlinearity. Recently, some improvements on uniform piecewise approximation method are proposed to partition the interval of the variable into segments of unequal lengths that is referred to as nonuniform segmentation. Figure 4.11 shows the uniform and

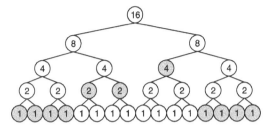

Figure 4.12 Illustration of nonuniform segmentation generated from uniform segments.

nonuniform segmentations for the evaluation of the function $\sqrt{x}$. Note that in the nonuniform segmentation, the segment density for region with sharper change (near $x = 0$) is larger than that for the region with smooth change (near $x = 1$). Nonuniform segmentation results in the reduction of total number of LUT entries, and hence the LUT area.

Figure 4.12 shows an example of a nonuniform segmentation obtained from uniform segments. Each of the 16 leaf nodes in the binary tree of Figure 4.12 represents an interval of unit length, which correspond to uniform segmentation where the input interval is partitioned into 16 segments of equal length. The gray nodes represent segments corresponding to nonuniform segmentation, where the nonuniform segments are generated by combining the segments corresponding to their child. As shown in Figure 4.12, the nonuniform segmentation requires 11 segments of size 1, 1, 1, 1, 2, 2, 4, 1, 1, 1, 1 units.

Although nonuniform segmentation helps in reducing the number of LUT entries with uniform segmentation scheme, it requires encoding of MSB to generate the address for accessing the LUT that requires additional hardware and involves additional delay. In order to achieve simpler address encoding, the size of nonuniform segments is usually taken to be power of two. In Reference [35], the lengths of nonuniform segments are either exponentially increasing or decreasing. In Reference [37], a generalized nonuniform segmentation scheme is presented that reduces the overhead of the area and delay in indexing the LUT.

4.4.2 Table-Addition (TA) Approach for Function Evaluation

TA methods use several LUTs along with some simple adders to calculate the values of function. Bipartite (BP) method is one of typical TA methods where the values of a function $f(x)$ are computed by adding values from two LUTs.

4.4.2.1 Bipartite Table-Addition Approach for Function Evaluation
In BP method, the n-bit fractional input x is split into three parts as

$$x = x_0 + x_1 2^{-n_0} + x_2 2^{-(n_0+n_1)} \tag{4.17}$$

where $n = n_0 + n_1$, and n is generally large, so that higher powers of 2^{-n} could be negligibly small compared to it. By Taylor series expansion of $f(x)$ at $(x_0 + x_1 2^{-n_0})$ we can have

$$f(x) = f(x_0 + x_1 2^{-n_0}) + x_2 2^{-(n_0+n_1)} \cdot f'(x_0 + x_1 2^{-n_0}) + \varepsilon_1 \qquad (4.18)$$

where ε_1 represents the higher terms in the expansion. If the derivative $f'(x_0 + x_1 2^{-n_0})$ in the above equation is approximated by $f'(x_0)$, which would introduce another error term ε_2, and the value of the function $f(x)$ can be obtained by adding two values $F_0(x_0, x_1)$ and $F_1(x_0, x_2)$ from two LUTs as follows:

$$f(x) = F_0(x_0, x_1) + F_1(x_0, x_2) + \varepsilon_1 + \varepsilon_2 \qquad (4.19)$$

where $F_0(x_0, x_1) = f(x_0 + x_1 2^{-n_0})$ and $F_1(x_0, x_2) + \varepsilon_1 = x_2 2^{-(n_0+n_1)} \cdot f'(x_0)$. When $\varepsilon_1 + \varepsilon_2$ is adequately small while considering the precession requirement of the application, then we can have

$$f(x) \approx F_0(x_0, x_1) + F_1(x_0, x_2) \qquad (4.20)$$

The architecture of the BP is shown in Figure 4.13a where the table of initial values (TIV) stores the initial values $F_0(x_0, x_1)$ of the function in each segment and the table of offset (TO) stores the offset values $F_1(x_0, x_2)$ relative to the initial values in TIV.

4.4.2.2 Accuracy of TA Method

In order to determine the bit width of each entry and the number of entries in the LUTs, we need to perform error analysis for BP in a similar way as has been done in PPA. The total error in BP method is sum of the approximation errors $\varepsilon_{apx} = \varepsilon_1 + \varepsilon_2$, the quantization errors $\varepsilon_q = \varepsilon_{q0} + \varepsilon_{q1}$, and the final rounding error ε_{rnd}. For faithfully rounded BP designs, the number of LUT entries and the bit widths should be appropriately selected so that total error is limited to 1 ulp, given by,

$$\varepsilon_{total} = \varepsilon_{apx} + \varepsilon_q + \varepsilon_{rnd} < 1\text{ulp}. \qquad (4.21)$$

As a rule of thumb, to achieve accuracy of n bits, the values of each of n_0, n_1, and n_2 are selected to be around $n/3$ to achieve accuracy of n fractional bits. Thus, the total number of entries in the two LUTs, TIV and TO, is about $2^{2n/3} + 2^{2n/3}$.

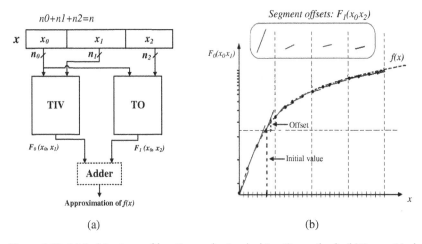

(a) (b)

Figure 4.13 (a) Architectures of function evaluation by bipartite methods. (b) Geometrical interpretation of the working of bipartite method.

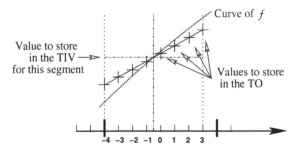

Figure 4.14 Symmetry of offset values with respect to the initial value at the midpoint.

Geometrical interpretation of BP is shown in Figure 4.13b for $n = 4$, $n_0 = 2$, $n_1 = n_2 = 1$. The initial input interval is divided into $2^{n_0} = 4$ subintervals of equal length based on the MSB part x_0 of the input x, and the value of function in each subinterval is approximated by adding the initial value $f(x_0 + x_1 2^{-n_0})$, and the offset value $x_2 2^{-(n_0+n_1)} \cdot f'(x_0)$ that share the same slope $f'(x_0)$ in the subinterval.

The values of a function stored in a TIV is the function value of the midpoint in each segment instead of the function values at the starting point. As shown in Figure 4.14, we can exploit the symmetry of the contents in the offset table to reduce the table size by half since the four offset values above the midpoint of a segment of initial value in Figure 4.14 are negative of the offset values below the midpoint. This approach is described as symmetric bipartite table

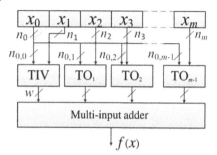

Figure 4.15 Architecture of multipartite (MP) design.

method (SBTM) in Reference [39], and has been adopted in the subsequent BP designs.

4.4.2.3 Multipartite Table-Addition Approach for Function Evaluation

The TA table method can be generalized further by decomposing the offset table TO into some smaller tables TO_j, $i = 1, 2, \ldots, m-1$, as shown in Figure 4.15, where m is the total number of tables, including one TIV and $m-1$ TOs. This method is called symmetric table addition method STAM in Reference [40], where $n_{0,0} = n_{0,1} = n_{0,2} = \cdots = n_{0,m-1} = n_0$. A more general and efficient method with $n_0 \geq n_{0,0} \geq n_{0,1} \geq n_{0,2} \geq \cdots \geq n_{0,m-1}$ is called multipartite (MP) in Reference [41]. MP methods can reduce the number of total table entries to order of $O(2^{n/2})$ compared with $O(2^{2n/3})$ in BP.

TA methods have been successfully applied to the design of low-precision (around 12-16-bit) communication systems [29]. But the major drawback is the large table size, especially in high-precision applications, making them suitable only for low-precision requirement. Properties of specific functions could be utilized in PPA as well as TA methods for identifying the input range, and to optimize the LUT size as discussed in Section 4.5 for the LUT-based evaluation of sigmoid function.

4.5 Applications

There are several other applications of LUT-based constant multiplications and evaluation of nonlinear functions [3–17]. The LUT-based approach can be used for efficient implementation of finite impulse response (FIR) filters that is discussed in detail in Reference [5]. The LUT-based computation of inner products with fixed vectors are popularly evaluated by using the principle of distributed arithmetic [18–20] that is discussed in a separate chapter. We discuss here three LUT-based implementation of cyclic convolution and orthogonal transforms, evaluation of reciprocals, and evaluation of sigmoid functions.

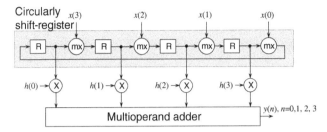

Figure 4.16 A structure for parallel implementation of circular convolution of length $N = 4$.

4.5.1 LUT-Based Implementation of Cyclic Convolution and Orthogonal Transforms

Cyclic convolution is used as a basic tool in digital signal and image processing applications [45]. It can be used for efficient computation of orthogonal transforms like DFT, DCT, and DHT. Cyclic convolution can be used for the implementation of linear convolution and also digital filters by overlap add and overlap save methods. Cyclic convolution of two sequences $\{h(n)\}$ and $\{x(n)\}$ for $n = 0, 1, ..., N - 1$ is defined as

$$y(n) = \sum_{k=0}^{N-1} h(k) \cdot x\big((n - k)_N\big) \qquad (4.22)$$

for $n = 0, 1, ..., N - 1$, where $(.)_N$ denotes modulo N operation.

In most applications, one of the sequences consists of constant values those are known *a priori*. The LUT-based constant multiplication scheme can be efficiently used for the implementation of cyclic convolution. For $N = 4$ we can write Eq. (4.22) as

$$y(0) = h(0) \cdot x(0) + h(1) \cdot x(3) + h(2) \cdot x(2) + h(3) \cdot x(1) \qquad (4.23)$$

$$y(1) = h(0) \cdot x(1) + h(1) \cdot x(0) + h(2) \cdot x(3) + h(3) \cdot x(2) \qquad (4.24)$$

$$y(2) = h(0) \cdot x(2) + h(1) \cdot x(1) + h(2) \cdot x(0) + h(3) \cdot x(3) \qquad (4.25)$$

$$y(3) = h(0) \cdot x(3) + h(1) \cdot x(2) + h(2) \cdot)x(1) + h(3) \cdot x(0) \qquad (4.26)$$

The 4-point cyclic convolution give by Eq. (4.23) can be realized by a structure shown in Figure 4.16. At the beginning of a period of 4 clock cycles, an input sequence $\{x(0), x(1), x(2), x(3)\}$ is fed in parallel to the circularly shift register of the structure and fed to the 4 multipliers. The product values are then added together by a multioperand adder to produce the first convolved output. During each of the next three subsequent clock cycles the input sequence $\{x(0), x(1), x(2), x(3)\}$ is circularly shifted and fed to the multipliers for generating the next three convolved output from the multioperand adder. Each of $h(i)$ for $i = 0, 1, 2$, and 3 of sequence $\{h(0), h(1), h(2), h(3)\}$ is a constant for different applications like computation of orthogonal transforms and block filtering. The multipliers in the convolution structure of Figure 4.16 can be implemented by the LUT-based constant multipliers. The optimization of algorithms and LUT-based architecture for the implementation of different orthogonal transforms are discussed in References [6–10].

4.5.2   LUT-Based Evaluation of Reciprocals and Division Operation

A straightforward LUT-based implementation of reciprocal using PPA method will require several tables for the storage of polynomial coefficients along with several multipliers and adders. Similarly, the direct implementation of bipartite or multipartite table-addition approach will have large LUT complexity. A fast algorithm for LUT-based computation of reciprocal and division using a small lookup table is presented in Reference [46]. The basic algorithm for the computation of reciprocal $1/X$ of a number of X (based on the algorithm of Reference [46]) is discussed in the following section.

4.5.2.1   Mathematical Formulation

Let X be a $2m$-bit positive number in fixed-point representation, given by,

$$X = 1 + 2^{-1}x_1 + 2^{-2}x_2 + \cdots + 2^{-(2m-1)}x_{(2m-1)} \tag{4.27}$$

where $x_i \in \{0, 1\}$ for all i. Note that any positive binary integer could be scaled by a suitable power of 2 integer to arrive at the representation given by Eq. (4.27).

Let us decompose X into two parts X_h and X_l, such that $X = X_h + X_l$. Let X_h contain the contribution for the $(m + 1)$ most significant bits of X and X_l is the contribution of $(m - 1)$ least significant bits of X, given by,

$$X_h = 1 + 2^{-1}x_1 + \cdots + 2^{-(m-1)}x_{(m-1)} + 2^{-m}x_m \tag{4.28}$$

$$X_l = 2^{-(m+1)}x_{(m+1)} + \cdots + 2^{-(2m-1)}x_{(2m-1)} \tag{4.29}$$

From Eq. (4.28), we can easily find that the minimum values of X_h and X_l are 1 and 0, respectively, while the maximum values of X_h and X_l are, respectively, $(X_h)_{\max} = (2 - 2^{-m})$ and $(X_l)_{\max} = (2^{-m} - 2^{-(2m-1)})$. We can write the reciprocal $1/X$ as

$$\frac{1}{X} = \frac{1}{X_h + X_l} = \frac{X_h - X_l}{X_h^2 - X_l^2} \approx \frac{X_h - X_l}{X_h^2} \tag{4.30}$$

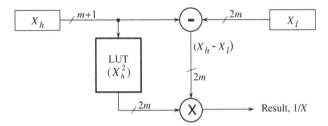

Figure 4.17 LUT-based structure for evaluation of reciprocal.

Note that the approximation of $(X_h^2 - X_l^2)$ to be X_h^2 in Eq. (4.30) can incorporate maximum fractional error less than 2^{-2m} (or 1/2 ulp). The error in the evaluation of the reciprocal, therefore, depends directly on the choice of decomposition of X into X_h and X_l. The less the width of X_h the more is the error in the result. We can however, find that LUT complexity increases exponentially with the width of X_h.

4.5.2.2   LUT-Based Structure for Evaluation of Reciprocal

Figure 4.17 shows the LUT-based structure for the evaluation of reciprocal of variable X using an algorithm based on Eq. (4.30). It consists of an LUT of 2^{m+1} entries to store the values of $1/X_h^2$ from which the values of $1/X_h^2$ corresponding to any m-bit values X_h can be retrieved. A subtractor is used to compute $(X_h - X_l)$ and the LUT output is multiplied with $(X_h - X_l)$ to obtain the desired result. We can notice that the bits of X_h and X_l are nonoverlapping. Therefore, the subtractor to compute $(X_h - X_l)$ could be realized by simple circuit consisting of an array of half adders. The main issue with the implementation of LUT-based circuit of Figure 4.17 is the complexity of LUT. For every one bit increase in the width of X_h, the LUT complexity is doubled. So it is important to have some approach for the reduction of LUT complexity.

4.5.2.3   Reduction of LUT Size

A simple lookup table for the computation of reciprocals for $m = 3$ is shown in Table 4.6. To reduce the size of the lookup table, the LUT entries could be normalized by appropriate scaling, such that the MSB of each entry is one. Therefore, the MSBs need not be stored in the LUT. The second column of the table consists of $\lfloor 1/X_h \rfloor$, which represents the scaled and truncated values of $1/X_h$ where the MSB is always 1 and consists of 5 fractional bits to the right of the radix point. Larger number of bits, however, could be taken to improve the

Table 4.6 A simple lookup table for computation of reciprocals (for $m = 3$).

X_h	$\lfloor 1/X_h \rfloor$	LUT entry
1.000	1.00000×1.00	00000
1.001	1.10010×0.10	10010
1.010	1.01000×0.10	01000
1.011	1.00001×0.10	00001
1.100	1.11000×0.01	11000
1.101	1.10000×0.01	10000
1.110	1.01001×0.01	01001
1.111	1.00100×0.01	00100

All the entries in the table are in binary representation.

accuracy of computation. The actual LUT entries are those five fractional bits shown in the third column of Table 4.6.

The 2-bit scale factor in the second column of this table could be stored in the LUT or derived by a simple logic circuit. If we take $X_h = 1 \cdot abc$, then the scale factor $s = s_0 \cdot s_1 s_2 s_3$, where $s_0 = \overline{a \cdot b \cdot c}$, $s_1 = \overline{s_0 + a}$, and $s_2 = \overline{s_1}$.

For division operation Y/X, we can multiply Y with the reciprocal Y. One can perform Booth encoding of Y to reduce the partial products for fast division and to reduce the area complexity.

4.5.3 LUT-Based Design for Evaluation of Sigmoid Function

Sigmoid activation functions are very often used in DSP applications involving artificial neural network (ANN). Let us consider here the evaluation of hyperbolic tangent sigmoid function, which is defined as

$$\tanh(x) = [e^x - e^{-x}]/[e^x + e^{-x}] \tag{4.31}$$

Direct computation of sigmoid by conventional arithmetic circuits using polynomial expansion involves large number of additions and multiplications, which contributes a major part to the critical path in hardware realization of an ANN system. LUT-based approach using PPA and TA technique are potential solutions to reduce the area and time complexity of implementation of sigmoid function. The delay of PPA is considerably large since it involves delay of multiplication and three-operand addition followed by table lookup. The TA method on the other hand, generally requires more than one large size table followed by additions. As discussed in the following, the symmetry property and saturation behavior of hyperbolic tangent can be used to substantially reduce the LUT size and delay for sigmoid evaluation.

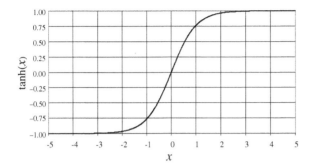

Figure 4.18 The hyperbolic tangent sigmoid function.

4.5.3.1   LUT Optimization Strategy using Properties of Sigmoid Function

As shown in Figure 4.18, the hyperbolic tangent has negative mirror symmetry about Y-axis, and the value of this function varies from -1 to $+1$ as x varies from $-\infty$ to $+\infty$. Besides, it varies almost linearly with x for small values of x. These behaviors of hyperbolic tangent relate to the following three basic properties:

$$\text{property 1:} \quad \tanh(-x) = -\tanh(x) \tag{4.32}$$

$$\text{property 2:} \quad \lim_{x \to 0} \tanh(x) = x \tag{4.33}$$

$$\text{property 3:} \quad \lim_{x \to \infty} \tanh(x) = 1 \text{ and } \lim_{x \to -\infty} \tanh(x) = -1 \tag{4.34}$$

Based on these properties, the LUT size could be reduced according to the following simple strategy:

- According to the property 1, we need to store the values of $\tanh(x)$ only for $x \geq 0$, while the values of the function for negative values of x could be evaluated by negating the LUT words stored for its corresponding positive values.
- According to the property 2, since $\tanh(x) \simeq x$ for small values of x, we can avoid storing the values of $\tanh(x)$ in this region. Instead of that we can obtain the values of $\tanh(x)$ directly from the values of x.
- According to the property 3, $\tanh(x)$ tends to ± 1, when $x \to \pm\infty$. The variation of $\tanh(x)$ is not significant for large values of $|x|$. For example, its variation for $|x| \geq 3$ is less than 0.00015, which is not significant for most of the ANN applications. Therefore, we can store $+1$ for all values of $x \geq 3$ if we do not need the error to be less than 0.00015.

As already discussed, based on the properties of sigmoid function, we may store the values of $\tanh(x)$ for $\delta \leq x \leq 3$, where δ is a limiting value that could

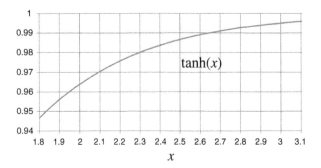

Figure 4.19 Saturating region of hyperbolic tangent sigmoid.

be derived from the accuracy requirement; such that $| \tanh(\delta) - \delta | < \epsilon$, for ϵ being the maximum allowable error. The value of δ depends on the accuracy requirement; and the actual range of $\tanh(x)$ to be stored in the LUT depends on the desired accuracy. In the following, we analyze the quantitative behavior of variation of $\tanh(x)$ to determine the range of values x for which we can assume $\tanh(x) \simeq x$, and not to store those values of $\tanh(x)$ in the LUTs. By expansion of exponentials in Eq. (4.31), and subsequent division we can find

$$\tanh(x) = x - \frac{x^3}{3} + \frac{2x^5}{15} - \frac{17x^7}{315} + \cdots \qquad (4.35)$$

For small values of x we can ignore the higher order terms, and approximate the expression of Eq. (4.31) as

$$\tanh(x) \simeq x - \frac{x^3}{3} + \frac{2x^5}{15} \qquad (4.36)$$

If ϵ is the maximum allowed error in assuming $\tanh(x) = x$, we can find maximum value of x for which we can assume $\tanh(x) = x$ using the following equation:

$$\frac{\delta^3}{3} - \frac{2\delta^5}{15} \leq \epsilon \qquad (4.37)$$

It can be seen from Figure 4.19 that for $x = 2, \tanh(x) > 0.96$, for $x = 2.4, \tanh(x) > 0.98$, and for $x = 2.7, \tanh(x) > 0.99$. Hence, we can approximate $\tanh(x) = 1$ for $x > 2.0, 2.4,$ or 2.7 if the maximum allowable errors are 0.04, 0.02, or 0.01, respectively. Based on the above observations, we store the values of $\tanh(x)$ for $\delta \leq x \leq \alpha$, where the value of delta is estimated according to Eq. (4.37), and $\alpha > 2.0, 2.4,$ or 2.7 for $\epsilon = 0.04, 0.02,$ or 0.01, respectively.

Table 4.7 LUT for hyperbolic tangent function using proposed compaction techniques.

LUT location	Limits of positive subdomain		Stored value	Maximum error
	X1	X2		
1	0.390625	0.453125	0.4049	0.0196
2	0.453125	0.515625	0.4558	0.0186
3	0.515625	0.578125	0.5038	0.0175
4	0.578125	0.640625	0.5490	0.0164
5	0.640625	0.703125	0.5911	0.0152
6	0.703125	0.78125	0.6348	0.0186
7	0.78125	0.859375	0.6791	0.0168
8	0.859375	0.9375	0.7190	0.0151
9	0.9375	1.046875	0.7609	0.0197
10	1.046875	1.171875	0.8057	0.0191
11	1.171875	1.328125	0.8493	0.0195
12	1.328125	1.53125	0.8916	0.0190
13	1.53125	1.859375	0.9329	0.0197
14	1.859375	2.90625	0.9740	0.0200
15	2.90625	—	1	0.0041

Maximum allowable error $= 0.02$, and address size $= 8$-bit. The input magnitude M satisfies the condition $I1 < I \leq I2$, where $I1$ and $I2$ are the binary equivalents of $X1$ and $X2$, respectively, in 8-bit representation.

4.5.3.2 Design Example for $\epsilon = 0.2$

One can store only one LUT word for two segments of argument values, out of which one segment consists of all positive values while the other consists of all negative values. Apart from that, the value stored for a given segment is the mean of the boundary values of the function for the positive set of argument values in the segment. Therefore, the difference between the maximum and minimum values of the function could be double of the maximum allowable error. The addressing scheme of the LUT for sigmoid evaluation for $\epsilon = 0.2$ is discussed in the following steps.

- *Determination of Lower and Upper Limits of LUT Input.*
  The lower limit of LUT address δ is found to be 0.390625 for $\epsilon = 0.2$ using Eq. (4.37). The upper limit of input x is found to 2.4 for this range of error.
- *Selection of Address Width.*
  By Matlab simulation, it is found that it is adequate to use 9-bit values of input x in 2's complement representation to have the difference $|\tanh(x_2) -$

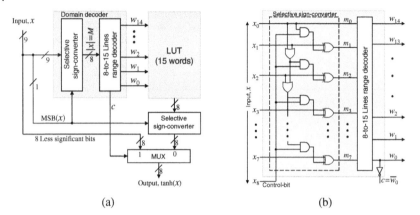

Figure 4.20 An LUT-based structure for evaluation of hyperbolic tangent sigmoid function. (a) The LUT. (b) The domain decoder circuit.

$\tanh(x_1)| < 0.02$, where x_1 and x_2 are two consecutive addresses/inputs (for $0.390625 < |x| \leq 2.4$). The bit width of $|x|$ is therefore, taken to be 8.

- *Selection of Domain Boundaries.*
  For all possible 8-bit positive values of x, in the whole range of $\tanh(x)$ for $0.390625 < x \leq 2.4$, is divided into n subdomains $R_i(x_{i1}, x_{i2})$ such that

$$\tanh(x_{i2}) - \tanh(x_{i1}) < 2\epsilon. \tag{4.38}$$

where $1 \leq i \leq n$, given that n is the smallest integer for which $x_{n2} \geq 2.4$. The address ranges $X_i \equiv [x_{i1}, x_{i2}]$ for $1 \leq i \leq n$ constitute the desired address ranges.

- *Content Assignment.*
  The average value of $\tanh(x_i)$ in the positive subdomain, that is, $[\tanh(x_{i2}) + \tanh(x_{i1})]/2$ is stored in the LUT location corresponding to positive subdomain specified by the interval $[x_{i1}, x_{i2}]$.

4.5.3.3 Implementation of Design Example for $\epsilon = 0.2$

An LUT for allowable error $\epsilon = 0.02$ is shown in Figure 4.20a. It is fed with the 9-bit input word x in 2's complement representation. The domain decoder (Figure 4.20b) generates the word-select signals to select one of the 15 words in the LUT that contains the desired value of sigmoid function. It consists of a selective sign-converter, which performs the 2's complement operation of the input word x, when the MSB of x is 1, that is, x is negative, otherwise the input is transferred unchanged to the output. The 8-bit magnitude of the input x, that is, $M = m_7 \, m_6 \, m_5 \, m_4 \, m_3 \, m_2 \, m_1 \, m_0$ generated by the selective sign-converter is fed as input to the 8-to-15 lines range decoder, which produces 15 word-select signals w_i for $1 \leq i \leq 15$ according to the logic expressions of Eqs. (4.39) and (4.40) using the range addressing technique [47].

$$w_1 = t_1, \text{ and } w_i = t_i \cdot \overline{t_{i-1}} \text{ for } 1 < i \leq 15 \text{ and} \tag{4.39}$$

$$t_1 = m_7 \cdot (m_6 + (m_5 \cdot m_4 \cdot m_3) \cdot (m_2 + (m_1 \cdot m_0)))$$
$$t_2 = m_7 + (m_6 \cdot m_5 \cdot m_4 \cdot m_3)$$
$$t_3 = m_7 + ((m_6 \cdot m_5) \cdot (m_4 + m_3 + m_2 + (m_1 \cdot m_0)))$$
$$t_4 = m_7 + (m_6 \cdot (m_5 + m_4) \cdot (m_3 + (m_2 \cdot m_1)))$$
$$t_5 = m_7 + (m_6 \cdot (m_5 + m_4 + (m_3 \cdot m_2)))$$
$$t_6 = m_7 + (m_6 \cdot (m_5 + m_4 + m_3 + m_2))$$
$$t_7 = (m_7 + m_6) + ((m_5 \cdot m_4 \cdot m_3 \cdot m_2) \cdot (m_1 + m_0))$$
$$t_8 = (m_7 + m_6) + (m_5 \cdot m_4 \cdot m_3)$$
$$t_9 = (m_7 + m_6) + ((m_5 \cdot m_4) \cdot (m_3 + m_2 + (m_1 \cdot m_0)))$$
$$t_{10} = (m_7 + m_6) + (m_5 \cdot (m_4 + m_3))$$
$$t_{11} = (m_7 + m_6) + (m_5 \cdot (m_4 + m_3) \cdot (m_2 + m_1))$$
$$t_{12} = (m_7 + m_6) + (m_5 \cdot (m_4 + m_3 + (m_2 \cdot m_1)))$$
$$t_{13} = (m_7 + m_6) + (m_5 \cdot (m_4 + m_3 + m_2 + m_1))$$
$$t_{14} = (m_7 + m_6 + m_5) + (m_4 \cdot m_3 \cdot m_2 \cdot m_1)$$
$$t_{15} = (m_7 + m_6 + m_5) + ((m_4 \cdot m_3) \cdot (m_2 + m_1)) \tag{4.40}$$

The logic expressions of Eqs. (4.39) and (4.40) are derived by simplification of Boolean expression of comparisons in order to map the input values to the subdomains in Table 4.7. The implementation of range-decoder based on logic expressions of Eqs. (4.39) and (4.40) can be simplified by sharing of logic subexpression. The subexpressions used to derive the word-select signals are listed in Table 4.8.

The LUT output is appended with a 0 at the MSB, and fed to the selective sign-conversion circuit. When the input x is negative, the LUT output is negated by this selective sign-conversion circuit. When the input x is less than 0.390625, the 2:1 MUX selects the input x; otherwise it selects the output of selective sign-converter using a control signal $c = \overline{w_0}$. The control signal c is also generated by the domain decoder.

4.6 Summary

LUT-based computing is very much close to the way in which human brain performs the computations, since we perform most of our manual calculations using the multiplication tables and addition tables that we have memorized.

Table 4.8 Logic expressions used to derive the word-select signals.

n	List of n-variable common logic expressions used in address generation
2	$(m_0 \cdot m_1), (m_1 \cdot m_2), (m_2 \cdot m_3), (m_3 \cdot m_4), (m_4 \cdot m_5), (m_5 \cdot m_6),$
	$(m_0 + m_1), (m_1 + m_2), (m_2 + m_3), (m_3 + m_4), (m_4 + m_5), (m_6 + m_7)$
3	$(m_5 + (m_6 + m_7)), (m_5 \cdot (m_3 + m_4)), (m_2 + (m_3 + m_4)), (m_3 + (m_1 \cdot m_2))$
	$(m_5 \cdot (m_3 \cdot m_4)), (m_2 + (m_1 \cdot m_0)), ((m_4 + m_5) \cdot m_6)$
4	$((m_1 + m_2) \cdot (m_3 \cdot m_4)), ((m_1 \cdot m_2) \cdot (m_3 \cdot m_4)), ((m_0 \cdot m_1) + (m_2 + m_3)),$
	$((m_1 \cdot m_2) + (m_3 + m_4)), ((m_2 \cdot m_3) \cdot (m_4 \cdot m_5)), ((m_2 + m_3) + (m_4 + m_5))$
	$((m_2 \cdot m_3) + (m_4 + m_5)), ((m_3 \cdot m_4) \cdot (m_5 \cdot m_6)), ((m_1 + m_2) + (m_3 + m_4))$
5	$m_5 \cdot ((m_1 + m_2) + (m_3 + m_4)), m_5 \cdot ((m_1 + m_2) \cdot (m_3 + m_4))$
	$m_5 \cdot ((m_1 \cdot m_2) + (m_3 + m_4)), m_6 \cdot ((m_2 + m_3) + (m_4 + m_5))$
	$(m_6 \cdot ((m_2 \cdot m_3) + (m_4 + m_5)), (m_0 \cdot m_1) + (m_2 + m_3 + m_4)$
6	$(m_4 \cdot m_5) \cdot ((m_0 \cdot m_1) + (m_2 + m_3)), (m_0 + m_1) \cdot ((m_2 \cdot m_3) \cdot (m_4 \cdot m_5))$
	$((m_1 \cdot m_2) + m_3) \cdot ((m_4 + m_5) \cdot m_6), ((m_0 \cdot m_1) + m_2) \cdot ((m_3 \cdot m_4) \cdot m_5)$
7	$(m_0 \cdot m_1) + (m_2 + m_3 + m_4) \cdot (m_5 \cdot m_6), m_6 + ((m_0 \cdot m_1) + m_2) \cdot ((m_3 \cdot m_4) \cdot m_5)$

Table-based computing is widely used in FPGA devices where Boolean functions are realized by LUT of 1-bit width. The LUT-based realization of logic function provides the desired reconfigurability to these devices. In this chapter, we have briefly discussed the basic concept and implementation of LUT-based system for constant multiplications and nonlinear function evaluation. Lookup tables could be used as constant multipliers in various DSP applications, e.g., implementation of FIR filters, computation of cyclic convolutions, and orthogonal transforms. The main problem with any LUT-based computing systems is the exponential increase in LUT size with the input word size. Odd multiple storage and antisymmetric product coding could be used for reduction of LUT size. Time multiplexed LUT-based designs could be used for reducing the table size further. Inner products with fixed vectors could be realized by LUT-based structures using the principle of distributed arithmetic, which is discussed in Chapter III.

Several DSP and multimedia applications require efficient computation of elementary arithmetic functions, for example, reciprocal $(1/x)$, square root $(\sqrt{x})$, logarithm $(\log_a x)$, exponential (a^x), and trigonometric functions such as $\sin x$, $\cos x$, $\tan x$, and $\tan^{-1} x$ of any given fixed-point variable x. LUT-based approach using lookup tables with some supporting arithmetic circuits like adders and/or multipliers are popularly used for the evaluation of non-linear functions. LUT-based function evaluation could make use of piecewise polynomial approximation method or table-addition method. Polynomial approximation methods involve multiplications followed by additions, and table-addition methods generally involve large size tables. For more efficient use of these LUT-based implementation, one can make use of the specific properties of the function to be evaluated. Three such examples are discussed in this chapter. An efficient LUT-based design is discussed for the evaluation of sigmoid function to be used in ANN applications, where LUT size is significantly reduced using the fundamental properties of hyperbolic tangent. We have also discussed a simple lookup table for the computation of reciprocal and division operation. We have discussed a simple implementation of cyclic convolution using LUT-based constant multipliers, as well.

References

1 B. Prince, Trends in scaled and nanotechnology memories, in *Proc. IEEE 2004 Conference on Custom Integrated Circuits*, 2005, p. 7.
2 K. Itoh, S. Kimura, and T. Sakata, VL SI memory technology: current status and future trends, in *Proc. 25th European Solid-State Circuits Conference. (ESSCIRC 99)*, 1999, pp. 3–10.
3 H.-R. Lee, C.-W. Jen, and C.-M. Liu, On the design automation of the memory-based VLSI architectures for FIR filters, *IEEE Trans. Consum. Electronics.*, vol. 39, no. 3, 1993, pp. 619–629.
4 D. J. Allred, H. Yoo, V. Krishnan, W. Huang, and D. V. Anderson, LMS adaptive filters using distributed arithmetic for high throughput, *IEEE Trans. Circuits Syst. I Regul. Pap.*, vol. 52, no. 7, 2005, pp. 1327–1337.
5 P. K. Meher, New approach to look-up-table design and memory-based realization of FIR digital filter, in *IEEE Trans. on Circuits Syst.*, vol. 57, no. 3, 2010, pp. 592–603.
6 J.-I. Guo, C.-M. Liu, and C.-W. Jen, The efficient memory-based VLSI array design for DFT and DCT, *IEEE Trans. Circuits Syst. II Analog Digit. Signal Process.*, vol. 39, no. 10, 1992, pp. 723–733.
7 H.-C. Chen, J.-I. Guo, T.-S. Chang, and C.-W. Jen, A memory-efficient realization of cyclic convolution and its application to discrete cosine transform, *IEEE Trans. Circuits Syst Video Technol.*, vol. 15, no. 3, 2005, pp. 445–453.
8 D. F. Chiper, M. N. S. Swamy, M. O. Ahmad, and T. Stouraitis, Systolic algorithms and a memory-based design approach for a unified architecture for the computation of DCT/DST/IDCT/IDST, *IEEE Trans. Circuits Syst.-I: Regul. Pap.*, vol. 52, no. 6, 2005, pp. 1125–1137.
9 P. K. Meher, Systolic designs for DCT using a low-complexity concurrent convolutional formulation, *IEEE Trans. Circuits Syst. Video Technol.*, vol. 16, no. 9, 2006, pp. 1041–1050.
10 P. K. Meher, J. C. Patra, and M. N. S. Swamy, High-throughput memory-based architecture for DHT using a New convolutional formulation, in *IEEE Trans. Circuits Syst. II*, vol. 54, no. 7, 2007, pp. 606–610.

11 V. Saichand, D. M. Nirmala, S. Arumugam, and N. Mohankumar, FPGA realization of activation function for artificial neural networks, in *Eighth Int. Conf. Intell. Syst. Des. Appl. (ISDA '2008)*, vol. 3, 2008, pp. 159–164.

12 C.-W. Lin and J.-S. Wang, A digital circuit design of hyperbolic tangent sigmoid function for neural networks, in *IEEE Int. Symp. Circuits Syst. (ISCAS'2008)*, 2008, pp. 856–859.

13 F. Piazza, A. Uncini, and M. Zenobi, Neural networks with digital LUT activation functions, in *Proc. Int. Joint Conf. Neural Netw. (IJCNN'1993)*, vol. 2, October 1993, pp. 1401–1404.

14 K. Leboeuf, A. H. Namin, R. Muscedere, H. Wu, and M. Ahmadi, High speed VLSI implementation of the hyperbolic tangent sigmoid function, in *Proc. Third Int. Conf. Converg. Hybrid Information Technol. (ICCIT '2008)*, vol. 1, 2008, pp. 1070–1073.

15 S. Vassiliadis, M. Zhang, and J. G. Delgado-Frias, Elementary function generators for neural-network emulators, *IEEE Trans. Neural Netw.*, vol. 11, no. 6, 2000, pp. 1438–1449.

16 P. K. Meher, An optimized lookup-table for the evaluation of sigmoid function for artificial neural networks, in *Proc. 18th IEEE/IFIP VLSI Syst. Chip Conf. (VLSI-SoC-2010)*, September 2010, pp. 91–95.

17 T. C. Chen, Automatic computation of exponentials, logarithms, ratios, and square roots, *IBM J. Res. Dev.*, vol. 16, 1972, pp. 380–388.

18 P. K. Meher, Unified systolic-like architecture for DCT and DST using distributed arithmetic, *IEEE Trans. Circuits Syst. I Regu. Pap.*, vol. 53, no. 5, 2006, pp. 2656–2663.

19 S. A. White, Applications of the distributed arithmetic to digital signal processing: a tutorial review, *IEEE ASSP Mag.*, vol. 6, no. 3, 1989, pp. 5–19.

20 P. K. Meher, S. Chandrasekaran, and A. Amira, FPGA realization of FIR filters by efficient and flexible systolization using distributed arithmetic, *IEEE Trans. Signal Process.*, vol. 56, no. 7, 2008, pp. 3009–3017.

21 S. Brown, , and J. Rose, Architecture of FPGAs and CPLDs: a tutorial, *EEE Des. Test Comput.*, vol. 13, 1996, pp. 42–57.

22 U. Farooq, Z. Marrakchi, and H. Mehrez, FPGA architectures an overview, *Tree-based Heterogeneous FPGA Architectures*, Springer, New York, 2012, 7–48.

23 P. K. Meher, New approach to LUT implementation and accumulation for memory-based multiplication, in *Proc. IEEE Int. Symp. Circuits Syst., (ISCAS '2009)*, 2009, pp. 453–456.

24 P. K. Meher, LUT optimization for memory-based computation, *IEEE Trans. Circuits Syst. Express Briefs*, vol. 57, no. 4, 2010, pp. 285–289.

25 B.-G. Nam, H. Kim, and H.-J. Yoo, Power and area-efficient unfied computation of vector and elementary functions for handheld 3D graphics systems, in *IEEE Trans. Comput.*, 2008, pp. 490–504.

26 D. De Caro, N. Petra, and A. G. M. Strollo, High-performance special function unit for programmable 3-D graphics processors, in *IEEE Trans. Circuits Syst., I*, 2009, pp. 1968–1978.

27 Y.-J. Kim, *et al.*, Homogeneous stream processors with embedded special function units for high-utilization programmable shaders, in *IEEE Trans. VLSI Syst.*, 2012, pp. 1691–1704.

28 D. De Caro, N. Petra, and A. G. M. Strollo, Direct digital frequency synthesizer using nonuniform picewise-linear approximation, in *IEEE Trans. Circuits Syst.-I*, 2011, pp. 2409–2419.

29 D. De Caro, N. Petra, and A. G. M. Strollo, Reducing lookup-table size in direct digital frequency synthesizers using optimized multipartite table method, in *IEEE Trans. Circuits Syst.-I*, 2008, pp. 2116–2127.

30 J.-M. Muller, *Elementary Functions: Algorithms and Implementation*. 2nd edition, Birkhauser, 2006.

31 B. Parhami, *Algorithms and Design Methods for Digital Computer Arithmetic*. 2nd edition, Oxford University Press, 2012.

32 A. G. M. Strollo, D. De Caro, and N. Petra, Elementary functions hardware implementation using constrained piecewise polynomial approximations, in *IEEE Trans. Comput.*, 2011, pp. 418–432.

33 D.-U. Lee, *et al.*, Hardware implementation trade-offs of polynomial approximations and interpolations, in *IEEE Trans. on Comput.*, 2008, pp. 686–701.

34 D.-U. Lee and J. D. Villasenor, Optimized custom precision function evaluation for embedded processors, in *IEEE Trans. on Comput.*, January 2009, pp. 46–59.

35 D.-U. Lee, R. C. C. Cheung, W. Luk, and J. D. Villasenor, Hierarchical segmentation for function evaluation, *IEEE Trans. VLSI Syst.*, 2009, pp. 103–116.

36 S.-F. Hsiao, H.-J. Ko, and C.-S. Wen, Two-level hardware function evaluation based on correction of normalized piecewise difference functions, in *IEEE Trans. Circuits Syst.-II*, 2012, pp. 292–296.

37 S.-F. Hsiao *et al.*, Design of hardware function evaluators using low-overhead non-uniform segmentation with addressing remapping, in *IEEE Trans. VLSI Syst.*, 2013, pp. 875–886.

38 J.-M. Muller, A few results on table-based methods, *Reliab. Comput.*, 1999, pp. 279–288.

39 J. Stine and M. J. Schulte, Symmetric bipartite tables for accurate function approximation, in *Proc. Intl. Symp. Comput. Arithmetic*, 1997, pp. 175–183.

40 J. Stine and M. J. Schulte, The symmetric table addition method for accurate function evaluation, in *J. VLSI Signal Process.*, vol. 21, no. 2, 1999, pp. 167–177.

41 F. de Dinechin and A. Tisserand, Multipartite table methods, in *IEEE Trans. Comput.*, vol. 54, no. 3, 2005, pp. 319–330.

42 J. Y. L. Low and C. C. Jong, A Memory-efficient tables-and-additions method for accurate computation of elementary functions, in *IEEE Trans. Comput.*, 2013, pp. 858–872.

43 D. Wang, J.-M. Muller, N. Brisebarre, and M. D. Ercegovac, (M, p, k)-friendly points: a table-based method to evaluate trigonometric functions, in *IEEE Trans. Circuits Syst. II*, vol. 61, no. 9, 2014, pp. 711–715.

44 S.-F. Hsiao, P.-C. Wei, and C.-P. Lin, An automatic hardware generator for special arithmetic functions using various ROM-based approximation approaches, in *Proc. Intl. Symp. Circuits Syst. (ISCAS)*, 2008, pp. 468–471.

45 D. G. Myers, *Digital Signal Processing: Efficient Convolution and Fourier Transform Techniques*, Prentice Hall, New York, 1990.

46 P. Hung, H. Fahmy, O. Mencer, and M. J. Flynn, Fast division algorithm with a small lookup table, in *Conf. Rec. Thirty-Third Asilomar Conf. Signals, Syst., and Comput.*, 1999, pp. 1465–1468.

47 R. Muscedere, V. Dimitrov, G. A. Jullien, and W. C. Miller, Efficient techniques for binary-to-multidigit multidimensional logarithmic number system conversion using range-addressable look-up tables, *IEEE Trans. Comput.*, vol. 54, no. 3, 2005, pp. 257–271.

5

CORDIC Circuits

Pramod Kumar Meher[1], Javier Valls[2], Tso-Bing Juang[3], K. Sridharan[4], and Koushik Maharatna[5]

[1] Independent Hardware Consultant

[2] Universitat Politecnica de Valencia, Instituto de Telecomunicaciones y Aplicaciones Multimedia, Valencia, Spain

[3] Department of Computer Science and Information Engineering, National Pingtung University, Pingtung, Taiwan

[4] Department of Electrical Engineering, Indian Institute of Technology Madras, Chennai, India

[5] School of Electronics and Computer Science, University of Southampton, Southampton, UK

5.1 Introduction

COordinate Rotation DIgital Computer is abbreviated as CORDIC. The key concept of CORDIC arithmetic is based on the simple and ancient principles of two-dimensional geometry. But the iterative formulation of a computational algorithm for its implementation was first described in 1959 by Volder [1,2] for the computation of trigonometric functions, multiplication and division. Not only a wide variety of applications of CORDIC have emerged over the years but also a lot of progress has been made in the area of algorithm design and development of architectures for high-performance and low-cost hardware solutions of those applications. CORDIC-based computing received increased attention in 1971, when Walther [3,4] showed that, by varying a few simple parameters, it could be used as a single algorithm for unified implementation of a wide range of elementary transcendental functions involving logarithms, exponentials, and square roots along with those suggested by Volder [1]. During the same time, Cochran [5] benchmarked various algorithms, and showed that CORDIC technique is a better choice for scientific calculator applications.

The popularity of CORDIC was very much enhanced thereafter primarily due to its potential for efficient and low-cost implementation of a large class

Arithmetic Circuits for DSP Applications, First Edition. Edited by Pramod Kumar Meher and Thanos Stouraitis.
© 2017 by The Institute of Electrical and Electronics Engineers, Inc. Published 2017 by John Wiley & Sons, Inc.

of applications that include: the generation of trigonometric, logarithmic, and transcendental elementary functions; complex number multiplication, eigenvalue computation, matrix inversion, solution of linear systems and singular value decomposition (SVD) for signal processing, image processing, and general scientific computation. Some other popular and upcoming applications are

(i) direct frequency synthesis, digital modulation and coding for speech/music synthesis, and communication;
(ii) direct and inverse kinematics computation for robot manipulation; and
(iii) planar and three-dimensional vector rotation for graphics and animation.

Although CORDIC may not be the fastest technique to perform these operations, it is attractive due to the simplicity of its hardware implementation, since the same iterative algorithm could be used for all these applications using the basic shift-add operations of the form $a \pm b \times 2^{-i}$.

Keeping the requirements and constraints of different application environments in view, the development of CORDIC algorithm and architecture has taken place for achieving high throughput rate and reduction of hardware-complexity as well as the latency of implementation. Some of the typical approaches for reduced-complexity implementation are focused on minimization of the complexity of scaling operation and the complexity of barrel-shifter in the CORDIC engine. Latency of implementation is an inherent drawback of the conventional CORDIC algorithm. Angle recoding schemes, mixed-grain rotation and higher radix CORDIC have been developed for reduced latency realization. Parallel and pipelined CORDIC have been suggested for high-throughput computation. The objective of this chapter is not to present a detailed survey of the developments of algorithms, architectures, and applications of CORDIC, which would require a few doctoral and masters level dissertations. Rather, we aim at providing the key developments in algorithms and architectures along with an overview of the major application areas and upcoming applications. We shall, however, discuss here the basic principles of CORDIC operations for the benefit of general readers.

The remainder of this chapter is organized as follows. In Section 5.2, we discuss the principles of CORDIC operation, covering the elementary ideas from coordinate transformation to rotation mode and vectoring mode operations followed by design of the basic CORDIC cell and multidimensional CORDIC. The key developments in CORDIC algorithms and architectures are discussed in Section 5.3, which covers the algorithms and architectures pertaining to high-radix CORDIC, angle recording, coarse-fine hybrid microrotations, redundant number representation, differential CORDIC, and pipeline implementation. In Section 5.4, we discuss the scaling and accuracy aspects including the scaling techniques, scaling-free CORDIC, quantization, and area–delay–accuracy trade-off. The applications of CORDIC to scientific computations, signal processing, communications, robotics, and graphics are discussed briefly in

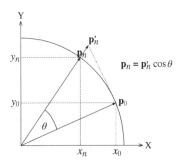

Figure 5.1 Rotation of vector on a two-dimensional plane.

Section 5.5. The conclusion along with future research directions are discussed in Section 5.6.

5.2 Basic CORDIC Techniques

In this section, we discuss the basic principle underlying the CORDIC-based computation, and present its iterative algorithm for different operating modes and planar coordinate systems. At the end of this section, we discuss the extension of two-dimensional rotation to multidimensional formulation.

5.2.1 The CORDIC Algorithm

As shown in Figure 5.1, the rotation of a two-dimensional vector $\mathbf{p}_0 = [x_0 \ y_0]$ through an angle θ, to obtain a rotated vector $\mathbf{p}_n = [x_n \ y_n]$ could be performed by the matrix product $\mathbf{p}_n = \mathbf{R}\mathbf{p}_0$, where $\mathbf{R}$ is the rotation matrix

$$\mathbf{R} = \begin{bmatrix} \cos\theta & -\sin\theta \\ \sin\theta & \cos\theta \end{bmatrix} \tag{5.1}$$

By factoring out the cosine term in Eq. (5.1), the rotation matrix $\mathbf{R}$ can be rewritten as

$$\mathbf{R} = \left[(1 + \tan^2\theta)^{-1/2}\right] \begin{bmatrix} 1 & -\tan\theta \\ \tan\theta & 1 \end{bmatrix} \tag{5.2}$$

and can be interpreted as a product of a scale-factor $K = [(1 + \tan^2\theta)^{-1/2}]$ with a pseudorotation matrix $\mathbf{R_c}$, given by

$$\mathbf{R_c} = \begin{bmatrix} 1 & -\tan\theta \\ \tan\theta & 1 \end{bmatrix} \tag{5.3}$$

The pseudorotation operation rotates the vector $\mathbf{p}_0$ by an angle θ and changes its magnitude by a factor $K = \cos\theta$, to produce a pseudo-rotated vector $\mathbf{p}'_n = \mathbf{R_c}\mathbf{p}_0$.

To achieve simplicity of hardware realization of the rotation, the key ideas used in CORDIC arithmetic are to (i) decompose the rotations into a sequence of elementary rotations through predefined angles that could be implemented with minimum hardware cost and (ii) avoid scaling, that might involve arithmetic operations such as square-root and division. The second idea is based on the fact the scale-factor contains only the magnitude information but no information about the angle of rotation.

5.2.1.1 Iterative Decomposition of Angle of Rotation

The CORDIC algorithm performs the rotation iteratively by breaking down the angle of rotation into a set of small predefined angles[1], $\alpha_i = \arctan(2^{-i})$, so that $\tan \alpha_i = 2^{-i}$ could be implemented in hardware by shifting through i bit locations. Instead of performing the rotation directly through an angle θ, CORDIC performs it by a certain number of microrotations through angle α_i, where

$$\theta = \sum_{i=0}^{n-1} \sigma_i \alpha_i, \quad \text{and } \sigma_i = \pm 1 \tag{5.4}$$

that satisfies the CORDIC convergence theorem [3]: $\alpha_i - \sum_{j=i+1}^{n-1} \alpha_j < \alpha_{n-1}$, $\forall i, i = 0, 1, \ldots, n-2$. But, the decomposition according to Eq. (5.4) could be used only for $-1.74329 \le \theta \le 1.74329$ (called the "convergence range") since $\sum_{i=0}^{\infty}(\alpha_i) = 1.743286\ldots$ Therefore, the angular decomposition of Eq. (5.4) is applicable for angles in the first and fourth quadrants. To obtain on-the-fly decomposition of angles into the discrete base α_i, one may otherwise use the nonrestoring decomposition [6]

$$\omega_0 = 0 \text{ and } \omega_{i+1} = \omega_i - \sigma_i \cdot \alpha_i \tag{5.5}$$

with $\sigma_i = 1$ if $\omega_i \ge 0$ and $\sigma_i = -1$ otherwise, where the rotation matrix for the ith iteration corresponding to the selected angle α_i is given by

$$\mathbf{R}(i) = K_i \begin{bmatrix} 1 & -\sigma_i 2^{-i} \\ \sigma_i 2^{-i} & 1 \end{bmatrix} \tag{5.6}$$

$K_i = 1/\sqrt{(1 + 2^{-2i})}$ being the scale factor, and the pseudorotation matrix

$$\mathbf{R}_c(i) = \begin{bmatrix} 1 & -\sigma_i 2^{-i} \\ \sigma_i 2^{-i} & 1 \end{bmatrix} \tag{5.7}$$

---

1 All angles are measured in radian unless otherwise stated.

Note that the pseudorotation matrix $\mathbf{R}_c(i)$ for the ith iteration alters the magnitude of the rotated vector by a scale factor $K_i = 1/\sqrt{(1 + 2^{-2i})}$ during the ith microrotation, which is independent of the value of σ_i (direction of microrotation) used in the angle decomposition.

5.2.1.2 Avoidance of Scaling

The other simplification performed by the Volder's algorithm [1] is to remove the scale factor $K_i = 1/\sqrt{(1 + 2^{-2i})}$ from Eq. (5.6). The removal of scaling from the iterative microrotations leads to a pseudorotated vector $\mathbf{p}'_n = \mathbf{R}_c \mathbf{p}_0$ instead of the desired rotated vector $\mathbf{p}_n = K\mathbf{R}_c \mathbf{p}_0$, where the scale factor K is given by

$$K = \prod_{i=0}^{n} K_i = \prod_{i=0}^{n} 1/\sqrt{(1 + 2^{-2i})} \qquad (5.8)$$

Since the scale factor of microrotations does not depend on the direction of microrotations and decreases monotonically, the final scale factor K converges to ~ 1.6467605. Therefore, instead of scaling during each microrotation, the magnitude of final output could be scaled by K. Therefore, the basic CORDIC iterations are obtained by applying the pseudorotation of a vector, $\mathbf{p}'_{i+1} = \mathbf{R}_c(i)\mathbf{p}_i$, together with the nonrestoring decomposition of the selected angles α_i, as follows:

$$x_{i+1} = x_i - \sigma_i \cdot 2^{-i} \cdot y_i \qquad (5.9a)$$
$$y_{i+1} = y_i + \sigma_i \cdot 2^{-i} \cdot x_i \qquad (5.9b)$$
$$\omega_{i+1} = \omega_i - \sigma_i \cdot \alpha_i \qquad (5.9c)$$

CORDIC iterations of Eq. (5.9) could be used in two operating modes, namely, the rotation mode (RM) and the vectoring mode (VM), which differ basically on how the directions of the microrotations are chosen. In the rotation mode, a vector $\mathbf{p}_0$ is rotated by an angle θ to obtain a new vector $\mathbf{p}'_n$. In this mode, the direction of each microrotation σ_i is determined by the sign of ω_i: if sign of ω_i is positive, then $\sigma_i = 1$ otherwise $\sigma_i = -1$. In the vectoring mode, the vector $\mathbf{p}_0$ is rotated toward the x-axis so that the y-component approaches zero. The sum of all angles of microrotations (output angle ω_n) is equal to the angle of rotation of vector $\mathbf{p}_0$, while output x'_n corresponds to its magnitude. In this operating mode, the decision about the direction of the microrotation depends on the sign of y_i: if it is positive then $\sigma_i = -1$ otherwise $\sigma_i = 1$. CORDIC iterations are easily implemented in both software and hardware. Figure 5.2 shows the basic hardware stage for a single CORDIC iteration. After each iteration the number of shifts is incremented by a pair of barrel-shifters. To have an n-bit output precision, $(n + 1)$ CORDIC iterations are needed. Note that it could be implemented by a simple selection operation in serial architectures like the one

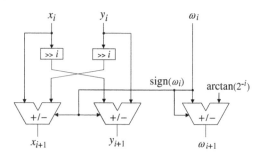

Figure 5.2 Hardware implementation of a CORDIC iteration.

proposed in the original work, or in fully parallel CORDIC architectures the shift operations could be hardwired, where no barrel shifters are involved.

Finally, to overcome the problem of the limited convergence range and, then to extend the CORDIC rotations to the complete range of $\pm\pi$, an extra iteration is required to be performed. This new iteration is shown in Eq. (5.10), which is required as an initial rotation through $\pm\pi/2$.

$$x_0 = -\sigma_{-i} \cdot y_{-i} \tag{5.10a}$$

$$y_0 = \sigma_{-i} \cdot x_{-i} \tag{5.10b}$$

$$\omega_0 = \omega_{-i} - \sigma_{-i} \cdot \alpha_{-i} \tag{5.10c}$$

where, $\alpha_{-i} = \pi/2$.

5.2.2 Generalization of the CORDIC Algorithm

In 1971, Walther found how CORDIC iterations could be modified to compute hyperbolic functions [3] and reformulated the CORDIC algorithm in to a generalized and unified form, which is suitable to perform rotations in circular, hyperbolic, and linear coordinate systems. The unified formulation includes a new variable m, which is assigned different values for different coordinate systems. The generalized CORDIC is formulated as follows:

$$x_{i+1} = x_i - m\sigma_i \cdot 2^{-i} \cdot y_i \tag{5.11a}$$

$$y_{i+1} = y_i + \sigma_i \cdot 2^{-i} \cdot x_i \tag{5.11b}$$

$$\omega_{i+1} = \omega_i - \sigma_i \cdot \alpha_i \tag{5.11c}$$

where

$$\sigma_i = \begin{cases} \text{sign}(\omega_i), & \text{for rotation mode} \\ \text{sign}(y_i), & \text{for vectoring mode} \end{cases}$$

For $m = 1, 0,$ or -1, and $\alpha_i = \tan^{-1}(2^{-i}), 2^{-i},$ or $\tanh^{-1}(2^{-i})$, the algorithm given by Eq. (5.11) works in circular, linear, or hyperbolic coordinate systems,

Table 5.1 Generalized CORDIC algorithm

m	Rotation mode	Vectoring mode
0	$x_n = K(x_o \cos\omega_0 - y_o \sin\omega_0)$	$x_n = K\sqrt{x_o^2 + y_0^2}$
	$y_n = K(y_o \cos\omega_0 + y_o \sin\omega_0)$	$y_n = 0$
	$\omega_n = 0$	$\omega_n = \omega_0 + \tan^{-1}(y_0/x_0)$
1	$x_n = x_o$	$x_n = x_o$
	$y_n = y_o + x_o\omega_0$	$y_n = 0$
	$\omega_n = 0$	$\omega_n = \omega_0 + (y_0/x_0)$
-1	$x_n = K_h(x_o \cosh\omega_0 - y_o \sinh\omega_0)$	$x_n = K_h\sqrt{x_o^2 - y_0^2}$
	$y_n = K_h(y_o \cosh\omega_0 + y_o \sinh\omega_0)$	$y_n = 0$
	$\omega_n = 0$	$\omega_n = \omega_0 + \tanh^{-1}(y_0/x_0)$

respectively. Table 5.1 summarizes the operations that can be performed in rotation and vectoring modes[2] in each of these coordinate systems. The convergence range of linear and hyperbolic CORDIC are obtained, as in the case of circular coordinate, by the sum of all α_i given by $C = \Sigma_{i=0}^{\infty}\alpha_i$. The hyperbolic CORDIC requires to execute iterations for $i = 4, 13, 40, \ldots$ twice to ensure convergence. Consequently, these repetitions must be considered while computing the scale-factor $K_h = \Pi(1 + 2^{-2i})^{-1/2}$, which converges to 0.8281.

5.2.3 Multidimensional CORDIC

The CORDIC algorithm was extended to higher dimensions using simple Householder reflection [7]. The Householder reflection matrix is defined as

$$\mathbf{H}_m = \mathbf{I}_m - 2.\frac{\mathbf{uu}^T}{\mathbf{uu}} \tag{5.12}$$

where $\mathbf{u}$ is an m-dimensional vector and $\mathbf{I}_m$ is the $m \times m$ identity matrix. The product $(\mathbf{H}_m\mathbf{v})$ reflects the m-dimensional vector $\mathbf{v}$ with respect to the hyperplane with normal $\mathbf{u}$ that passes through the origin. Basically, the Householder-

---

2 In the rotation mode, the components of a vector resulting due to rotation of a vector through a given angle are derived, while in the vectoring mode the magnitude as well as the phase angle of a vector are estimated from the component values. The rotation and vectoring modes are also known as the vector rotation mode and the angle accumulation mode, respectively.

based CORDIC performs the vectoring operation of an m-dimensional vector to one of the axes.

For the sake of clarity, we consider here the case of 3D vector $P_0 = [x_0 \ y_0 \ z_0]$ projected on to the x-axis in the Euclidean space. The rotation matrix for 3D case, corresponding to the ith iteration, $\mathbf{R}_{H3}(i)$, is given by the product of two simple Householder reflections as

$$\mathbf{R}_{H3}(i) = \left[\mathbf{I}_3 - 2.\frac{e_1 e_1^{\mathrm{T}}}{e_1^{\mathrm{T}} e_1}\right] \cdot \left[\mathbf{I}_3 - 2.\frac{\mathbf{u}_i \mathbf{u}_i^{\mathrm{T}}}{\mathbf{u}_i^{\mathrm{T}} \mathbf{u}_i}\right] \tag{5.13}$$

where $e_1 = [1 \ 0 \ 0]^{\mathrm{T}}$, and $\mathbf{u}_i = [1 \ \sigma_{yi} t_i \ \sigma_{zi} t_i]^{\mathrm{T}}$ with $t_i = \tan \alpha_i = 2^{-i}$, and $\sigma_{yi} = \mathrm{sign}(x_i y_i)$ and $\sigma_{zi} = \mathrm{sign}(x_i z_i)$ being the directions of microrotations.

One can write the ith rotation matrix in terms of the pseudorotation matrix as $\mathbf{R}_{H3}(i) = K_{Hi} \mathbf{R}_{HC3}(i)$, where $K_{Hi} = 1/\sqrt{(1 + 2^{-2i+1})}$ is the scale-factor and $\mathbf{R}_{HC3}(i)$ is the pseudorotation matrix which could be expressed as function of the shifting and decision variables as

$$\begin{bmatrix} 1 & \sigma_{yi} 2^{-i+1} & \sigma_{zi} 2^{-i+1} \\ -\sigma_{yi} 2^{-i+1} & 1 - 2^{-2i+1} & -\sigma_{yi} \sigma_{zi} 2^{-2i+1} \\ -\sigma_{zi} 2^{-i+1} & -\sigma_{zi} \sigma_{yi} 2^{-2i+1} & 1 - 2^{-2i+1} \end{bmatrix} \tag{5.14}$$

Therefore, the ith iteration of 3D Housholder CORDIC rotation results $\mathbf{p}_{i+1} = \mathbf{R}_{HC3}(i)\mathbf{p}_i$, and, the vector is projected to x-axis, such that after n iterations x_n gives the length of the vector scaled by $\Pi_{i=0}^{n}(K_{Hi})$ with (n-1) bit precision [8].

5.3 Advanced CORDIC Algorithms and Architectures

CORDIC computation is inherently sequential due to two main bottlenecks: (1) the microrotation for any iteration is performed on the intermediate vector computed by the previous iteration, and (2) the $(i + 1)$th iteration could be started only after the completion of the ith iteration, since the value of σ_{i+1} which is required to start the $(i + 1)$th iteration is known only after the completion of the ith iteration. To alleviate the second bottleneck some attempts have been made for evaluation of σ_i values corresponding to small microrotation angles [9,10]. However, the CORDIC iterations cannot still be performed in parallel due to the first bottleneck. A partial parallelization has been realized in Reference [11] by combining a pair of conventional CORDIC iterations into a single merged iteration that provides better area-delay efficiency. But the accuracy is slightly affected by such merging and cannot be extended to a higher number of conventional CORDIC iterations since the induced error becomes unacceptable [11]. Parallel realization of CORDIC iterations to handle the first bottleneck by direct unfolding of microrotation is possible, but that would result in increase in computational complexity and the advantage of simplicity of CORDIC algorithm gets degraded [12,13]. Although no popular architectures are known to us for fully parallel implementation of CORDIC, different forms of pipelined implementation of CORDIC have however been proposed for improving the computational throughput [14].

Since the CORDIC algorithm exhibits linear-rate convergence, it requires $(n + 1)$ iterations to have n-bit precision of the output. Overall latency of the computation thus amounts to product the word length and the CORDIC iteration period. The speed of CORDIC operations is therefore constrained either by the precision requirement (iteration count) or the duration of the clock period. The duration of clock period on the other hand mainly depends on the large carry propagation time for the addition/subtraction during each microrotation. It is a straightforward choice to use fast adders for reducing the iteration period at the expense of large silicon area. Use of carry-save adder is a good option to reduce the iteration period and overall latency [15]. Timmermann et al. have suggested a method of truncation of CORDIC algorithm after $(n + 1)/2$ iterations (for n-bit precision), where the last iteration performs a single rotation for implementing the remaining angle. It lowers the the latency time but involves one multiplication or division, respectively, in the rotation or vectoring mode [9].

To handle latency bottlenecks, various techniques have been developed and reported in the literature. Most of the well known algorithms could be grouped under, high-radix CORDIC, the angle recoding method, hybrid microrotation scheme, redundant CORDIC, and differential CORDIC which we discuss briefly in the following subsections.

5.3.1 High-Radix CORDIC Algorithm

The radix-4 CORDIC algorithm [16] is given by

$$x_{i+1} = x_i - \sigma_i \cdot 4^{-i} \cdot y_i \tag{5.15a}$$

$$y_{i+1} = y_i + \sigma_i \cdot 4^{-i} \cdot x_i \tag{5.15b}$$

$$\omega_{i+1} = \omega_i - \sigma_i \cdot \alpha_i \tag{5.15c}$$

where $\sigma_i \in \{-2, -1, 0, 1, 2\}$ and the elementary angles $\alpha_i = \arctan(\sigma_i 4^{-i})$. The scale factor for the ith iteration $K_i = 1/\sqrt{(1 + \sigma_i^2 4^{-2i})}$. In order to preserve the norm of the vector the output of microrotations is required to be scaled by a factor

$$K = 1/\prod_i \sqrt{(1 + \sigma_i^2 4^{-2i})} \tag{5.16}$$

To have n-bit output precision, the radix-4 CORDIC algorithm requires $n/2$ microrotations, which is half that of radix-2 algorithm. However, it requires more computation time for each iteration and involves more hardware compared to the radix-2 CORDIC to select the value of σ_i out of five different possibilities. Moreover, the scalefactor, given by Eq. (5.16), also varies with the rotation angles since it depends on σ_i which could have any of the five different values. Some techniques have therefore been suggested for scale factor compensation through iterative shift-add operations [16,17]. A high-radix CORDIC algorithm in vectoring mode is also suggested in Reference [18], which can be used for reduced latency operation at the cost of larger size tables for storing the elementary angles and prescaling factors than the radix-2 and radix-4 implementation.

5.3.2 Angle Recoding Methods

The purpose of angle recoding (AR) is to reduce the number of CORDIC iterations by encoding the angle of rotation as a linear combination of a set of selected elementary angles of microrotations. AR methods are well-suited for many signal processing and image processing applications where the rotation angle is known *a priori*, such as when performing the discrete orthogonal transforms like discrete Fourier transform (DFT), the discrete cosine transform (DCT), and so on.

5.3.2.1 Elementary Angle Set Recoding

In the conventional CORDIC, any given rotation angle is expressed as a linear combination of n values of elementary angles that belong to the set $S = \left\{ (\sigma \cdot \arctan(2^{-r})) : \sigma \in \{-1, 1\}, r \in \{1, 2, \ldots, n-1\} \right\}$ in order to obtain an n-bit value as $\theta = \sum_{i=0}^{n-1} [\sigma_i \cdot \arctan(2^{-i})]$. However, in AR methods, this constraint is relaxed by adding zeros to the linear combination to obtain the desired angle using relatively fewer terms of the form $(\sigma \cdot \arctan 2^{-r})$ for $\sigma \in \{1, 0, -1\}$. The elementary angle set (EAS) used by AR scheme is given by $S_{EAS} = \left\{ (\sigma \arctan 2^{-r}) : \sigma \in \{-1, 0, 1\}, r \in \{1, 2, \ldots, n-1\} \right\}$. One of the simplest form of the angle recoding method based on the greedy algorithm proposed by Hu and Naganathan [19] tries to represent the remaining angle using the closest elementary angle $\pm \arctan 2^{-i}$. The angle recoding algorithm of Reference [19] is briefly stated in Table 5.2. Using this recoding scheme, the total number of iterations could be reduced by at least 50% keeping the same n-bit accuracy unchanged. A similar method of angle recoding in vectoring mode called as the backward angle recoding is suggested in References [20] .

Table 5.2 Angle recoding algorithm

---

initialize: $\theta_0 = \theta$, $\sigma_i = 0$ for $0 \le i \le (n-1)$ and $k = 0$.

repeat until $|\theta_k| < \arctan(2^{-n+1})$ do:

1. choose i_k, $0 \le i_k \le (n-1)$ such that

$$||\theta_k| - \arctan(2^{-i_k})| = \min_{0 \le i \le (n-1)} ||\theta_k| - \arctan(2^{-i})|.$$

2. $\theta_{k+1} = \theta_k - \sigma_{i_k} \arctan(2^{-i_k})$, where $\sigma_{i_k} = \mathrm{sign}(\theta_k)$.

5.3.2.2 Extended Elementary Angle Set Recoding

Wu *et al.* [21] have suggested an AR scheme based on an extended elementary angle set (EEAS), that provides a more flexible way of decomposing the target rotation angle. In the EEAS approach, the set S_{EEAS} of the elementary angle set is extended further to $S_{\mathrm{EEAS}} = \{(\arctan(\sigma_1 \cdot 2^{-r_1} + \sigma_2 \cdot 2^{-r_2})) : \sigma_1, \sigma_2 \in \{-1, 0, 1\}$ and $r_1, r_2 \in \{1, 2, ..., n-1\}\}$. EEAS has better recoding efficiency in terms of the number of iterations and can yield better error performance than the AR scheme based on EAS. The pseudorotation for ith microrotations based on EEAS scheme is given by

$$x_{i+1} = x_i - [\sigma_1(i) \cdot 2^{-r_1(i)} + \sigma_2(i) \cdot 2^{-r_2(i)}]y_i$$
$$y_{i+1} = y_i + [\sigma_1(i) \cdot 2^{-r_1(i)} + \sigma_2(i) \cdot 2^{-r_2(i)}]x_i \tag{5.17}$$

The pseudorotated vector $\begin{bmatrix} x_{R_m} & y_{R_m} \end{bmatrix}$, obtained after R_m (the required number of microrotations) iterations, according to Eq. (5.17), needs to be scaled by a factor $K = \Pi(K_i)$, where $K_i = [1 + (\sigma_1(i) \cdot 2^{-r_1(i)} + \sigma_2(i) \cdot 2^{-r_2(i)})^2]^{-1/2}$ to produce the rotated vector. For reducing the scaling approximation and for a more flexible implementation of scaling, similar to the EEAS scheme for the microrotation phase, a method has also been suggested in References [21], as given below

$$\widetilde{x}_{i+1} = \widetilde{x}_i + [k_1(i) \cdot 2^{-s_1(i)} + k_2(i) \cdot 2^{-s_2(i)}]\widetilde{y}_i$$
$$\widetilde{y}_{i+1} = \widetilde{y}_i + [k_1(i) \cdot 2^{-s_1(i)} + k_2(i) \cdot 2^{-s_2(i)}]\widetilde{x}_i \tag{5.18}$$

where $\widetilde{x}_0 = x_{R_m}$ and $\widetilde{y}_0 = x_{R_m} \cdot k_1, k_2 \in \{-1, 0, 1\}$ and $q_1, q_2 \in \{1, 2, ..., n-1\}\}$.

The iterations for microrotation phase as well as the scaling phase could be implemented in the same architecture to reduce the hardware cost, as shown in Figure 5.3.

5.3.2.3 Parallel Angle Recoding

The AR methods [19,21] could be used to reduce the number of iterations by more than 50%, when the angle of rotation is known in advance. However, for

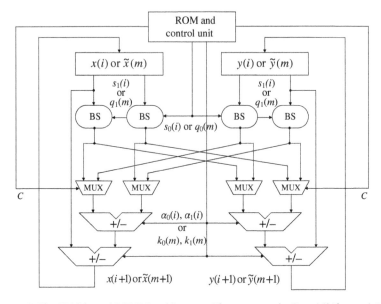

Figure 5.3 The EEAS-based CORDIC architecture. BS represents the Barrel Shifter and C denotes the control signals for the microrotations.

unknown rotation angles, their hardware implementation involves more cycle time than the conventional implementation, which results in a reduction in overall efficacy of the algorithm. To reduce the cycle time of CORDIC iterations in such cases, a parallel angle selection scheme is suggested in Reference [22], which can be used in conjunction with the AR method, to gain the advantages of the reduction in iteration count, without further increase in the cycle time. The parallel AR scheme in Reference [22] is based on dynamic angle selection, where the elementary angles α_i can be tested in parallel and the direction for the microrotations can be determined quickly to minimize the iteration period. During each iteration, the residual angle ω, is passed to a set of n adder-subtractor units that compute $\Delta_i = (\omega - \sigma_i \cdot \alpha_i)$ for each elementary angle $\alpha_i = \arctan 2^{-i}$ in parallel and the differences Δ_i for $0 \leq i \leq n$ are then fed to a binary tree-like structure to compare them against each other to find the smallest difference. The $\sigma_i \cdot \alpha_i$ corresponding to the smallest difference $(\Delta_i)_{\min}$ is used as the angle of microrotation. The architecture for parallel angle recoding of Reference [22] is shown in Figure 5.4.

The parallel AR reduces the overall latency at the cost of high hardware complexity of add/subtract-compare unit. For actual implementation, it is required to find a space-time trade-off and look at the relative performance in comparison with other approaches as well. The AR schemes based on EAS and

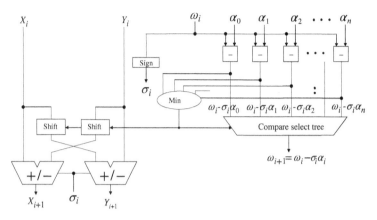

Figure 5.4 The architecture for parallel angle recoding.

EEAS however are useful for those cases where the angle of rotation is known in advance.

5.3.3 Hybrid or Coarse-Fine Rotation CORDIC

Based on the radix-2 decomposition, any rotation angle θ with n-bit precision could be expressed as a linear combination of angles from the set $\{2^{-i} : i \in \{1, 2, ..., n-1\}\}$, given by $\sum_{i=0}^{n-1} b_i 2^{-i}$, where $b_i \in \{0, 1\}$, explicitly specifies whether there is need of a microrotation or not. But, radix-2 decomposition is not used in the conventional CORDIC because that would not lead to simplicity of hardware realization. Instead, arctangents of the corresponding values of radix-2-based set are used as the elementary angle set with a view to implement the CORDIC operations only by shift-add operations. The key idea underlying the coarse-fine angular decomposition is that for the fine values of $\alpha_j = \arctan(2^{-j})$, (i.e., when $j > \lceil n/3 \rceil - 1$), $\tan(2^{-j})$ could be replaced by 2^{-j} in the radix set for expansion of θ, since $\tan(2^{-j}) \approx 2^{-j}$ when j is sufficiently large.

5.3.3.1 Coarse-Fine Angular Decomposition

In the coarse-fine angular decomposition, the elementary angle set contains the arctangents of power-of-two for more-significant part while the less significant part contains the power-of-two values, such that the radix set is given by $S = S_1 \bigcup S_2$, where $S_1 = \{ \arctan 2^{-i} : i \in \{1, 2, ..., p-1\} \}$ and $S_2 = \{ 2^{-j} : j \in \{p, p+1, .., n-1\} \}$, and j is assumed to be sufficiently large such that $\tan(2^{-j}) \to 2^{-j}$ [10]. For the hybrid decomposition scheme, the rotation angle could be partitioned into two terms expressed as

$$\theta = \theta_M + \theta_L \tag{5.19}$$

where θ_M and θ_L are said to be the coarse and fine subangles, respectively, given by

$$\theta_M = \sum_{i=1}^{p-1} \sigma_i \arctan 2^{-i} \ \text{for} \ \sigma_i \in \{1, -1\} \tag{5.20}$$

$$\theta_L = \sum_{i=p}^{n-1} d_i 2^{-i}, \ \text{for} \ d_i \in \{0, 1\} \tag{5.21}$$

A combination of coarse and fine microrotations are used in hybrid CORDIC operations in two cascaded stages. Coarse rotations are performed in stage-1 to have an intermediate vector

$$\begin{bmatrix} x_M \\ y_M \end{bmatrix} = \begin{bmatrix} 1 & -\tan(\theta_M) \\ \tan(\theta_M) & 1 \end{bmatrix} \begin{bmatrix} x_0 \\ y_0 \end{bmatrix} \tag{5.22}$$

and fine rotations are performed on the output of stage-1 to obtain the rotated output

$$\begin{bmatrix} x_n \\ y_n \end{bmatrix} = \begin{bmatrix} 1 & -\tan(\theta_L) \\ \tan(\theta_L) & 1 \end{bmatrix} \begin{bmatrix} x_M \\ y_M \end{bmatrix} \tag{5.23}$$

5.3.3.2 Implementation of Hybrid CORDIC

To derive the efficiency of hybrid CORDIC, the coarse and fine rotations are performed by separate circuits as shown in Figure 5.5. The coarse rotation phase is performed by the CORDIC processor-I and the fine rotation phase is performed by CORDIC processor-II.

To have fast implementation, processor-I performs a pair of ROM look-up operations followed by addition to realize the rotation through angle θ_H. Since θ_L could be expressed as a linear combination of angles of small enough magnitude 2^{-j}, where $\tan(2^{-j}) \rightarrow 2^{-j}$, the computation of fine rotation phase can be realized by a sequence of shift-and-add operations. For implementation of the fine rotation phase, no computations are involved to decide the direction of microrotation, since the need of a microrotation is explicit in the radix-2 representation of θ_L. The radix-2 representation could also be recoded to express $\theta_L = \sum \tilde{b}_i 2^{-i}$, where $\tilde{b}_i \in \{-1, 1\}$ as shown in Reference [9]. Since the direction of microrotations are explicit in such a representation of θ_L, it would be possible to implement the fine rotation phase in parallel for low-latency realization.

The hybrid decomposition could be used for reducing the latency by ROM-based realization of coarse operation. This can also be used for reducing the hardware complexity of fine rotation phase since there is no need to find the direction of microrotation. Several options are however possible for the implementation of these two stages. A form of hybrid CORDIC is suggested in Reference [23] for very high precision CORDIC rotation where the ROM size is reduced to nearly $n \times 2^{n/5}$ bits. The coarse rotations could be implemented as conventional CORDIC through shift-add operations of microrotations if the latency is tolerable.

5.3.3.3 Shift-Add Implementation of Coarse Rotation

Using the symmetry properties of the sine and cosine functions in different quadrants, the rotation through any arbitrary angle θ could be mapped from the full range $[0, 2\pi]$ to the first half the first quadrant $[0, \pi/4]$. The coarse-fine partition could be applied thereafter for reducing the number of microrotations necessary for fine rotations. To implement the course rotations through shift-add operations the coarse subangle θ_M is represented in References [24,25] in terms of elementary rotations of the form $(\arctan 2^{-k})$ as

$$\theta_M = \sum_{i=2}^{n/3} 2^{-i} = \sum_{i=2}^{n/3} (\arctan 2^{-i} + \theta_{Li}) \tag{5.24}$$

where, θ_{Li} is a correction term.

Using Eq. (5.24) on Eq. (5.19), one can find $\theta = \sum\limits_{i=2}^{n/3} \arctan 2^{-i} + \widetilde{\theta}_L$, where

$$\widetilde{\theta}_L = \theta_L + \sum_{i=2}^{n/3} \theta_{Li} \tag{5.25}$$

It is shown [25] that, based on the above decompositions using radix-2 representation, both coarse and fine rotations could be implemented by a sequence of shift-and-add operations in CORDIC iterations without ROM lookup table or the real multiplication operation. One such implementation is shown in Figure 5.6. Processor-I performs CORDIC operations like that of conventional CORDIC for nearly the first one-third of the iterations and the residual angle as well as the intermediate rotated vector is passed to the processor-II. Processor-II can perform the fine rotation in one of the possible ways as in case of the circuit of Figure 5.5.

The coarse-fine rotation approach in some modified forms has been applied for reduced-latency implementation of sine and cosine generation [24–28], high-speed and high-precision rotation [24,26], and conversion of rectangular to polar coordinates and vice versa [29,30].

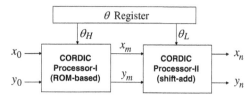

Figure 5.5 The architecture for a hybrid CORDIC algorithm [10].

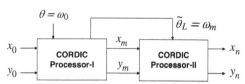

Figure 5.6 The shift-add architecture for a hybrid CORDIC algorithm.

5.3.3.4 Parallel CORDIC-Based on Coarse-Fine Decomposition

In Reference [31], the authors have proposed two angle recoding techniques for parallel detection of direction of microrotations, namely, the binary to bipolar recoding (BBR) and microrotation angle recoding (MAR) to be used for the coarse part of the input angle θ_H. BBR is used to obtain the polarity of each bit in the radix-2 representation of θ_H to determine the rotation direction. MAR is used to decompose each positional binary weight 2^{-i}, $\forall i, i = 1, 2, ..., m-1$ into a linear combination of arctangent terms. It is further shown in Reference [32] that the rotation direction can be decided once the input angle is known to enable parallel computation of the microrotations. Although the CORDIC rotation can be executed in parallel according to Reference [32], the method for decomposition of each positional binary weight produces many extra stages of microrotation, especially when the bit-width of input angle increases. A more efficient recoding scheme has been proposed in Reference [33] for the reduction of number of microrotations to be employed in parallel CORDIC rotations.

5.3.4 Redundant Number-Based CORDIC Implementation

Addition/subtraction operations are faster in the redundant number system, since unlike the binary system, it does not involve carry propagation. The use of redundant number system is therefore another way to speedup the CORDIC iterations. A CORDIC implementation based on the redundant number system called as redundant CORDIC was proposed by Ercegovac and Lang and applied to matrix triangularization and singular value decomposition [34]. Rotation mode redundant CORDIC has been found to result in fast implementation of sinusoidal function generation, unitary matrix transformation, angle calculation and rotation [34–38]. Although redundant CORDIC can achieve a fast carry-free computation, the direction of the microrotation (the sign factor σ_i) cannot be determined directly unlike the case of the conventional CORDIC, since the redundant number system allows a choice $\sigma_i = 0$ along with the conventional choices 1 and −1 such that $\sigma_i \in \{-1, 0, 1\}$. Therefore, it requires a different formulation for selection of $\sigma_i = 0$, which is different for binary signed-digit representation and carry-save implementation. In radix-2 signed-digit representation, assuming $-(\Sigma_{k=i}^{\infty} \arctan 2^{-i}) \leq \omega_i \leq (\Sigma_{k=i}^{\infty} \arctan 2^{-i})$, it is shown that [6]

$$\sigma_i = \begin{cases} -1 & \text{if } \tilde{\omega}_i < 0 \\ 0 & \text{if } \tilde{\omega}_i = 0 \\ 1 & \text{if } \tilde{\omega}_i > 0 \end{cases} \tag{5.26}$$

where $\tilde{\omega}_i$ is the value of $2^j \omega_j$ truncated after the first fractional digit. Similarly, for carry-save implementation, it is

$$\sigma_i = \begin{cases} -1 & \text{if } \tilde{\omega}_i < -1/2 \\ 0 & \text{if } \tilde{\omega}_i = 1/2 \\ 1 & \text{if } \tilde{\omega}_i > 1/2 \end{cases} \tag{5.27}$$

It can be noted from Eqs. (5.26) and (5.27), that in some of the iterations no rotations are performed, so that the scale factor becomes a variable which depends on the angle of rotation. Since the redundant CORDIC of Reference [34] uses nonconstant scale factor, Takagi *et al.* [35] have proposed the double-rotation method and correcting-rotation method to keep the value of scale factor constant. In double rotation method, in each iteration two microrotations are performed, such that when $\sigma_i = 0$, one positive and one negative microrotations are performed, and when $\sigma_i = +1$ or -1, respectively, two positive or two negative microrotations are performed. The scale factor is retained constant in this case since the number of microrotations is fixed for any rotation angle but it doubles the iteration count. The correcting-rotation method examines the sign of $\tilde{\omega}_i$ constituted by some most significant digits of ω_i, and if $\tilde{\omega}_i \neq 0$ then σ_i is taken to be sign($\tilde{\omega}_i$) and σ_i is taken to be +1 otherwise. It is shown that the error occurring in this algorithm could be corrected by repetition of the iterations for $i = p, 2p, 3p...$, and so on, where p is the size of $\tilde{\omega}_i$. The branching CORDIC was proposed in Reference [36] for fast online implementation for redundant CORDIC with a constant scale factor. The main drawback of this method, however, is its necessity of performing two conventional CORDIC iterations in parallel, which consumes more silicon area than classical methods [39]. The work proposed in Reference [34] has also been extended to the vectoring mode [37], and correcting operations are included further to keep the scaling factor constant so as to eliminate the hardware for scaling.

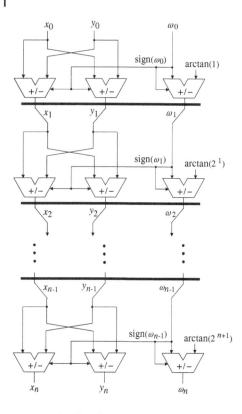

Figure 5.7 Pipelined architecture for conventional CORDIC.

5.3.5 Pipelined CORDIC Architecture

Since the CORDIC iterations are identical, it is very much convenient to map them into pipelined architectures. The main emphasis in efficient pipelined implementation lies with the minimization of the critical path. The earliest pipelined architecture that we find was suggested by Deprettere *et al.* [14]. Pipelined CORDIC circuits have been used thereafter for high-throughput implementation of sinusoidal wave generation, fixed and adaptive filters, discrete orthogonal transforms and other signal processing applications [40–44]. A generic architecture of pipelined CORDIC circuit is shown in Figure 5.7. It consists of n stages of CORDIC units where each of the pipelined stages consists of a basic CORDIC engine of the kind shown in Figure 5.2. Since the number of shifts to be performed by the shifters at different stages is fixed (shift operation through i-bit positions is performed at the ith stage) in case of pipelined CORDIC the shift operations could be hardwired with adders and therefore shifters are eliminated in the pipelined implementation. The critical-path of pipelined CORDIC thus amounts to the time required by the add/subtract operations in each of the stages. When three adders are used in each stage as shown in Figure 5.7, the critical-path amounts to $T_{ADD} + T_{MUX} + T_{2C}$, where T_{ADD}, T_{MUX} and T_{2C} are the time required for addition, 2:1 multiplexing and 2's complement operation, respectively. For known and constant angle rotations the sign of microrotations could be predetermined, and the need of multiplexing could be avoided for reducing the critical path. The latency of computation, thus, depends primarily on the time required for an addition. Since there is very little room for reducing the critical path in the pipelined implementation of conventional CORDIC, digit-on-line pipelined CORDIC circuits based on the differential CORDIC (D-CORDIC) algorithm have been suggested to achieve higher throughput and lower pipeline latency.

5.3.6 Differential CORDIC Algorithm

D-CORDIC algorithm is equivalent to the usual CORDIC in terms of accuracy as well as convergence, but it provides faster and more efficient redundant number-based implementation of both rotation mode and vectoring mode CORDIC. It introduces some temporary variables corresponding to the CORDIC variables x, y, and ω, that generically defined as

$$\hat{g}_{i+1} = \text{sign}(g_i) \cdot g_{i+1} \tag{5.28}$$

which implies that $|\hat{g}_{i+1}| = |g_{i+1}|$ and $\text{sign}(g_{i+1}) = \text{sign}(g_i) \cdot \text{sign}(\hat{g}_{i+1})$. The signs of g_i are, therefore, considered as being differentially encoded signs of $\hat{g}_i$ in the differential CORDIC algorithm [45]. The rotation and vectoring mode D-CORDIC algorithms are outlined in Table 5.3.

D-CORDIC algorithm is suitable for efficient pipelined implementation, which is utilized by Ercegovac and Lang [34] using online arithmetic based on redundant number system. Since the output data in the redundant

Table 5.3 Differential CORDIC algorithm.

Mode	Algorithm
Rotation mode	$x_{i+1} = x_i - \text{sign}(\omega_i) \cdot 2^{-i} \cdot y_i$
	$y_{i+1} = y_i + \text{sign}(\omega_i) \cdot 2^{-i} \cdot x_i$
	$|\hat{\omega}_{i+1}| = ||\hat{\omega}_i| - \alpha_i|$
	$\text{sign}(\omega_{i+1}) = \text{sign}(\omega_i) \cdot \text{sign}(\hat{\omega}_{i+1})$
Vectoring mode	$\hat{x}_{i+1} = \hat{x}_i + |\hat{y}_i| \cdot 2^{-i}$
	$\omega_{i+1} = \omega_i + \text{sign}(y_i) \cdot \alpha_i$
	$\text{sign}(y_{i+1}) = \text{sign}(y_i) \cdot \text{sign}(\hat{y}_{i+1})$
	$|\hat{y}_{i+1}| = ||\hat{y}_i| - \hat{x}_i \cdot 2^{-i}|$

online arithmetic can be available in the most significant digit-first (MSD-first) fashion, the successive iterations could be implemented by a set of cascaded stages, where processing time between the successive stages is overlapped with a single-digit time-skew, that results in a significant reduction in overall latency of computation. Moreover, in some redundant number representations, the absolute values and sign of the output are easily determined, for example, in binary signed-digit (BSD) representation, the sign of a number corresponds to the sign of the first nonzero MSD, and negation of the number can be performed just by flipping signs of nonzero digits. A two-dimensional systolic D-CORDIC architecture is derived in Reference [46] where phase accumulation is performed for direct digital frequency synthesis in the digit-level pipelining framework.

5.4 Scaling, Quantization, and Accuracy Issues

As discussed in Section 5.2.1, scaling is a necessary operation associated with the implementation of CORDIC algorithm. Scaling in CORDIC could be of two types: (i) constant factor scaling and (ii) variable factor scaling. In case of variable factor scaling, the scale factor changes with the rotation angle. It arises mainly because some of the iterations of convental CORDIC are ignored (and that varies with the angle of rotation), as in the case of higher-radix CORDIC and most of the optimized CORDIC algorithms. The techniques for scaling compensation for each such algorithms have been studied extensively for minimizing the scaling overhead. In case of conventional CORDIC, as given by Eq. (5.8), after sufficiently large number of iterations, the scale factor K converges to ~ 1.6467605, which leads to constant factor scaling since the scale factor remains the same for all the angle of rotations. Constant factor scaling could be efficiently implemented in a dedicated scaling unit designed by canonical signed digit (CSD)-based technique [47] and common subexpression elimination (CSE) approach [48,49]. When the sum of the output of more than one independent CORDIC operations are to be evaluated, one can perform only one scaling of the output sum [50] in the case of constant factor scaling. In the following subsections, we briefly discuss some interesting developments on implementation of online scaling and realization of scaling-free CORDIC. Besides, we outline here the sources of error that may arise in a CORDIC design and their impact on implementation.

5.4.1 Implementation of Mixed-Scaling Rotation

Dewilde *et al.* [51] have suggested the online scaling where shift-add operations for scaling and microrotations are interleaved in the same circuit. This approach has been used in Reference [52] and improved further in Reference [53]. In the mixed-scaling rotation (MSR) approach, pioneered by Wu *et al.* [54–56], the microrotation and scaling phases are merged into a unified vector rotational model to minimize the overhead of the scaling operation [54–56]. The MSR-CORDIC can be applied to DSP applications, in which the rotation angles are usually known *a priori*, for example, the twiddle factor in fast Fourier transform (FFT) and kernel components in other sinusoidal transforms. It is shown in Reference [55] that the MSR technique can significantly reduce the total iteration count so as to improve the speed performance and enhance the signal-to-quantization-noise ratio (SQNR) performance by controlling the internal dynamic range. The MSR-CORDIC scheme has been applied to a variable-length FFT processor design [29], and found to result in significant hardware reduction in the implementation of twiddle-factor multiplications. Although the interleaved scaling and MSR-CORDIC provide hardware reduction, they also lead to the reduction of throughput. For high-throughput implementation, one should implement the microrotations and scaling in two separate pipelined stages.

5.4.2 Low-Complexity Scaling

When the elementary angles pertaining to a rotation are "sufficiently small," defined by $\sin(\alpha_i) \cong \alpha_i = 2^{-i}$, and the rotations are only in one direction, the CORDIC rotation is given by the representation [57]

$$
\begin{bmatrix} x_n \\ y_n \end{bmatrix} = \prod_{i=p}^{n-1} \begin{bmatrix} 1 - 2^{-(2i+1)} & 2^{-i} \\ -2^{-i} & 1 - 2^{-(2i+1)} \end{bmatrix} \begin{bmatrix} x_i \\ y_i \end{bmatrix}
\tag{5.29}
$$

and $\omega_{i+1} = \omega_i - 2^{-i}$, (considering clockwise microrotations only), where x_i and y_i are the components of the vector after the ith microrotation, n is the input word length and $p = \lfloor (n - 2.585)/3 \rfloor$. The formulation of Eq. (5.29) performs the "actual" rotation where the norm of the vector is preserved at every microrotation.

However, the problem with this formulation is that the overall range of angles for which it can be used is very small because for 16-bit word length, the largest such angle is $\theta = \Sigma_{i=4}^{15} \alpha_i = \pm 7.16$, which obviously is quite small compared to the entire coordinate space. To overcome this problem, argument reduction is performed through "domain folding" [58] by mapping the target rotation angles into the range $[0 \ \pi/8]$. Besides, the elementary rotations are carried out in an adaptive manner to enhance the rate of convergence so as to force the approximation error of final angle below a specified limit [59]. But the domain-folding in some cases, involves a rotation through $\pi/4$ which demands a scaling by a factor of $1/\sqrt{2}$. Besides, the target range $[0 \ \pi/8]$ is still much larger than the range of convergence of the scaling-free realization.

The formulation of Eq. (5.29), therefore, could be effectively used when a rotation through $\pi/4$ is not required and angles of rotations could be folded to the range $-7.16°$ to $7.16°$. Generalized algorithms and their corresponding architectures to perform the scale-factor compensation in parallel with the CORDIC iterations for both rotation and vectoring modes are proposed in Reference [60], where the compensation overhead is reduced to a couple of iterations. It is shown in Reference [61] that since the scale factor is known in advance, one can perform the minimal recoding of the bits of scaling factor and implement the multiplication thereafter by a Wallace tree. It is a good solution of low-latency scaling, particularly for pipelined CORDIC architectures.

5.4.3 Quantization and Numerical Accuracy

Errors in CORDIC are mainly of two types: (i) the angle approximation error that originates from quantization of rotation angle represented by a linear combination of finite numbers of elementary angles; and (ii) the finite wordlength of the datapath resulting in the rounding/truncation of output that increases cumulatively through the successive iterations of microrotations. A third source of error that also comes into the picture results from the scaling of pseudorotated outputs. The scaling error is, however, also due to the use of finite word length in the scaling circuitry and is predominantly a rounding/truncation error. A detailed discussion on rounding error due to fixed and floating point implementations is available in Reference [62]. In his earlier work, Walther [3] concluded that the errors in the CORDIC output are bounded and $\log_2 n$ extra bits are required in the datapaths to take care of the errors. Hu [62] has provided more precise error bounds due to the angle approximation error for different CORDIC modes for fixed point as well as floating point implementations. The error bound resulting for fixed point representation of arctangents is further analyzed by Kota and Cavallaro [63] and its impact on practical implementation has been discussed.

5.4.4 Area–Delay–Accuracy Trade-off

Area, accuracy, and latency of CORDIC algorithm depend mainly on the iteration count and its implementation. To achieve n-bit accuracy, if fixed point arithmetic is applied, the word length of x and y data path is $(n + 2 + \log 2(n))$ and for the computation of the angle θ, it is $(n + \log 2(n))$ [45,63]. The hardware requirement, therefore, increases accordingly with the desired accuracy. Floating-point implementation naturally gives higher accuracy than its fixed-point counterpart, but at the cost of more complex hardware. To minimize the angle approximation error, the smallest elementary angle α_{n-1} needs to be as small as possible [62]. This consequently demands more number of right-shifts

Table 5.4 Computations using CORDIC algorithm in different configurations.

Operation	Configuration	Initialization	Output	Postprocessing and remarks
$\cos\theta,\ \sin\theta,\ \tan\theta$	CC-RM	$x_0 = 1$ $y_0 = 0$ and $\omega_0 = \theta$	$x_n = \cos\theta$ $y_n = \sin\theta$	$\tan\theta = (\sin\theta / \cos\theta)$
$\cosh\theta,\ \sinh\theta$ $\tanh\theta,\ \exp(\theta)$	HC-RM	$x_0 = 1$ $y_0 = 0$ and $\omega_0 = \theta$	$x_n = \cosh\theta$ $y_n = \sinh\theta$	$\tanh\theta = (\cosh\theta / \sinh\theta)$ $\exp(\theta) = (\cosh\theta + \sinh\theta)$
$\ln(a),\ \sqrt{a}$	HC-VM	$x_0 = a + 1$ $y_0 = a - 1$ and $\omega_0 = 0$	$x_n = \sqrt{a}$ $\omega_n = \frac{1}{2}\ln(a)$	$\ln(a) = 2\omega_n$
Arctan (a)	CC-VM	$x_0 = a$ $y_0 = 1$ and $\omega_0 = 0$	$\omega_n = \arctan(a)$	No pre or postprocessing
Division (b/a)	LC-VM	$x_0 = a$ $y_0 = b$ and $\omega_0 = 0$	$\omega_n = b/a$	No pre or postprocessing
Polar-to-rectangular	CC-RM	$x_0 = R$ $y_0 = 0$ and $\omega_0 = \theta$	$x_n = R\cos\theta$ $y_n = R\sin\theta$	No pre or postprocessing
Rectangular-to-polar $\tan^{-1}(b/a)$ and $\sqrt{a^2 + b^2}$	CC-VM	$x_0 = a$ $y_0 = b$ and $\omega_0 = 0$	$x_n = \sqrt{a^2 + b^2}$ $\omega_n = \arctan(b/a)$	No pre or postprocessing

The computation of $\tan\theta$ and $\tanh\theta$ require one division operation, while $\exp(\theta)$ and $\ln(a)$ require one addition and one shift, respectively, for postprocessing. The computation of $\sqrt{a}$ and $\ln(a)$ require one increment and one decrement for preprocessing as shown in column 3, for "initialization."

and more hardware for the barrel-shifters and adders. Besides, to have better angle approximation, more number of iterations are required which increases the latency. The additional accuracy resulting from floating-point implementation or better angle approximation may not, however, be necessary in many applications. Thus, there is a need for trade-off between hardware-cost, latency, and numerical accuracy subject to a particular application. Therefore, the designer has to check how much numerical accuracy is needed along with area and speed constraints for the particular application; and can accordingly decide on fixed or floating-point implementation and should set the word length and optimal number of iterations.

5.5 Applications of CORDIC

CORDIC technique is basically applied for rotation of a vector in circular, hyperbolic, or linear coordinate systems, which in turn could also be used for generation of sinusoidal waveform, multiplication and division operations, and evaluation of angle of rotation, trigonometric functions, logarithms, exponentials, and square-root [6,64,65]. Table 5.4 shows some elementary functions and operations that can be directly implemented by CORDIC. The table also indicates whether the coordinate system is circular (CC), linear (LC), or hyperbolic (HC), and whether the CORDIC operates in rotation mode (RM) or vectoring mode (VM), the initialization of the CORDIC and the necessary pre or postprocessing step to perform the operation. The scale factors are, however, obviated in Table 5.4 for simplicity of presentation. In this section, we discuss how CORDIC is used for some basic matrix problems like QR decomposition and singular-value decomposition. Moreover, we make a brief presentation on the applications of CORDIC to signal and image processing, digital communication, robotics, and 3D graphics.

5.5.1 Matrix Computation

5.5.1.1 QR Decomposition

QR decomposition of a matrix can be performed through Givens rotation [66] that selectively introduces zeros into the matrix. Givens rotation is an orthogonal transformation of the form

$$\begin{bmatrix} c & c \\ -s & c \end{bmatrix} \cdot \begin{bmatrix} a \\ b \end{bmatrix} = \begin{bmatrix} r \\ 0 \end{bmatrix} \tag{5.30}$$

where $\theta = \tan^{-1}(b/a)$, $c = \cos\theta$, $s = \sin\theta$, and $r = \sqrt{a^2 + b^2}$. The QR decomposition requires two types of iterative operations to obtain an upper-triangular matrix using orthogonal transformations. Those are: (i) to calculate the Givens rotation angle and (ii) to apply the calculated angle of rotation to the rest of the rows. Circular coordinate CORDIC is a good choice to implement both these Givens rotations, where the first operation is performed by a VM CORDIC and the second one is performed by an RM CORDIC. The CORDIC-based QR decomposition can be implemented in VLSI with suitable area–time trade-off using a systolic triangular array, a linear array, or a single CORDIC processor that is reconfigurable for rotation and vectoring modes of operations. Detailed explanations of these architectures are available in References [64,67].

5.5.1.2 Singular Value Decomposition and Eigenvalue Estimation

Singular value decomposition of a matrix $\mathbf{M}$ is given by $\mathbf{M} = \mathbf{U}\boldsymbol{\Sigma}\mathbf{V}^T$, where $\mathbf{U}$ and $\mathbf{V}$ are orthogonal matrices and $\boldsymbol{\Sigma}$ is a diagonal matrix of singular values. For CORDIC-based implementation of SVD, it is decomposed into 2×2 SVD problems, and solved iteratively. To solve each 2×2 SVD problem, two-sided Givens rotation is applied to each of the 2×2 matrices to nullify the off-diagonal elements, as described in the following:

$$\begin{bmatrix} \cos\theta_l & -\sin\theta_l \\ \sin\theta_l & \cos\theta_l \end{bmatrix} \mathbf{M} \begin{bmatrix} \cos\theta_r & -\sin\theta_r \\ \sin\theta_r & \cos\theta_r \end{bmatrix} = \begin{bmatrix} \psi_1 & 0 \\ 0 & \psi_2 \end{bmatrix} \tag{5.31}$$

where $\mathbf{M}$ is a 2×2 input matrix to be decomposed, and θ_l and θ_r are, respectively, the left and right rotation angles, calculated from the elements of $\mathbf{M}$ using the following two relations:

$$\begin{aligned} \theta_l + \theta_r &= \arctan([c + b]/[d - a]) \\ \theta_l - \theta_r &= \arctan([c - b]/[d + a]) \end{aligned} \text{ for } \mathbf{M} = \begin{bmatrix} a & b \\ c & d \end{bmatrix} \tag{5.32}$$

CORDIC-based architectures for SVD using this method were developed by Cavallaro and Luk [68]. A simplified design of array processor for the particular case ($c = b$, i.e., $\theta_l = \theta_r$) was developed further by Delosme [69] for the symmetric Eigenvalue problem. In a relatively recent paper [70], Liu *et al.* have proposed an application-specific instruction set processor (ASIP) for the real-time implementation of QR decomposition and SVD where circular coordinate CORDIC is used for efficient implementation of both these functions.

5.5.2 Signal Processing and Image Processing Applications

CORDIC techniques have a wide range of DSP applications including fixed/adaptive filtering [8], and the computation of discrete sinusoidal transforms such as the DFT [50,52,71,72], discrete Hartley transform (DHT) [53,73,74], discrete cosine transform (DCT) [75–78], discrete sine transform (DST) [76–78], and chirp Z transform (CZT) [79]. The DFT, DHT, and DCT

[80] of an N-point input sequence $\{x(l)$ for $l = 0, 1, \cdots, N - 1\}$, in general, are given by

$$X(k) = \sum_{l=0}^{N-1} C(k, l) \cdot x(l), \quad \text{for} \quad k = 0, 1, \cdots, N - 1 \tag{5.33}$$

where the transform kernel matrix is defined as

$$C(k, l) = \begin{cases} \cos(2\pi kl/N) - j \sin(2\pi kl/N), & \text{for DFT} \\ \cos(2\pi kl/N) + \sin(2\pi kl/N), & \text{for DHT} \\ \cos(\pi k(2l + 1)/2N), & \text{for DCT} \end{cases}$$

The input sequence for the DFT is, in general, complex and the computation of Eq. (5.33) can be partitioned into blocks of form: $[\mathrm{Re}[x(l)] \cdot \cos(2\pi kl/N) \pm \mathrm{Im}[x(l)] \cdot \sin(2\pi kl/N)]$, which is in the same form of the output of RM-CORDIC, for $\theta = 2\pi kl/N$. In case of DHT, similarly the computation can also be transformed into a $N/2$ computations of the form $[x(l) \cdot \cos(2\pi kl/N) \pm [x(N - l)] \cdot \sin(2\pi kl/N)]$ to be implemented efficiently by RM-CORDIC units. These features of DFT and DHT are used to design parallel and pipelined architectures for the computation of these two transforms [50,52,53,71–74]. It is shown that [76,77] by simple input–output modification, one can transform the DCT and DST kernels into the DHT form to compute then by rotation mode CORDIC. Similarly in Reference [79], CZT is represented by a DFT-like kernel by simple preprocessing and postprocessing operations, and implemented through CORDIC rotations. The CORDIC technique has also been used in many image processing operations like spatial domain image enhancement for contrast stretching, logarithmic transformation and power-law transformation, image rotation, and Hough transform for line detection [81,82]. CORDIC implementation of some of these applications are discussed in References [83,84]. Several other signal processing applications are discussed in detail in Reference [64], which we do not intend to repeat here.

5.5.3 Applications to Communication

CORDIC algorithm can be used for efficient implementation of various functional modules in a digital communication system [85]. Most applications of CORDIC in communications use the circular coordinate system in one or both CORDIC operating modes. The RM-CORDIC is mainly used to generate mixed signals, while the VM-CORDIC is mainly used to estimate phase and frequency parameters. We briefly outline here some of the important communication applications.

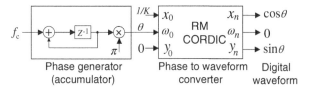

Figure 5.8 A CORDIC-based direct digital synthesizer. $\theta = \pi \cdot \left[\sum f_c\right]$.

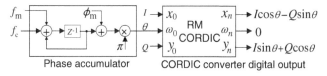

Figure 5.9 A generic scheme to use RM CORDIC for digital modulation. I and Q are, respectively, the in-phase and quadrature signals to be modulated. $x_n = I \cdot \cos\theta - Q \cdot \sin\theta$, $y_n = I \cdot \sin\theta + Q \cdot \cos\theta$ and $\theta = \left[\left(\sum(f_c + f_m)\right) + \phi_m\right]\pi$.

5.5.3.1 Direct Digital Synthesis

Direct digital synthesis is the process of generating sinusoidal waveforms directly in the digital domain. A direct digital synthesizer (DDS) (as shown in Figure 5.8) consists of a phase accumulator and a phase-to-waveform converter [86,87]. The phase-generation circuit increments the phase according to $\left[\sum f_c\right] \cdot \pi$, where f_c is the normalized carrier frequency in every cycle and feeds the phase information to the phase-to-waveform converter. The phase-to-waveform converter could be realized by an RM-CORDIC [88,89], as shown in Figure 5.8. The cosine and sine waveforms are obtained, respectively, by the CORDIC outputs x_n and y_n.

5.5.3.2 Analog and Digital Modulation

A generic scheme to use CORDIC in RM for digital modulation is shown in Figure 5.9, where the phase-generation unit of Figure 5.8 is changed to generate the phase according to $\left[\left(\sum(f_c + f_m)\right) + \phi_m\right] \cdot \pi$. f_c and f_m are the normalized carrier and the modulating frequencies, respectively, and ϕ_m is the phase of modulating component. By suitable selection of the parameters f_c, f_m and ϕ_m and the CORDIC inputs x_0 and y_0, the generic scheme of Figure 5.9 it could be used for digital realization of analog amplitude modulation (AM), phase modulation (PM), and frequency modulation (FM), as well as the digital modulations, for example, amplitude-shift keying (ASK), phase-shift keying (PSK), and frequency-shift keying (FSK) modulators. It could also be used for the up/down converters for quadrature amplitude modulators (QAM) and full mixers for complex signals or phase and frequency corrector circuits for synchronization [85].

5.5.3.3 Other Communication Applications

By operating the CORDIC in vectoring mode, one can compute the magnitude and the angle of an input vector. The magnitude computation can be used for envelope-detection in an AM receiver or to detect FSK signal if it is placed after mark or space filters [90]. The angle computation in VM CORDIC, on the other hand, can be used to detect FM and FSK signals and to estimate phase and frequency parameters [91]. A single VM-CORDIC can be used to perform these computations for the implementation of a slicer for a high-order constellation like the 32-APSK used in DVB-S2.

CORDIC circuits operating in both modes are also required in digital receivers for the synchronization stage to perform a phase or frequency estimation followed by a correction stage. This can be done by using two different CORDIC units, to meet the high-speed requirement in Costas loop for phase recovery in a QAM modulation [92,93]. On the other hand the burst-based communication system that needs a preamble for synchronization purposes, for example, in case of IEEE 802.11a WLAN-OFDM receivers, can use a single CORDIC unit configurable for both operating modes since the estimation and correction are not performed simultaneously [94,95]. Apart from these, the CORDIC-based QR decomposition has been used in multiinput–multioutput (MIMO) systems to implement V-BLAST detectors [96–98], and to implement a recursive least square (RLS) adaptive antenna beamformer [67,99,100].

5.5.4 Applications of CORDIC to Robotics and Graphics

Two of the key problems where CORDIC provides area and power-efficient solutions are (i) direct kinematics and (ii) inverse kinematics of serial robot manipulators. How CORDIC is applied in these applications is discussed below.

5.5.4.1 Direct Kinematics Solution (DKS) for Serial Robot Manipulators

A robot manipulator consists of a sequence of links, connected typically by either revolute or prismatic joints. For an N-degrees-of-freedom manipulator, there are N joint-link pairs with link 0 being the supporting base and the last link is attached with a tool. The joints and links are numbered outwardly from the base. The coordinates of the points on the ith link represented by (x_i, y_i, z_i) change successively for $i \in \{0, 1, \dots, N - 1\}$ due to successive rotations and translations of the links. The translation operations are realized by simple additions of coordinate values while the new coordinates of any point due to rotation are computed by RM-CORDIC circuits.

5.5.4.2 Inverse Kinematics for Robot Manipulators

The inverse kinematics problem involves determination of joint variables for a desired position and orientation for the tool. The CORDIC approach is valuable to find the inverse kinematic solution when a closed form solution is possible (when, in particular, the desired tool tip position is within the robot's work envelope and when joint angle limits are not violated). The authors in Reference [101] present a maximum pipelined CORDIC-based architecture for efficient computation of the inverse kinematics solution. It is also shown [101,102] that up to 25 CORDIC processors are required for the computation of the entire inverse kinematics solution for a six-link PUMA-type robotic arm. Apart from implementation of rotation operations, CORDIC is used in the evaluation of trigonometric functions and square root expressions involved in the inverse kinematics problems [103].

5.5.4.3 CORDIC for Other Robotics Applications

CORDIC has also been applied to robot control [104,105], where CORDIC circuits serve as the functional units of a programmable CPU coprocessor. Another application of CORDIC is for kinematics of redundant manipulators [106]. It is shown in Reference [106] that the case of inverse kinematics can be implemented efficiently in parallel by computing pseudoinverse through singular value decomposition. Collision detection is another area where CORDIC has been applied to robotics [107]. A CORDIC-based highly parallel solution for collision detection between a robot manipulator and multiple obstacles in the workspace is suggested in Reference [107]. The collision detection problem is formulated as one that involves a number of coordinate transformations. CORDIC-based processing elements are used to efficiently perform the coordinate transformations by shift-add operations.

5.5.4.4 CORDIC for 3D Graphics

The processing in graphics such as 3D vector rotation, lighting, and vector interpolation are computation-intensive and are geometric in nature. CORDIC architecture is therefore a natural candidate for cost-effective implementation of these geometric computations in graphics. A systematic formulation to represent 3D computer graphics operations in terms of CORDIC-type primitives is provided in Reference [108]. An efficient stream processor based on CORDIC-type modules to implement the graphic operations is also suggested in Reference [108]. 3D vector interpolation is also an important function in graphics, which is required for good-quality shading [109] for graphic rendering. It is shown that the variable-precision capability of CORDIC engine could be utilized to realize a power-aware implementation of the 3D vector interpolator [110].

5.6 Conclusions

The beauty of CORDIC is its potential for unified solution for a large set of computational tasks involving the evaluation of trigonometric and transcendental functions, calculation of multiplication, division, square root and logarithm, solution of linear systems, QR-decomposition, and SVD. Moreover, CORDIC is implemented by a simple hardware through repeated shift-add operations. These features of CORDIC have made it an attractive choice for a wide variety of applications. Several algorithms and architectures have been developed over the years to speed-up the CORDIC by reducing its iteration counts and through its pipelined implementation. Moreover, its applications in several diverse areas including signal processing, image processing, communication, robotics, and graphics apart from general scientific and technical computations have been explored. Latency of computation, however, continues to be the major drawback of the CORDIC algorithm, since we do not have efficient algorithms for its parallel implementation. But, CORDIC on the other hand is inherently suitable for pipelined designs, due to its iterative behavior, and small cycle time compared with the conventional arithmetic. For high-throughput applications, efficient pipelined-architectures with multiple-CORDIC units could be developed to take the advantage of pipelineability of CORDIC, because the digital hardware is getting cheaper along with the progressive device-scaling. Research on fast implementation of shift-accumulation operation, exploration of new number systems for CORDIC, optimization of CORDIC for constant rotation have scope for further reduction of its latency. Another way to use CORDIC efficiently, is to transform the computational algorithm into independent segments, and to implement the individual segments by different CORDIC processors. With enhancement of its throughput and reduction of latency, it is expected that CORDIC would be useful for many high-speed and real-time applications. The area-delay-accuracy trade-off for different advanced algorithms may be investigated in detail and compared with in future work.

References

1 J. E. Volder, The CORDIC trigonometric computing technique, *IRE Trans. Electron. Comput.*, vol. 8, 1959, pp. 330–334.

2 J. E. Volder, The birth of CORDIC, *J. VLSI Signal Process.*, vol. 25, 2000, pp. 101–105.

3 J. S. Walther, A unified algorithm for elementary functions, in *Proc. 38th Spring Joint Comput. Conf.*, Atlantic City, NJ, 1971, pp. 379–385.

4 J. S. Walther, The story of unified CORDIC, *J. VLSI Signal Process.*, vol. 25, no. 2, 2000, pp. 107–112.

5 D. S. Cochran, Algorithms and accuracy in the HP-35, *HP J.*, vol. 23, no. 10, 1972, pp. 10–11.

6 J.-M. Muller, *Elementary Functions: Algorithms and Implementation.* Birkhauser, Boston, MA, 2006.

7 S.-F. Hsiao and J.-M. Delosme, Householder CORDIC algorithms, *IEEE Trans. Comput.*, vol. 44, no. 8, 1995, pp. 990–1001.

8 E. Antelo, J. Villalba, and E. L. Zapata, A low-latency pipelined 2D and 3D CORDIC processors, *IEEE Trans. Comput.*, vol. 57, no. 3, 2008, pp. 404–417.

9 D. Timmermann, H. Hahn, and B. J. Hosticka, Low latency time CORDIC algorithms, *IEEE Trans. Comput.*, vol. 41, no. 8, 1992, pp. 1010–1015.

10 S. Wang, V. Piuri, and J. E. E. Swartzlander, Hybrid CORDIC algorithms, *IEEE Trans. Comput.*, vol. 46, no. 11, 1997, pp. 1202–1207.

11 S. Wang and E. E. Swartzlander Jr., Merged CORDIC algorithm, in *IEEE Int. Symp. Circuits Syst., (ISCAS '95)*, vol. 3, 1995, pp. 1988–1991.

12 B. Gisuthan and T. Srikanthan, Pipelining flat CORDIC based trigonometric function generators, *Microelectron. J.*, vol. 33, 2002, pp. 77–89.

13 S. Suchitra, S. Sukthankar, T. Srikanthan, and C. T. Clarke, Elimination of sign precomputation in flat CORDIC, in *IEEE Int. Symp. Circuits Syst., (ISCAS '05)*, vol. 4, 2005, pp. 3319–3322.

14 E. Deprettere, P. Dewilde, and R. Udo, Pipelined CORDIC architectures for fast VLSI filtering and array processing, in *IEEE Int. Conf. Acoust., Speech, Signal Process., (ICASSP '84)*, vol. 9, 1984, pp. 250–253.

15 H. Kunemund, S. Soldner, S. Wohlleben, and T. Noll, CORDIC processor with carry save architecture, in *Proc. 16th Euro. Solid State Circuits Conf. (ESSCIRC '90)*, 1990, pp. 193–196.

16 E. Antelo, J. Villalba, J. D. Bruguera, and E. L. Zapatai, High performance rotation architectures based on the radix-4 CORDIC algorithm, *IEEE Trans. Comput.*, vol. 46, no. 8, 1997, pp. 855–870.

17 P. R. Rao and I. Chakrabarti, High-performance compensation technique for the radix-4 CORDIC algorithm, *IEEE Proc. Comput.Digit. Techniq.*, vol. 149, no. 5, 2002, pp. 219–228.

18 E. Antelo, T. Lang, and J. D. Bruguera, Very-high radix circular CORDIC: vectoring and unified rotation/vectoring, *IEEE Trans. Comput.*, vol. 49, no. 7, 2000, pp. 727–739.

19 Y. H. Hu and S. Naganathan, An angle recoding method for CORDIC algorithm implementation, *IEEE Trans. Comput.*, vol. 42, no. 1, 1993, pp. 99–102.

20 Y. H. Hu and H. H. M. Chern, A novel implementation of CORDIC algorithm using backward angle recoding (BAR), *IEEE Trans. Comput.*, vol. 45, no. 12, 1996, pp. 1370–1378.

21 C.-S. Wu, A.-Y. Wu, and C.-H. Lin, A high-performance/low-latency vector rotational CORDIC architecture based on extended elementary angle set and trellis-based searching schemes, *IEEE Trans. Circuits Syst. II: Analog Digit. Signal Process.*, vol. 50, no. 9, 2003, pp. 589–601.

22 T. K. Rodrigues and E. E. Swartzlander, Adaptive CORDIC: Using parallel angle recoding to accelerate CORDIC rotations, in *Fortieth Asilomar Conf. Signals, Syst. Comput., (ACSSC '06)*, 2006, pp. 323–327.

23 M. Kuhlmann and K. K. Parhi, P-CORDIC: a precomputation based rotation CORDIC algorithm, *EURASIP J. Appl. Signal Process.*, vol. 2002, no. 9, 2002, pp. 936–943.

24 D. Fu and A. N. Willson Jr., A high-speed processor for digital sine/cosine generation and angle rotation, in *Conf. Rec. 32nd Asilomar Conf. Signals, Syst. Comput.*, vol. 1, 1998, pp. 177–181.

25 C.-Y. Chen and W.-C. Liu, Architecture for CORDIC algorithm realization without ROM lookup tables, in *Proc. 2003 Int. Symp. Circuits Syst., (ISCAS '03)*, vol. 4, 2003, pp. 544–547.

26 D. Fu and A. N. Willson Jr., A two-stage angle-rotation architecture and its error analysis for efficient digital mixer implementation, *IEEE Trans. Circuits Syst. I: Regul. Pap.*, vol. 53, no. 3, 2006, pp. 604–614.

27 S. Ravichandran and V. Asari, Implementation of unidirectional CORDIC algorithm using precomputed rotation bits, in *45th Midwest Symp. Circuits Syst., (MWSCAS-2002)*, vol. 3, 2002, pp. 453–456.

28 C.-Y. Chen and C.-Y. Lin, High-resolution architecture for CORDIC algorithm realization, in *Proc. Int. Conf. Commun., Circuits Syst., (ICCCS'06)*, vol. 1, 2006, pp. 579–582.

29 D. D. Hwang, D. Fu, and A. N. Willson Jr., A 400-MHz processor for the conversion of rectangular to polar coordinates in 0.25-/mu-m CMOS, *IEEE J. Solid State Circuits*, vol. 38, no. 10, 2003, pp. 1771–1775.

30 S.-W. Lee, K.-S. Kwon, and I.-C. Park, Pipelined cartesian-to-polar coordinate conversion based on SRT division, *IEEE Trans. Circuits Syst. II: Express Briefs*, vol. 54, no. 8, 2007, pp. 680–684.

31 S.-F. Hsiao, Y.-H. Hu, and T.-B. Juang, A memory-efficient and high-speed sine/cosine generator based on parallel CORDIC rotations, *IEEE Signal Process. Lett.*, vol. 11, no. 2, 2004.

32 T.-B. Juang, S.-F. Hsiao, and M.-Y. Tsai, Para-CORDIC: parallel CORDIC rotation algorithm, *IEEE Trans. Circuits Syst. I: Regul. Pap.*, vol. 51, no. 8, 2004.

33 T.-B. Juang, Area/delay efficient recoding methods for parallel CORDIC rotations, in *IEEE Asia Pac. Conf. Circuits Syst., (APCCAS'06)*, 2006, pp. 1539–1542.

34 M. D. Ercegovac and T. Lang, Redundant and on-line CORDIC: application to matrix triangularization and SVD, *IEEE Trans. Comput.*, vol. 39, no. 6, 1990, pp. 725–740.

35 N.Takagi, T.Asada, and S.Yajima, Redundant CORDIC methods with a constant scale factor for sine and cosine computation, *IEEE Trans. Comput.*, vol. 40, no. 9, 1991, pp. 989–995.

36 J. Duprat and J.-M. Muller, The CORDIC algorithm: new results for fast vlsi implementation, *IEEE Trans. Comput.*, vol. 42, no. 2, 1993, pp. 168–178.

37 J.-A. Lee and T. Lang, Constant-factor redundant CORDIC for angle calculation and rotation, *IEEE Trans. Comput.*, vol. 41, no. 8, 1992, pp. 1016–1025.

38 N. D. Hemkumar and J. R. Cavallaro, Redundant and on-line CORDIC for unitary transformations, *IEEE Trans. Comput.*, vol. 43, no. 8, 1994, pp. 941–954.

39 J. Valls, M. Kuhlmann, and K. K. Parhi, Evaluation of CORDIC algorithms for FPGA design, *J. VLSI Signal Process. Syst.*, vol. 32, no. 3, 2002, pp. 207–222.

40 D.E.Metafas and C.E.Goutis, A floating point pipeline CORDIC processor with extended operation set, in *IEEE Int. Symp. Circuits Syst., (ISCAS'91)*, vol. 5, 1991, pp. 3066–3069.

41 Z. Feng and P. Kornerup, High speed DCT/IDCT using a pipelined CORDIC algorithm, in *12th Symp. Comput. Arithmetic*, 1995, pp. 180–187.

42 M. Jun, K. K. Parhi, G. J. Hekstra, and E. F. Deprettere, Efficient implementations of pipelined CORDIC based iir digital filters using fast orthonormal μ-rotations, *IEEE Trans. Signal Process.*, vol. 48, no. 9, 2000.

43 M. Chakraborty, A. S. Dhar, and M. H. Lee, A trigonometric formulation of the lms algorithm for realization on pipelined CORDIC, *IEEE Trans. Circuits Syst. II: Express Briefs*, vol. 52, no. 9, 2005.

44 E. I. Garcia, R. Cumplido, and M. Arias, Pipelined CORDIC design on FPGA for a digital sine and cosine waves generator, in *Int. Conf. Electrical Electron. Eng., (ICEEE'06)*, 2006, pp. 1–4.

45 H. Dawid and H. Meyr, The differential CORDIC algorithm: constant scale factor redundant implementation without correcting iterations, *IEEE Trans. Comput.*, vol. 45, no. 3, 1996, pp. 307–318.

46 C. Y. Kang and E. E. Swartzlander, Jr., Digit-pipelined direct digital frequency synthesis based on differential CORDIC, *IEEE Trans. Circuits Syst. I: Regul. Pap.*, vol. 53, no. 5, 2006, pp. 1035–1044.

47 R. I. Hartley, Subexpression sharing in filters using canonic signed digit multipliers, *IEEE Trans Circuits Syst. II: Analog Digit. Signal Process.*, vol. 43, no. 10, 1996, pp. 677–688.

48 O. Gustafsson, A. G. Dempster, K. Johansson, M. D. Macleod, and L. Wanhammar, Simplified design of constant coefficient multipliers, *J. Circuits, Syst. Signal Process.*, vol. 25, no. 2, 2006, pp. 225–251.

49 G. Gilbert, D. Al-Khalili, and C. Rozon, Optimized distributed processing of scaling factor in CORDIC, in *the 3rd Int. IEEE-NEWCAS Conf.*, 2005, pp. 35–38.

50 A. M. Despain, Fourier transform computers using CORDIC iterations, *IEEE Trans. Comput.*, vol. 23, no. 10, 1974, pp. 993–1001.

51 P. Dewilde, E. F. Deprettere, and R. Nouta, Parallel and pipelined VLSI implementation of signal processing algorithms, in *VLSI and Modern Signal Processing*, (eds S. Y. Kung, H. J. Whitehouse, and T. Kailath), Prentice-Hall, Englewood Cliffs, NJ, 1995.

52 K. J. Jones, High-throughput, reduced hardware systolic solution to prime factor discrete Fourier transform algorithm, *IEE Proc. Comput. Digit. Tech.*, vol. 137, no. 3, 1990, pp. 191–196.

53 P. K. Meher, J. K. Satapathy, and G. Panda, Efficient systolic solution for a new prime factor discrete Hartley transform algorithm, *IEE Proc. Circuits, Devices Syst.*, vol. 140, no. 2, 1993, pp. 135–139.

54 Z.-X. Lin and A.-Y. Wu, Mixed-scaling-rotation CORDIC (MSR-CORDIC) algorithm and architecture for scaling-free high-performance rotational operations, in *IEEE Int. Conf. Acoust., Speech Signal Process., (ICASSP '03)*, vol. 2, 2003, pp. 653–656.

55 C.-H. Lin and A.-Y. Wu, Mixed-scaling-rotation CORDIC (MSR-CORDIC) algorithm and architecture for high-performance vector rotational DSP applications, *IEEE Trans. Circuits Syst. I: Regul. Pap.*, vol. 52, no. 11, 2005, pp. 2385–2396.

56 C.-L. Yu, T.-H. Yu, and A.-Y. Wu, On the fixed-point properties of mixed-scaling-rotation CORDIC algorithm, in *IEEE Workshop Signal Process. Syst.*, 2007, pp. 430–435.

57 A. S. Dhar and S. Banerjee, An array architecture for fast computation of discrete hartley transform, *IEEE Trans. Circuits Syst.*, vol. 38, no. 9, 1991, pp. 1095–1098.

58 K. Maharatna, A. Troya, S. Banerjee, and E. Grass, New virtually scaling free adaptive CORDIC rotator, *IEEE Proc. Comput. Digit. Tech.*, vol. 151, no. 6, 2004, pp. 448–456.

59 K. Maharatna, S. Banerjee, E. Grass, M. Krstic, and A. Troya, Modified virtually scaling free adaptive CORDIC rotator algorithm and architecture, *IEEE Trans. Circuits Syst. Video Technol.*, vol. 15, no. 11, 2005, pp. 1463–1474.

60 J. Villalba, T. Lang, and E. Zapata, Parallel compensation of scale factor for the CORDIC algorithm, *J. VLSI Signal Process.*, vol. 19, no. 3, 1998, pp. 227–241.

61 D. Timmermann, H. Hahn, B. J. Hosticka, and B. Rix, A new addition scheme and fast scaling factor compensation methods for CORDIC algorithms, *Integr. VLSI J.*, vol. 11, no. 1, 1991, pp. 85–100.

62 Y. H. Hu, The quantization effects of the CORDIC algorithm, *IEEE Trans. Signal Process.*, vol. 40, no. 4, 1992, pp. 834–844.

63 K. Kota and J. R. Cavallaro, Numerical accuracy and hardware tradeoffs for CORDIC arithmetic for special-purpose processors, *IEEE Trans. Comput.*, vol. 42, no. 7, 1993, pp. 769–779.

64 Y. H. Hu, CORDIC-based VLSI architectures for digital signal processing, *IEEE Signal Process. Mag.*, vol. 9, no. 3, 1992, pp. 16–35.

65 F. Angarita, A. Perez-Pascual, T. Sansaloni, and J. Vails, Efficient FPGA implementation of cordic algorithm for circular and linear coordinates, in *Int. Conf. Field Program. Logic Appl.*, 2005, pp. 535–538.

66 G. H. Golub and C. F. Van Loan, *Matrix Computations*, 3rd ed. Johns Hopkins University Press, Baltimore, MD, 1996.

67 G. Lightbody, R. Woods, and R. Walke, Design of a parameterizable silicon intellectual property core for QR-based RLS filtering, *IEEE Trans. Very Large Scale Integr. (VLSI) Syst.*, vol. 11, no. 4, 2003, pp. 659–678.

68 J. R. Cavallaro and F. T. Luk, CORDIC arithmetic for a SVD processor, *J. Parallel Distrib. Comput.*, vol. 5, 1988, pp. 271–290.

69 J. M. Delosme, A processor for two-dimensional symmetric eigenvalue and singular value arrays, in *IEEE 21st Asilomar Conf. Circuits, Syst. Comput.*, 1987, pp. 217–221.

70 Z. Liu, K. Dickson, and J. V. McCanny, Application-specific instruction set processor for SoC implementation of modern signal processing algorithms, *IEEE Trans. Circuits Syst. I: Regul. Pap.*, vol. 52, no. 4, 2005, pp. 755–765.

71 K. J. Jones, 2D systolic solution to discrete Fourier transform, *IEEE Proc. Comput. Digit. Tech.*, vol. 136, no. 3, 1989, pp. 211–216.

72 T.-Y. Sung, Memory-efficient and high-speed split-radix FFT/IFFT processor based on pipelined CORDIC rotations, *IEEE Proc. Vision, Image Signal Proc.*, vol. 153, no. 4, 2006, pp. 405–410.

73 L.-W. Chang and S.-W. Lee, Systolic arrays for the discrete Hartley transform, *IEEE Trans. Signal Proc.*, vol. 39, no. 11, 1991 pp. 2411–2418.

74 P. K. Meher and G. Panda, Novel recursive algorithm and highly compact semisystolic architecture for high throughput computation of 2D DHT, *Electron. Lett.*, vol. 29, no. 10, 1993, pp. 883–885.

75 W.-H. Chen, C. H. Smith, and S. C. Fralick, A fast computational algorithm for the discrete cosine transform, *IEEE Trans. Commun.*, vol. 25, no. 9, 1977, pp. 1004–1009.

76 B. Das and S. Banerjee, Unified CORDIC-based chip to realise DFT/DHT/DCT/DST, *IEEE Proc. Comput. Digit. Tech.*, vol. 149, no. 4, 2002, pp. 121–127.

77 J.-H. Hsiao, L.-G. Ghen, T.-D. Chiueh, and C.-T. Chen, High throughput CORDIC-based systolic array design for the discrete cosine transform, *IEEE Trans. Circuits Syst. Video Technol.*, vol. 5, no. 3, 1995, pp. 218–225.

78 D. C. Kar and V. V. B. Rao, A CORDIC-based unified systolic architecture for sliding window applications of discrete transforms, *IEEE Trans. Signal Process.*, vol. 44, no. 2, 1996, pp. 441–444.

79 Y. H. Hu and S. Naganathan, A novel implementation of chirp Z-transform using a CORDIC processor, *IEEE Trans. Acoust., Speech, Signal Process.*, vol. 38, no. 2, 1990, pp. 352–354.

80 J. G. Proakis and D. G. Manolakis, *Digital Signal Processing: Principles, Algorithms and Applications.* Prentice-Hall, Upper Saddle River, NJ, 1996.

81 R. C. Gonzalez, *Digital Image Processing*, 3rd ed. Prentice Hall, Upper Saddle River, NJ, 2008.

82 N. Guil, J. Villalba, and E. L. Zapata, A fast Hough transform for segment detection, *IEEE Trans. Image Process.*, vol. 4, no. 11, 1995, pp. 1541–1548.

83 S. M. Bhandakar and H. Yu, VLSI implementation of real-time image rotation, in *Int. Conf. Image Process.*, vol. 2, 1996, pp. 1015–1018.

84 S. Sathyanarayana, S. R. Kumar, and S. Thambipillai, Unified CORDIC based processor for image processing, in *15th Int. Conf. Digit. Signal Process.*, 2007, pp. 343–346.

85 J. Valls, T. Sansaloni, A. Perez-Pascual, V. Torres, and V. Almenar, The use of CORDIC in software defined radios: a tutorial, *IEEE Commun. Mag.*, vol. 44, no. 9, 2006.

86 L. Cordesses, Direct digital synthesis: a tool for periodic wave generation (part 1), *IEEE Signal Process. Mag.*, vol. 21, no. 4, 2004.

87 J. Vankka, *Digital Synthesizers and Transmitters for Software Radio*. Springer, Dordrecht, 2005.

88 J. Vankka, Methods of mapping from phase to sine amplitude in direct digital synthesis, in *50th IEEE Int. Frequency Control Symp.*, 1996, pp. 942–950.

89 F. Cardells-Tormo and J. Valls-Coquillat, Optimisation of direct digital frequency synthesisers based on CORDIC, *Electron. Lett.*, vol. 37, no. 21, 2001.

90 M. E. Frerking, *Digital Signal Processing in Communication Systems*. Van Nostrand Reinhold, New York, 1994.

91 H. Meyr, M. Moeneclaey, and S. A. Fechtel, *Digital Communication Receivers : Synchronization, Channel Estimation, and Signal Processing*. John Wiley and Sons, Inc., New York, 1998.

92 F. Cardells, J. Valls, V. Almenar, and V. Torres, Efficient FPGA-based QPSK demodulation loops: Application to the DVB standard, *Lect. Notes Comput. Sci.*, vol. 2438, 2002, pp. 102–111.

93 C. Dick, F. Harris, and M. Rice, FPGA implementation of carrier synchronization for QAM receivers, *J. VLSI Signal Process.*, vol. 36, 2004, pp. 57–71.

94 M. J. Canet, F. Vicedo, V. Almenar, and J. Valls, FPGA implementation of an IF transceiver for OFDM-based WLAN, in *IEEE Workshop Signal Process. Syst., (SIPS'04)*, 2004, pp. 227–232.

95 A. Troya, K. Maharatna, M. Krstic, E. Grass, U. Jagdhold, and R. Kraemer, Low-power VLSI implementation of the inner receiver for OFDM-based WLAN systems, *IEEE Trans. Circuits Syst. I: Regul. Pap.*, vol. 55, no. 2, 2008, pp. 672–686.

96 Z. Guo and P. Nilsson, A VLSI architecture of the square root algorithm for V-BLAST detection, *J. VLSI Signal Process. Syst.*, vol. 44, no. 3, 2006, pp. 219–230.

97 Z. Khan, T. Arslan, J. S. Thompson, and A. T. Erdogan, Analysis and implementation of multiple-input, multiple-output VBLAST receiver from area and power efficiency perspective, *IEEE Trans. Very Large Scale Integr. (VLSI) Syst.*, vol. 14, no. 11, 2006, pp. 1281–1286.

98 F. Sobhanmanesh and S. Nooshabadi, Parametric minimum hardware QR-factoriser architecture for V-BLAST detection, *IEEE Proc. -Circuits, Devices Syst.*, vol. 153, no. 5, 2006, pp. 433–441.

99 C. M. Rader, VLSI systolic arrays for adaptive nulling [radar], *IEEE Signal Process. Mag.*, vol. 13, no. 4, 1996, pp. 29–49.

100 R. Hamill, J. V. McCanny, and R. L. Walke, Online CORDIC algorithm and VLSI architecture for implementing QR-array process., *IEEE Trans. Signal Process.*, vol. 48, no. 2, 2000, pp. 592–598.

101 C. Lee and P. Chang, A maximum pipelined CORDIC architecture for inverse kinematic position computation, *IEEE Trans. Robot. Autom.*, vol. RA-3, no. 5, 1987, pp. 445–458.

102 R. Harber, J. Li, X. Hu, and S. Bass, The application of bit-serial CORDIC computational units to the design of inverse kinematics processors, in *Proc. IEEE Int. Conf. Robot. Autom.*, 1988, pp. 1152–1157.

103 C. Krieger and B. Hosticka, Inverse kinematics computations with modified CORDIC iterations, *IEEE Proc. Comput. Digit. Tech.*, vol. 143, 1996, pp. 87–92.

104 Y. Wang and S. Butner, A new architecture for robot control, in *Proc. IEEE Int. Conf. Robot. Autom.*, 1987, pp. 664–670.

105 Y. Wang and S. Butner, RIPS: a platform for experimental real-time sensory-based robot control, *IEEE Trans. Syst., Man, Cybern.*, vol. 19, 1989, pp. 853–860.

106 I. Walker and J. Cavallaro, Parallel VLSI architectures for real-time kinematics of redundant robots, in *Proc. of IEEE Int. Conf. Robot. Autom.*, 1993, pp. 870–877.

107 M. Kameyama, T. Amada, and T. Higuchi, Highly parallel collision detection processor for intelligent robots, *IEEE J. Solid State Circuits*, vol. 27, 1992, pp. 300–306.

108 T. Lang and E. Antelo, High-throughput CORDIC-based geometry operations for 3D computer graphics, *IEEE Trans. Comput.*, vol. 54, no. 3, 2005, pp. 347–361.

109 B. Phong, Illumination for computer generated pictures, *Commun. ACM*, vol. 18, no. 6, 1975, pp. 311–317.

110 J. Euh, J. Chittamuru, and W. Busrleson, CORDIC vector interpolator for power-aware 3D computer graphics, in *IEEE Workshop Signal Process. Syst. (SIPS '02)*, 2002, pp. 240–245.

6

RNS-Based Arithmetic Circuits and Applications

P.V. Ananda Mohan

Centre for Development of Advanced Computing, Bangalore, India

6.1   Introduction

The origin of residue number system (RNS) dates back to more than 1700 years, which relates to a puzzle written by a Chinese scholar Sun Tzu in the book *Suan-ching* in the third century. The puzzle is written in the form of a verse to find a unique positive integer that leaves the remainders 2, 3, and 2, respectively, when we divide by a set of relatively prime integers 7, 5, and 3. Sun Tzu had formulated a method to use these remainders (or residues) to find the unique integer, which is well known now as the Chinese remainder theorem (CRT). Since the CRT provides a method of determination of such a unique integer from the set of residues, it is possible to represent integers by the set of residues generated by a set of relatively prime integers. The theory of modular congruence and RNS was developed systematically by Carl Friedrich Gauss in his famous Latin book *Disquisitiones Arithmeticae* that was published on 1801. The first paper on RNS in english language is found to be one written by Harvey L. Garner that appeared in *IRE Transactions on Electronic Computers* in June 1959 [1]. Szabo and Tanaka [2] thereafter have extensively popularized this approach. Since then several papers and books on RNS-based computing have appeared in literature [3–7]. RNS now is a well developed computing technique and popularly used in several digital signal processing, communication, and security applications.

A residue number system is defined by a set of mutually prime integers $\{m_i$, for $i = 1, 2, \ldots, n\}$ called as the moduli set such that $\gcd(m_i, m_j) = 1$, where gcd refers to the greatest common divisor of m_i and m_j. The, product

Arithmetic Circuits for DSP Applications, First Edition. Edited by Pramod Kumar Meher and Thanos Stouraitis.
© 2017 by The Institute of Electrical and Electronics Engineers, Inc. Published 2017 by John Wiley & Sons, Inc.

of all the moduli, $M = m_1 \ldots m_n$ is the dynamic range of the RNS of unsigned numbers, that is, any unsigned numbers between 0 and $M-1$ can be uniquely represented by the residues. Alternatively, numbers between $-M/2$ to $\frac{M}{2} - 1$ when M is even and $-\frac{M-1}{2}$ to $\frac{M-1}{2}$ when M is odd can be uniquely represented by the residues corresponding to the moduli set. To understand working of the number representation in RNS and arithmetic operations in RNS domain, let us consider the moduli set $\{3, 5, 7\}$, that is, $m_1 = 3$, $m_2 = 5$, and $m_3 = 7$. We can have $M = m_1 \times m_2 \times m_3 = 3 \times 5 \times 7 = 105$. We can find that any unsigned number between 0 and 104 can be uniquely represented by the residues corresponding to moduli set (m_1, m_2, m_3). To represent the number $A = 52$ in the residue domain, we need the residues of the number with respect to the moduli 3, 5, and 7, which we can find to be $a_1 = 52 \bmod 3 = 1$, $a_2 = 52 \bmod 5 = 2$, and $a_3 = 52 \bmod 7 = 3$. Therefore, we can represent the decimal number 52 in residue domain by $(1, 2, 3)$ in the RNS. Similarly, the residue domain representation of any other number $B = 16$ can be found to be $(1, 1, 2)$. We can add the numbers $A = (1, 2, 3)$ and $B = (1, 1, 2)$ in the residue domain to have $A + B = (1 + 1) \bmod 3$, $(2 + 1) \bmod 5$, and $(3 + 2) \bmod 7 = (2, 2, 5)$. The residue domain representation $(2, 2, 5)$ can be converted to decimal integer using Chinese remainder theorem or other methods to be discussed later. We can find that the sum of $A = 52$ and $B = 16$ in decimal as 68.

If $\{x_1, \ldots, x_n\}$, $\{y_1, \ldots, y_n\}$, and $\{z_1, \ldots, z_n\}$ are RNS representations of the integers X, Y, and Z, where $Z = X$ o Y, then $z_i = z_i$ o y_i for $i = 1, \ldots, n$, where o is an arithmetic operation like addition, subtraction, and multiplication.

For RNS-based computation, we need to do the following:

(i) Convert the decimal or binary number to RNS representation by evaluating the residues.
(ii) Perform all arithmetic operations in residue domain.
(iii) Convert back the result in residue representation to decimal or binary representation.

Accordingly, an RNS-based processor consists of (i) binary to RNS convertor, (ii) Residue domain processors, and (iii) the RNS to binary converter.

The front-end of a RNS-based processor (see Figure 6.1) is a binary to RNS converter known as forward converter, which produces k residues corresponding to k moduli pertaining to each of the binary input operands. The k sets of output of the forward converter are processed by k residue processor blocks. Each residue processor block contains the required number of arithmetic circuits and produce an output in residue domain. The last stage in the RNS-based processor, known as reverse converter, converts the k residues of residue processing blocks to a conventional binary number. This process known as reverse conversion is very important and needs to be hardware-efficient and time-efficient.

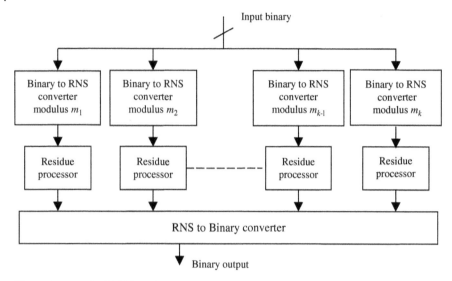

Figure 6.1 A typical RNS-based processor.

There are several advantages of RNS: (1) from Figure 6.1 we can see that the processing in different residue processors are realized in parallel and there is no carry propagation from one residue processor to the others. (2) A large decimal integer can be represented by several parallel operations on smaller numbers in residue domain. The word size used in each of the residue processor is significantly small compared with the size of input of the binary operands. The various residue processors need smaller word length and hence the additions and multiplications can be performed faster. Of course, all operations in the residue processors are modulo operations. (3) The radix-based systems, for example, the decimal or the binary number systems provide a polynomial representation of numbers, where every digit relates to some magnitude and has definite weight or place value. This is not true for the digits in RNS, and due to this reason the digits in RNS representation do not contain the magnitude information of the numbers directly. This is a disadvantage of the RNS to be used for general purpose computation but on the other hand provides better fault-tolerance compared to conventional radix-based number systems.

This chapter deals with various issues involved in the design of RNS-based processors. In Section 2, modulo addition and subtraction are considered and in Section 3, modulo multiplication and squaring are described. In Section 4, the problem of binary to RNS conversion is addressed. In Section 5, RNS to binary conversion is discussed whereas the topic of scaling and base extension are considered in Section 6. Magnitude comparison and sign detection are reviewed in Section 7 followed by error correction in Section 8. In Section 9 some practical applications of RNS are described.

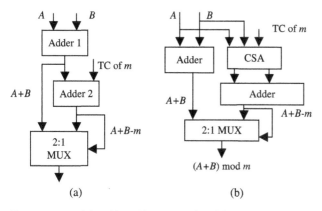

Figure 6.2 Modulo adders: (a) sequential and (b) parallel.

6.2 Modulo Addition and Subtraction

We discuss the design of arithmetic circuits of modulo adders and modulo subtractor in the following. The modulo addition of two operands A and B can be implemented using the architectures of Figure 6.2a and b [8,9].

For modulo addition, $A + B$ is computed first in Adder 1 and then m_i is subtracted from the result to compute $(A + B - m)$ by Adder 2. A 2 : 1 multiplexer is used to select either $(A + B)$ if $(A + B - m)$ is negative, or to select $(A + B - m)$ otherwise. Thus, the computation time of a modulo adder is that of one n-bit addition, one $(n + 1)$-bit addition, and the delay of the 2 : 1 multiplexer. In the architecture of Figure 6.2b, on the otherhand, both $(A + B)$ and $(A + B - m)$ are computed in parallel and one of these is selected using a 2 : 1 multiplexer depending on the sign of $(A + B - m)$. Note that a carry-save-adder (CSA) stage is needed for computing $(A + B - m)$, which is followed by a carry-propagate-adder (CPA). Thus, the area complexity of structure of Figure 6.2b is more than that of Figure 6.2a by one adder, but the addition time is less than the other by one almost one addition time.

Subtraction is similar to the addition operation wherein $(A - B)$ and $(A - B + m)$ are computed sequentially or in parallel following architectures similar to Figure 6.2a and b. The computation time of modulo subtraction is almost the same as that of modulo adder.

6.2.1 Modulo $(2^n - 1)$ Adders

We can see that finding the modulo 2^n sum of two integers is quite simple. Only n LSBs of the integer need to be retained while the rest of the most significant bits could be ignored. Suppose we want to find $(7 + 6) \bmod 2^3$, which is equal to 5. The binary representation of 13 is $(1101)_2$. The three LSBs of $(1101)_2$ is $(101)_2$, which equal to decimal 5. We can find that it is also simple to find the modulo addition operations in the residue domain if the moduli are $(2^n - 1)$, and $(2^n + 1)$. Among different suitable RNS moduli, the powers-of-two and nearly power-of-two moduli sets, for example, 2^n, $(2^n - 1)$, and $(2^n + 1)$ are therefore the most popular choice for RNS implementation of signal processing applications. For any two integers $0 < A, B < 2^n - 1$, we can easily find that

$$(A + B) \bmod (2^n - 1) = (A + B) - (2^n - 1)$$
$$= (A + B + 1) \bmod 2^n \text{ if } X = A + B \text{ is} \geq 2^n \qquad (6.1)$$
$$= X \text{ otherwise}$$

Equation (6.1) provides correct numerical values for all values of $(A + B)$ except for $(A + B) = 2^n - 1$, when it produces $= 2^n - 1$ which could be considered as an alternate form of $|0|\bmod (2^n - 1)$. Note that zero is assumed to have two representations $(0\ldots0$ and $1\ldots.1)$. The condition $A + B \geq 2^n$ is equivalent to $C_{out} = 1$, where C_{out} is the output carry of the addition $(A + B)$. Equation (6.1) therefore could be written as

$$(A + B) \bmod (2^n - 1) = (A + B + C_{out}) \bmod 2^n \qquad (6.2)$$

There could be various straightforward solutions to compute Eq. (6.2) as discussed in the following:

(a) The addition of Eq. (6.2) is carried out in two clock cycles. In the first clock cycle $(A + B)$ is computed where the input carry is taken to be 0. The carry-out of the first clock cycle is added in the second clock cycle.
(b) The computation of $A + B$ is performed by an adder, and the result $(A + B)$ is conditionally incremented if the carry out of the addition $(A + B)$ is 1.
(c) Two adders are used to compute the two possible results $A + B + 1$ and $A + B$. If the carry-out of $A + B$ is 1 then the result $A + B + 1$ is selected while the result $A + B$ is selected if the carry-out of $A + B$ is 0.

The solution (a) requires two clock cycles where duration of each clock cycle is at least one addition time and involves only one adder circuit. The solution (b) requires one addition time and one increment time to perform the modulo addition. It requires one adder circuit and one incrementer. Solution (c) requires two adders and one 2:1 multiplexer but the computation needs one addition time and the delay of a multiplexer to select the correct result.

The modulo $(2^n - 1)$ addition of Eq. (6.2) with double representation of zero $(1\ldots.1$ and $0\ldots.0)$ can be implemented by an n-bit end-around-carry parallel-prefix adder with $C_{in} = C_{out}$. In an end-around-carry adder, the carry-out of the adder is fed back as input carry bit, such that the final sum W is

$$W = A + B + C_{out} \tag{6.3}$$

The computation of Eq. (6.3) by a conventional adder could lead to undesirable race condition. To avoid that various possible techniques are discussed in the literature.

Efstathiou *et al.* [10] have described a mod $(2^n - 1)$ adder. In this design, the output carry C_{out} resulting from the addition assuming the input carry to be zero is taken into account in reformulating the equations to compute the sum. The authors have considered one and two level CLAs in this approach.

Zimmerman [11] has suggested using an additional level for adding end-around-carry for realizing a mod $(2^n - 1)$ adder (see Figure 6.3a) which needs extra hardware and more over, this carry has a large fan out thus making it slower. Kalampoukas *et al.* [12] have considered modulo $(2^n - 1)$ adders using the idea of carry recirculation at each prefix level as shown in Figure 6.3b. Here, no extra level of adders will be required. In addition, the fan out requirement of the carry output is also removed. These architectures are very fast while consuming large area.

Efstathiou *et al.* [13] have suggested design using small length adder blocks in select-prefix configuration. The word length of these small adders can be chosen so as to minimize delay. Patel and Boussakta [14] have suggested first calculation of $A + B + 1$ assuming carry in of "1." Subsequently, a conditional decrement can be performed. They also ensure a single representation of zero in the result.

6.2.2 Modulo $(2^n + 1)$ Adders

Diminished-1 arithmetic [15] is very often used for modulo $(2^n + 1)$ addition of two numbers the A and B. In diminished-1 representation A is represented as A' $= A - 1$. Since 0 cannot be represented by non-negative value in diminished-1 representation, 0 is processed separately. The numbers corresponding to A in the range 1 to 2^n are represented by A' in the range of 0 to $2^n - 1$. $A = 0$, is identified by a zero indication bit $= 1$ in a separate channel. Let $A + B = W$. Hence, in diminished-1 notation this can be written as $(A' + 1) + (B' + 1) = W' + 1$, which implies that $A' + B' + 1 = W'$. Modulo $(2^n + 1)$ addition can be formulated in diminished-1 notation as follows:

$$\begin{aligned} W' \bmod (2^n + 1) &= A' + B' + 1 - (2^n + 1) \\ &= (A' + B') \bmod 2^n \text{ if } A' + B' + 1 \geq 2^n \\ &\quad (A' + B' + 1) \text{ otherwise} \end{aligned} \tag{6.4}$$

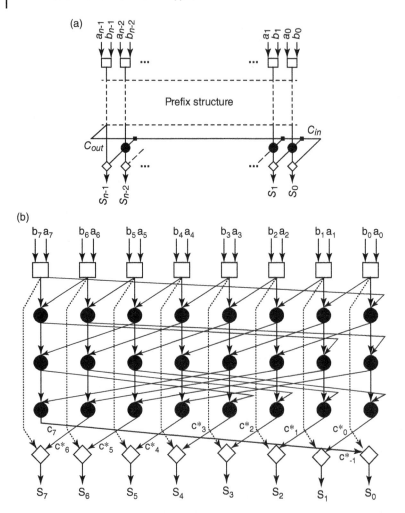

Figure 6.3 Modulo $(2^n - 1)$ adder architectures due to (a) Zimmermann (Adapted with permission from Reference [11]. Copyright 1999, IEEE.) and (b) modulo $(2^8 - 1)$ adder due to Kalampoukas *et al.* (Adapted with permission from Reference [12]. Copyright 2000, IEEE.)

To calculate W' mod $(2^n + 1)$ we need to increment $A' + B'$ if $A' + B' + 1 < 2^n$, that is, when $C_{out} = 0$. As in case of modulo $(2^n - 1)$ addition, the modulo $(2^n + 1)$ addition can be performed by end-around-carry parallel-prefix adder with $C_{in} = \text{NOT}(C_{out})$. We can thus write Eq. (6.4) as

$$(A' + B' + 1) \bmod (2^n + 1) = (A' + B' + \text{NOT}(C_{out})) \bmod 2^n \qquad (6.5)$$

Several mod $(2^n + 1)$ adders have been proposed in literature. The techniques described in the case of mod $(2^n - 1)$ adders have been extended to mod $(2^n + 1)$ adders as well. These are as follows:

(a) CLA implementation derived by associating the re-entering carry equation with those producing the carries of the modulo addition [18], (b) using parallel-prefix adders and using recirculation of the carries in each level of parallel-prefix structure [16], and (c) using select-prefix blocks [13].

Efstathiou *et al.* [16] have suggested for mod $(2^n + 1)$ addition in the case of normal numbers computing $(A + B + 2^n - 1)$ which is a $(n + 2)$–bit number. If MSB of the result is zero, we need to add $(2^n + 1)$ mod $(2^n + 1)$. If MSB = 1, the LSBs yield the answer. Vergos *et al* [17] have suggested adders mod $(2^n + 1)$ for both weighted and diminished-1 operands. For weighted operands, the diminished-1 adder shall be preceded by a carry save adder stage.

The computation carried out is

$$(A + B)_{2^n+1} = ((A_n + B_n + D + 1)_{2^n+1} + 1)_{2^n+1}$$

Where A_n, B_n are the n-bit LSBs of A and B and $D = 2^n - 4 + 2\overline{c_{n+1}} + \overline{S_n}$, where c_{n+1} and s_n are the sum and carry of addition of a_n and b_n. Vergos *et al.* have presented two mod $(2^n + 1)$ adder architectures [18] for diminished-1 numbers.

Juang *et al.* [19] have suggested weighted modulo $(2^n + 1)$ adder for addition of $(n + 1)$-bit inputs A and B using a diminished-1 adder with additional correction schemes. They compute $\left(A + B\right)_{2^n+1})_{2^n} = (A + B - (2^n + 1))_{2^n}$ if $A + B > 2^n$ and $(A + B - (2^n + 1))_{2^n} + 1$ otherwise.

6.3 Modulo Multiplication and Modulo Squaring

The modulo multiplication $(A \times B)$ mod C can be carried out by first multiplying A with B and then dividing the result with C to obtain the remainder. The quotient obtained in this process is not of interest for us. The word length of the product is $2n$ bits for n-bit input operands A and B, while the word length of C is n bits. Since the division is a time and area consuming operation, the modulo multipliers are usually realized in an integrated manner where the partial product addition and modulo reduction are performed each step. This will ensure that the word length is increased by at most 2 bits. An example will illustrate the algorithm for modulo multiplication [20].

Consider 13×17 mod 19. We start with MSB of 13 and in each step, the operation $E_{i+1} = (2E_i + a_iB)$ mod C is computed with $E_{-1} = 0$. The following illustrates the procedure:

$$E_0 = (2 \times 0 + 1 \times 17) \bmod 19 = 17 \; b_3 = 1$$
$$E_1 = (2 \times 17 + 1 \times 17) \bmod 19 = 13 \; b_2 = 1$$
$$E_2 = (2 \times 13 + 0 \times 17) \bmod 19 = 7 \; b_1 = 0$$
$$E_3 = (2 \times 7 + 1 \times 17) \bmod 19 = 12 \; b_0 = 1$$

Note, however, that each step involves a left shift (appending LSB of zero to realize multiplication by 2), a conditional addition depending on b_i value to allow A or not and a modulo C reduction. The operand in the bracket $<3C$ meaning that the word length is 2 bits more and the modulo reduction needs to be performed by subtracting C or $2C$ or none from the value. The method can be extended to higher radix (considering 2 or more bits of B at each step) as well, with some expense in hardware but reducing the modulo multiplication time.

Hence, usually ROMs are used to realize the multiplication in RNS for small moduli. On the other hand, recent work has focused on powers of two related moduli set since modulo multiplication can be simpler using the periodic property of moduli. Mod 2^n multiplication is simple. Only n LSBs of the product need to be obtained. An array multiplier can be used for instance omitting the full adders in the n MSB positions.

6.3.1 Multipliers for General Moduli

The residue number multiplication can be carried out by several methods. Two popular techniques used are based on index calculus and quarter-square multiplication. Soderstrand and Vernia [21] have suggested multipliers based on index calculus. Here, the residues are expressed as exponents of a chosen base modulo p. As an illustration, choosing base 2 and $p = 11$, since $2^8 \bmod 11 = 3$, and $2^4 \bmod 11 = 5$, corresponding to the inputs 3 and 5, the indices are 8 and 4, respectively. The indices corresponding to the inputs for a chosen base are read from the LUTs and added mod $(p - 1)$ and then using another LUT, the actual product mod p can be obtained. Note that the multiplication modulo p is isomorphic to modulo $(p - 1)$ addition of the indices. Note further that zero detection logic is required if the input is zero.

In the quarter-square technique [21], $XY \bmod m_i$ is found as

$$(XY)_{m_i} = \left(\frac{\left((X+Y)^2_{m_i} + (X+Y)^2_{m_i} \right)_{m_i}}{4} \right)_{m_i} \tag{6.6}$$

```
      0 1 1 0 0              0 1 1 0 0
        0 1 1 0 0              1 1 0 0
      0 1 1 0 0                1 0 0
    0 0 0 0 0                  0 0
  0 1 1 0 0                    0
                             0 0 0 1 1
                             0 0 1 1 0
                             0 1 1 0 0
                         1 0 0 1 0 0 1
                           0 1 0 0 1
                           0 0 1 1 0
                           0 1 1 1 1
```

Figure 6.4 Stouraitis *et al.* technique for computing $(AX + B)$ mod m_j. (Reproduced with permission from Reference [22]. Copyright 1993, IEEE.)

Thus, using look-up tables, both terms in the numerator can be found and added mod m_j. The division by 4 is also carried out using a ROM. These designs are well suited for small moduli.

Designs based on combinational logic will be needed for larger word length (160 bits–2048 bits) moduli for application in cryptography-authentication and key exchange algorithms.

Stouraitis *et al.* [22] have proposed full adder-based implementation of $(AX + B)$ mod m_j. The usual CSA-based partial product (PP) addition is modified by rewriting the MSBs in bit positions above b bits as new words 2^x mod m_j where b is the number of bits used to represent A, B, X, and m_j, These words are next added with the words with b least significant bits to yield a smaller length word ($<2b$ bits). This word is again reduced by rewriting MSBs as new words as before. In two to three iterations, the result obtained will be of length b bits. Note that one modulo subtraction may be needed at the end to obtain the final result. As an example consider (12×23) mod $29 = 15$. The bits in the positions to the left of five bits need to be rewritten: for example, 2^5 mod $29 = 3$, 2^6 mod $29 = 6$, 2^7 mod $2 = 12$. In this case, in general, 4 bits can be 1 in 2^5 position, 3 bits can be one in 2^6 position and 2 can be 1 in 2^7 position and 1 bit can be one in 2^8 position. The sum of the rewritten bit matrix can be seen to be 73 in the first step while the actual product is 276. In the next step, it will reduce further (see Figure 6.4).

6.3.2 Multipliers mod ($2^n - 1$)

The multiplication $(A \times B)$mod $(2^n - 1)$ involves addition of modified partial products obtained by circularly rotating left by i bits the bits of the input word A and adding in case b_i is 1. The resulting n words need to be added in a CSA with EAC followed by a CPA with EAC in the last stage or using a Wallace Tree followed by a CPA with EAC [11,23]. Consider the following example.

Compute $AB \bmod M = 1011 \times 1101 \bmod 1111$

1	0	1	1	PP_0	b_0A
0	0	0	0	PP_1	b_1A Rotated left by 1bit
1	1	1	0	PP_2	b_2A Rotated left by 2bits
0	1	0	1	SUM	
0	1	0	**1**	CARRY	Bold is used for EACbit
1	1	0	1	PP_3	b3A rotated left by 3 bits
1	1	0	1	SUM	
1	0	1	0	CARRY	
1	0	0	0	Final	EAC adder

Efstathiou *et al* [24] have pointed out that while using Booth's algorithm in the case of even n for a modulus of $2^n - 1$, the number of partial products can be only $n/2$. The most significant digit can be seen to be a_{n-1} that corresponds to $a_{n-1}2^n \bmod (2^n - 1) = a_{n-1}$. Thus, in place of a_{-1} (which is zero) in the first digit, we can use a_{n-1}. Accordingly, the first digit will be $(a_{n-1} + a_0 - 2a_1)$.

Radix-8 Booth encoded mod $(2^n - 1)$ multipliers have been described by Muralidharan and Chang [25]. They have suggested implementing the hard multiples needed $+3X$ and $-3X$ in partially redundant form by using Bevick's method [26]. The addition of a big word is considered as addition of several small words thus allowing trade-off of delay of the multiplier. This technique has been extended to the case using parallel-prefix adders for generating hard multiples also. The number of partial products in Radix-8 Booth encoding are $\left\lfloor \frac{n}{3} \right\rfloor + 1$.

6.3.3 Multipliers mod ($2^n + 1$)

The multiplication mod $(2^n + 1)$ is slightly involved [27–38]. Here we use the property that $2^{k+i} \bmod (2^k + 1) = -2^i$. Note however, instead of adding two's complement of the MSB field of the partial product, we add the one's complement and add a correction term of 1 for each of these terms in the end. Whenever carry vector is computed, the MSB is inverted and put as LSB. An example $(11 \times 13 \bmod 17)$ will illustrate this idea.

Example 6.1 Compute $1011 \times 1101 \bmod 10001$.

The quarter square multiplier [27] needs only $2 \times 2^n \times n$ bits of ROM. The index calculus technique needs $3 \times 2^n \times n$ bits of ROM [27]. Another method uses a $(n + 1)$-bit $\times(n + 1)$-bit multiplier followed by a dedicated correction unit after subtraction of MSBs from LSBs using a mod 2^n adder and then to

1 0 1 1	1 0 1 1	PP0
0 0 0 0	0 0 0 1	PP1 added 1
1 0 1 1	1 1 0 1	PP2 added 3
	0 1 1 1	SUM
	0 0 1 0	CARRY with MSB inverted and put as LSB; added 1
1 0 1 1	1 0 1 0	PP3 Added 7
	1 1 1 1	SUM
	0 1 0 1	CARRY
	0 0 1 1	Correction term added
	1 0 0 1	SUM
	1 1 1 1	CARRY with MSB inverted and put as LSB;added 1
	0 1 1 1	Result of mod 17 adder

reduce the result mod $(2^n + 1)$ using another modulo 2^n adder. This follows the well-known Low–High lemma [28] stated as

$$(AB)\bmod (2^n + 1) = ((AB)\bmod (2^n) - (AB)\, div\, (2^n))\bmod (2^n + 1)\ \text{if}$$
$$(AB)\bmod (2^n) \geq (AB)\, div\, (2^n)) \tag{6.7}$$
$$= (AB)\bmod (2^n) - (AB)\, div\, (2^n) + 2^n + 1 \quad \text{if} \quad (AB)\bmod (2^n) < (AB)\, div\, (2^n))$$

Note that when $A_n = 0$, $B_n = 0$, the result of multiplication denoting $P = AB = 2^n P_H + P_L$ is $(AB)\bmod (2^n + 1) = (P_L - P_H)$. Thus, one's complement of P_H and 1 need to be added to P_L. If carry is 1, the result is n LSBs. If carry is zero, then add 1 to the result. When $A = 2^n$ or $B = 2^n$, the above procedure is followed. On the other hand, if $A_n = B_n = 1$, the LSB is 1 since $2^{2n}\bmod (2^n + 1) = 1$ and the procedure is same as before.

We also have for multiplication [11,29], corresponding to $C = AB$, where A and B are given in diminished-1 form as A' and B',

$$C\bmod (2^n + 1) = (AB)\bmod (2^n + 1) = (A' + 1)(B' + 1)\bmod (2^{n'} + 1)$$
$$= (A'B' + A' + B' + 1)\bmod (2^n + 1) = (C' + 1)\bmod (2^n + 1)$$
$$\text{or}\ C' = (A'B' + A' + B')\bmod (2^n + 1) \tag{6.8}$$

Thus, multiplication of $A' \times B'$ followed by addition of A' and B' will be required to obtain the diminished-1 number. If either operand is zero or both are zero, the answer is zero.

In Wang *et al.* [29] using diminished-1 representation for both operand A and B, the $(n + 2)$ partial products including the correction term generated using a

$$
\begin{array}{lcccccccc}
PP_0 = & a_0b_7 & a_0b_6 & a_0b_5 & a_0b_4 & a_0b_3 & a_0b_2 & a_0b_3 & \overline{a_0b_0} \\
PP_1 = & a_1b_6 & a_1b_5 & a_1b_4 & a_1b_3 & a_1b_2 & a_1b_1 & a_1b_0 & \overline{a_1b_7} \\
PP_2 = & a_2b_5 & a_2b_4 & a_2b_3 & a_2b_2 & a_2b_1 & a_2b_0 & a_2b_7 & \overline{a_2b_6} \\
PP_3 = & a_3b_4 & a_3b_3 & a_3b_2 & a_3b_1 & a_3b_0 & a_3b_7 & a_3b_6 & \overline{a_3b_5} \\
PP_4 = & a_4b_3 & a_4b_2 & a_4b_1 & a_4b_0 & a_4b_7 & a_4b_6 & a_4b_5 & \overline{a_4b_4} \\
PP_5 = & a_5b_2 & a_5b_1 & \overline{a_5b_0} & a_5b_7 & a_5b_6 & a_5b_5 & a_5b_4 & \overline{a_5b_3} \\
PP_6 = & a_6b_1 & a_6b_0 & \overline{a_6b_7} & a_6b_6 & a_6b_5 & a_6b_4 & a_6b_3 & a_6b_2 \\
PP_7 = & a_7b_0 & a_7b_7 & a_7b_6 & a_7b_5 & a_7b_4 & a_7b_3 & a_7b_2 & a_7b_1 \\
PP_8 = & a_7 & a_6 & a_5 & a_4 & a_3 & a_2 & a_1 & a_0 \\
PP_9 = & b_7 & b_6 & b_5 & b_4 & b_3 & b_2 & b_1 & b_0 \\
PP_{10} = & 0 & 0 & 0 & 0 & 0 & 0 & 0 & 0
\end{array}
$$

Figure 6.5 Diminished-1 mod ($2^8 + 1$) multiplier (Adapted with permission from Reference [30]. Copyright 1998, IEEE.)

counter were accumulated in an adder tree. The correction term is shown to be a constant in Reference [30].

Efstathiou *et al.* [30] have described a mod ($2^n + 1$) multiplier for diminished-1 representation for both the operands *A* and *B*. They rewrite the partial products by inverting the MSBs beyond the *n*th bit position and show that a correction factor of 0 will be required (see Figure 6.5). They compute $AB + A + B$. Note that an all-zero carry vector is also added. All the ($n + 3$) number of *n*-bit words are added using a CSA with inverted end-around-carry being added as LSB. The final diminished-1 adder uses a parallel-prefix architecture with carry being inverted and recirculated at each prefix level.

Radix-4 Booth encoding was used by Ma [31] in reducing the number of partial products to $n/2$ but two additional modulo ($2^n + 1$) CSAs were needed to reduce the final sum of the partial products. Sousa and Chaves [32] have used modified Radix-4 Booth Recoding and Wallace tree addition. In case of *n* even, $n/2$ partial products, a correction term and "1" are added whereas for *n* odd, ($n + 1$)/2 partial products, a correction term and a constant "2" are added. The correction term needs a complex combinational logic circuit. Chen *et al.* design [33] needed ($n/2$) partial products, a constant and a correction term was simplified.

Muralidharan and Chang [34] have reduced the partial products to $\left\lfloor \frac{n}{3} \right\rfloor + 6$ using radix-8 Booth encoding and customized hard multiple generators and a simple correction term. Designs that accept one input in weighted form and another in diminished-1 form also have been described [35] using radix-4 Booth Recoding needing $n/2$ partial products for *n* even and ($n + 1$)/2 partial products for *n* odd except for one correction term. The multiplier can receive full inputs and avoids ($n + 1$)- bit circuits.

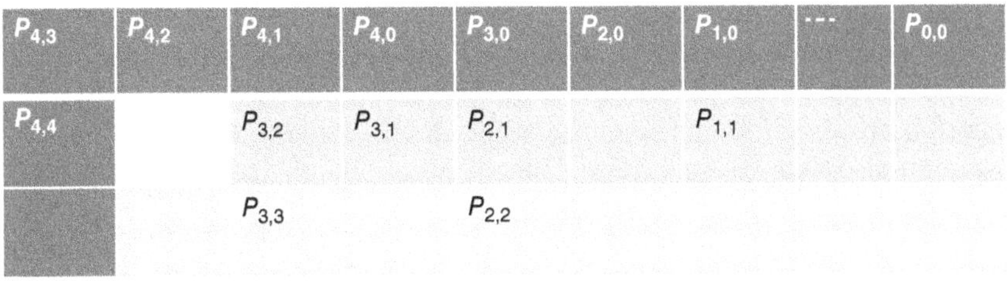

$P_{4,3}$	$P_{4,2}$	$P_{4,1}$	$P_{4,0}$	$P_{3,0}$	$P_{2,0}$	$P_{1,0}$	$\cdots$	$P_{0,0}$
$P_{4,4}$		$P_{3,2}$	$P_{3,1}$	$P_{2,1}$		$P_{1,1}$		
		$P_{3,3}$		$P_{2,2}$				

Figure 6.6 Bit matrix of a squarer.

Wrzyszcz and Milford [37] have observed that the array of bits in the partial products can be reduced to $n \times n$ for weighted representation of input numbers. This array is followed by three n-bit parallel adders and a final row of multiplexers. Vergos and Efstathiou [38] have extended this technique leading to $(n + 1)$ partial products with inverted EAC CSA tree and a final adder.

Multimodulus radix-4 and radix-8 Booth encoded multipliers also have been explored [36]. These share the hardware resources for mod 2^n, $2^n - 1$ and $2^n + 1$. Radix-4 is used for multipliers mod $(2^n - 1)$, whereas radix-8 is used for mod $(2^n + 1)$ multipliers. While these increase the delay of the multiplier, the area saving can be considerable. These use multiplexers to select the bits of the input operand or inverted bits or zero depending on the modulus chosen. Booth encoding was used for mod 2^n multipliers also.

6.3.4 Modulo Squarers

In many applications, for example, adaptive filtering and image processing squaring mod m is of interest. The use of a modulo multiplier may not be economical in such applications. Piestrak [39] has suggested using custom designs for realizing compact low area squarers mod m. By writing the partial products and noting that two identical bits in any column can be written in the next column as a single entry, the product matrix can be simplified. As an illustration for a 5-bit squarer, we obtain the matrix shown in Figure 6.6.

Note that $P_{ij} = a_i\, a_j$. Next, by using the periodic property of moduli, based on $2^n \bmod (2^n - 1) = 1$ or $2^n \bmod (2^n + 1) = -1$, the upper bits can be mapped into lower bit positions appropriately for finding mod $(2^n - 1)$ or $(2^n + 1)$. In the case of moduli having half period, the bits in the upper half need to be inverted and added and a correction term will be needed as explained before. The various words can be added using a CSA chain followed by a modulo adder. Note that serial implementation is possible for which the reader is referred to the excellent tutorial review [40].

Modulo $(2^n - 1)$ squarers have been described in which the partial product bits in the higher bit positions are mapped to LSB positions first. The partial product array can be simplified by writing two identical bits in a column as a single bit in the next left column as explained before. Further simplifications can be done to reduce the number of bits in each column. Next using EAC CSA tree followed by CPA with EAC, the result can be obtained [41]. Booth encoded squarers also have been considered that have used Booth folding encoding [42].

Modulo $(2^n + 1)$ squarers [43] for the case of diminished-1 representation compute the diminished-1 output as $Q^* = ((A^*)^2 + 2A^*)_{2^n+1}$. Designs for weighted representation also have been described [44], which map the higher n bits in the partial products in LSB positions after inversion and suitable correction word is also added. Multimoduli squarer architectures for 2^n, $2^n - 1$, $2^n + 1$ [45,46] also have been described that share the partial product bits and use multiplexers to select the appropriate inputs to form the partial products. They cater in the case of mod $(2^n + 1)$ for weighted as well as diminished-1 representations. Architectures that perform multiplication or compute sum of squares mod $(2^n - 1)$ and mod $(2^n + 1)$ have been suggested [47]. In the case of diminished-1 representation, we need to compute $P^* = (A^*B^* + A^* + B^*) \bmod (2^n + 1)$ in case of multiplication and $SS^* = ((A^*)^2 + (B^*)^2 + 2A^* + 2B^* + 1) \bmod (2^n + 1)$ in the case of sum of squares.

6.4 Forward (binary to RNS) Conversion

The given binary number needs to be converted to RNS. Straightforward method is to use a divider to obtain the residue while ignoring the quotient obtained. But, as is well-known, division is a complicated process. As such, alternative techniques to obtain residue easily by a sequential technique have been investigated. An example will illustrate the procedure.

Example 6.2. Find the residue of 892 mod 19.
Expressing 892 in binary form, we have 11 0111 1100. (We can start with the fifth bit from the right since 12 mod 19 is 12 itself). We know the consecutive powers of two mod 19 as 1, 2, 4, 8, 16, 13, 7, 14, 9, 18. Thus, we can add the residues corresponding to 2^i mod 19 for the ith bit if it is "1." This yields (4 + 8 + 16 + 13 + 7 + 9 + 18) mod 19 = 18. Note that at each step, when a new residue corresponding to a new power of 2 is added, modulo reduction can be done to avoid unnecessary growth of the result: (((4 + 8) mod 19 + 16) mod 19 + 13) mod 19, and so on.   □

Note, however, that certain simplifications can be made by noting the periodic property of residues of 2^k mod m_i [48–53]. By Fermat's theorem, $2^{18} = 1$ mod 19. Hence, we have $2^{18k+i} = 2^i$ mod 19. Thus, if T_i is period of the modulus m_i (i.e., $2^{T_i} = 1$ mod m_i), all T_i bit words in the given binary number can be added first in a carry save adder with EAC (end around carry) to first obtain a T_i bit word from which using the procedure described above the residue mod m_i can be obtained.

Example 6.3. Find the residue of $(0001\ 0100\ 1110\ 1101\ 1011\ 0001\ 0100\ 1110$ $1101\ 1011)_2$ mod $19 = (89887166171)_{10}$ mod 19.

Thus, the three 18 bit words (here $T_i = 18$) can be added with EAC to obtain

				0001
01	0011	1011	0110	1100
01	0100	1110	1101	1011
10	1000	1010	0100	1000

$\square$

166,472. The conversion procedure presented in example 6.2 can be used to obtain the residue of this 18 bit number. In short, the periodic property of 2^k mod m_i can be used to simplify computation.

Another simplification is possible for moduli satisfying the property $2^{(m_i-1)/2}$ mod $m_i = 1$. Considering modulus 19 again, we observe that $2^9 = -1$ mod 19. Thus, the residues in the upper half of a period are opposite in sign to those in the lower half. This property can be used to reduce the CSA word length to $(m_i - 1)/2$.

An example will illustrate this idea.

Example 6.4. Find the residue of $(0001\ 0100\ 1110\ 1101\ 1011\ 0001\ 0100\ 1110$ $1101\ 1011)_2$ mod $19 = (89887166171)_{10}$ mod 19.

Dividing the given word into 9 bit fields starting from LSB, we have

$W_4 =$		0001
$W_3 = 0$	1001	1101
$W_2 = 1$	0110	1100
$W_1 = 0$	1010	0111
$W_0 = 0$	1101	1011

$\square$

Thus, adding together alternate fields in separate CSAs, that is, adding W_0, W_2, W_4 we get $S_o = 10\ 0100\ 1000$ and adding W_1, W_3 we have $S_e = 1\ 0100\ 0100$. Subtracting S_e from S_o, we have $S = 0001\ 0000\ 0100$. (Here, subtraction is two's complement addition of S_o with S_e). Note that the word lengths of S_o and S_e are more than n-bits depending on the number of n-bit fields in the given binary number. The residue of this 9-bit word can be found easily using the technique described earlier as 13.

Interesting variations for implementation have been described in literature where 2^k mod m_i are precomputed and stored and added sequentially depending on the input bit value [48]. In another technique, the residue of 2^x mod m_i is computed from 2^{x-1} mod m_i using a doubler and modulo subtractor [52]. The doubling can be simply achieved means by appending a LSB of zero.

Interestingly, for special moduli of the form $2^k - 1$ and $2^k + 1$, the second stage of binary to RNS conversion of a smaller length word of either T_i or $T_i/2$ bits can altogether be avoided. For moduli of the form $2^k - 1$, since $T_i = k$, the input word can be divided into k bit fields all of which can be added in a CSA with EAC to yield the final residue. On the other hand, for moduli of the form $2^k + 1$, since $T_i = 2\,k$, all odd k-bit fields can be added mod 2^k and all even k-bit fields can be added mod $(2k + 1)$ to obtain S_o and S_e and one final adder gives $(S_o - S_e)$ mod $(2^k + 1)$.

Example 6.5. As an illustration, $(892)_{10}$ mod $15 = (0011\ 0111\ 1100)_2$ mod $15 = (3 + 7 + 12)$ mod $15 = 7$ and 892 mod $17 = (3-7 + 12) = 8$.

Recently, some variations of this approach have also been considered in literature. Low and Chang [53] have suggested rewriting the residues with large Hamming weight as sum of two words one with a small Hamming weight and a correction term. This will help in reducing the number of bits to be added for computing the residue mod m_i. For example, $x_{15}2^{15}$ mod 29 where x_{15} is a bit in the given input binary word can be written as $x_{15}2^{15}$ mod $29 = 27 + 2\overline{x_{15}}$. Thus, we need to add 2 or 0 having Hamming weight of 1 instead of 4 in case of 27. Note that all correction terms can be added as a single correction term. For the modulus 29, we can rewrite residues 11, 13, 15, 19, 21, 23, 25, 27, 28 which have 3 or more "one" bits using words with less number of "ones." □

Premkumar [54] and Premkumar *et al.* [55] have suggested a technique known as modular exponentiation. Instead of storing 2^i mod m_i, they use combinational logic to obtain the residue.

In the case of moduli $(2^n \pm k)$ [56], multipliers will be required to obtain the residues corresponding to various n bit fields. As an example, for a $4n$-bit input, $W_3 2^{3n} + W_2 2^{2n} + W_1 2^n + W_0$ mod $(2^n + k)$ reduces to $(-W_3 k^3 + W_2 k^2 - W_1 k + W_0)$ mod $(2^n + k)$ which needs three multipliers with prestored values k^3, k^2, and k followed by addition of the four terms and reduction mod $(2^n + k)$.

Recent trend has been to combine the hardware for various moduli by exploiting the commonality in the computation [57]. For example, for the moduli $(2^n - 1)$ and $(2^n + 1)$ reduction of a $4n$-bit word $W = W_3 2^{3n} + W_2 2^{2n} + W_1 2^n +$

W_0 can be carried out by noting the redundancy in computations: $W \bmod (2^n - 1) = ((W_3 + W_1) + (W_0 + W_2)) \bmod (2^n - 1)$ and $W \bmod (2^n + 1) = ((W_3 + W_1) - (W_0 + W_2)) \bmod (2^n + 1)$. Hence, the adders for computing $W_3 + W_1$ and $W_2 + W_0$ can be shared. Another technique [58] exploits the common factors in the periods of moduli. As an illustration for moduli 3, 5, 17, residue computation for an input 32-bit number can be by first reducing the 32-bit word to a 8-bit word mod 255. In a second stage, the 8-bit word can be reduced mod15 and 17 by considering the two 4-bit words. In a third stage, the mod 3 and 5 computation can be carried out.

6.5 RNS to Binary Conversion

There are two basic classical approaches to converting a number from RNS to binary form. These are based on Chinese remainder theorem (CRT) and mixed radix conversion (MRC) [2].

6.5.1 CRT-Based RNS to Binary Conversion

The binary number X corresponding to given residues $(x_1, x_2, x_3, \ldots, x_k)$ in the RNS $\{m_1, m_2, m_3, \ldots m_k\}$ can be derived using CRT as

$$X = \left(\left(x_1 \left(\frac{1}{M_1} \right)_{m_1} \right)_{m_1} M_1 + \left(x_2 \left(\frac{1}{M_2} \right)_{m_2} \right)_{m_2} M_2 + \cdots \left(x_k \left(\frac{1}{M_k} \right)_{m_k} \right)_{m_k} M_k \right) \bmod M$$

$$(6.9)$$

where $m_i = M/m_i$ for $i = 1, 2, \ldots k$ and $M = m_1 m_2 \ldots m_k$. The quantities $(\frac{1}{M_j})_{m_j}$ are known as the multiplicative inverses of $M_j \bmod m_j$. The summation in Eq. (6.9) is less than kM and the reduction mod M to obtain X is a cumbersome process. The advantage of the CRT is that the weighting of the residues x_i can be done in parallel and results summed, followed by reduction mod M.

CRT can be efficiently used in case of the three and four moduli sets $\{2^n - 1, 2^n, 2^n + 1\}$, $\{2^n - 1, 2^n, 2^n + 1, 2^{2n} + 1\}$, and $\{2^{2n} - 1, 2^n, 2^{2n} + 1\}, \{2^n - 1, 2^n, 2^{n+1} - 1\}$, $\{2^n - 1, 2^n, 2^{n-1} - 1\}$ since n bits of the decoded number are available directly as residue corresponding to modulus 2^n and the modulo reduction needed in the end with respect to the product of remaining moduli can be efficiently implemented in the case of the first three moduli sets.

We will consider the RNS to binary conversion for the moduli set $\{2^n - 1, 2^n, 2^n + 1\}$ based on CRT in detail next in view of the immense attention paid in literature [59–70]. For this moduli set, we have $M = 2^n(2^{2n} - 1)$, $M_1 = 2^n(2^n + 1)$, $M_2 = 2^{2n} - 1$, $M_3 = 2^n(2^n - 1)$ and $(1/M_1) \bmod (2^n - 1) = 2^{n-1}$, $(1/M_2) \bmod$

$(2^n) = -1$, $(1/M_3) \bmod (2^n + 1) = (2^{n-1} + 1)$. Thus, we can obtain using CRT [61] for the given residues (x_1, x_2, x_3),

$$X = [2^n(2^n + 1)2^{n-1}x_1 - (2^{2n} - 1)x_2 + 2^n(2^n - 1)(2^{n-1} + 1)x_3] \bmod (2^n(2^{2n} - 1))$$
$$= Y2^n + x_2 \tag{6.10}$$

Since we know the LSBs of X as x_2, we can obtain the $2n$ MSBs of X by computing

$$Y = \left(\frac{X - x_2}{2^n}\right) \bmod (2^{2n} - 1) \tag{6.11}$$

From Eqs (6.10) and (6.11), we have

$$Y = [(2^n + 1)2^{n-1}x_1 - 2^n x_2 + (2^n + 1)2^{n-1}x_3 - x_3] \bmod (2^{2n} - 1) \tag{6.12}$$

Interestingly, the computation of Y involves summing of four $2n$-bit words, which can be easily found by bit manipulations of x_1, x_2, and x_3 (rotations and one's complimenting) due to the mod $(2^{2n} - 1)$ operations involved [61,65]. We consider the first term $A_1 = [(2^n + 1)2^{n-1}x_1] \bmod (2^{2n-1}) = (2^{2n-1}x_1 + 2^{n-1}x1) \bmod (2^{2n} - 1)$ for instance. Writing x_1 as the n bit number $x_{1,n-1}x_{1,n-2}x_{1,n-3} \cdots x_{1,2}x_{1,1}x_{1,0}$, we have

$$A_1 = x_{1,0}x_{1,n-1}x_{1,n-2}x_{1,n-3} \cdots x_{1,2}x_{1,1}x_{1,0}x_{1,n-1}x_{1,n-2}x_{1,n-3} \cdots x_{1,2}x_{1,1}$$

The second term $A_2 = (-2^n x_2) \bmod (2^{2n} - 1)$ can be written as the $2n$-bit word

$$A_2 = \bar{x}_{2,n-1}\bar{x}_{2,n-2} \cdots \bar{x}_{2,2}\bar{x}_{2,1}\bar{x}_{2,0} \underbrace{111 \ldots 111}_{n\,\text{bits}}$$

where bars indicate one's complement of bits.

The third term is slightly involved since x_3 is $(n + 1)$-bit wide. We use the fact that when $x_{3,n}$ is 1, invariably $x_{3,0}$ is 0. Proceeding in a similar manner as before, $A_3 = ((2^n + 1)2^{n-1}x_3 - x_3) \bmod (2^{2n} - 1)$ can be obtained as the sum of the two words

$$A_{31} = (x_{3,n} + x_{3,0})x_{3,n-1} \cdots x_{3,2}x_{3,1}(x_{3,n} + x_{3,0})_{3,n-1} \cdots x_{3,1} \text{ and}$$

$$A_{32} = \underbrace{1 \quad 1 \ldots \quad 1}_{(n-1)\text{bits}} \quad \overline{x_{3,n}x_{3,n-1}} \quad \overline{x_{3,n-2} \cdots x_{3,0}}$$

Dhurkadas [66] suggested rewriting the three words A_2, A_{31}, and A_{32} to yield the two words

$$B_2 = \bar{x}_{2,n-1} \quad \bar{x}_{2,n-2} \cdots \bar{x}_{2,1} \bar{x}_{2,0} \overline{x_{3,n} + x_{3,n-1}} \bar{x}_{3,n-2} \cdots \bar{x}_{3,0}$$

$$B_3 = (x_{3,n} + x_{3,0}) x_{3,n-1} \cdots x_{3,2} x_{3,1} \quad x_{3,0} \quad x_{3,n-1} \cdots x_{3,1}$$

Thus, the three words A_1, B_2, and B_3 need to be summed in a carry-save-adder with end around carry (to take care of mod ($2^{2n} - 1$) operation) and the resulting sum and carry vectors need to be added using a CPA with end-around-carry.

Several improvements have been made in the past two decades by examining the bit structure of the three operands, by using n-bit CPAs in place of $2n$-bit CPA to reduce the addition time [67–77]. CRT can be applied to other RNS systems having three or more moduli. The various operands to be summed can be easily obtained by bit manipulations (rotation of word, bit inversions) but the final summation and modulo reduction can be very involved. Thus, three moduli system described above, is believed to be attractive.

Soderstrand *et al.* [78] have suggested the computation of weighted sum in Eq. (6.9) scaled by $2/M$. Each of these scaled values can be represented as words with one integer bit and several fractional bits and all the scaled values can be added. The multiple of 2 in the integer part of the resulting sum can be discarded. The integer part will convey information about the sign of the number. However, Vu [79] has pointed out that the precision of these fractions shall be proper so as to yield a definite indication of sign. Considering the CRT expression given in Eq. (6.9), and multiplying both sides by $2/M$, we obtain the scaled value

$$X_s = \sum_{i=1}^{j} \frac{2}{m_i} \left(\left(\frac{1}{M_i} \right)_{m_i} x_i \right)_{m_i} = \sum_{i=1}^{j} u_i \tag{6.13}$$

Considering that the error due to finite bit representation in u_i as e_i, such that $0 \le e_i \le 2^{-t}$ where $t + 1$ bits are used to represent each u_i, it can be shown that the total error $e < 2/M$ for M even and $e < 1/M$ for M odd or $n2^{-t} \le 2/M$ for M even and $n2^{-t} \le 1/M$ for M odd.

An example will illustrate the idea of sign detection. Consider the moduli set {11, 13, 15, 16} and the two cases corresponding to the real numbers +1 and −1. The residues are (1,1,1,1) and (10,12,14,15), respectively. The Eq. (6.13) becomes in the first case

$$X_s = \frac{16}{11} + \frac{2}{13} + \frac{4}{15} + \frac{2}{16}$$

Adding the first two terms and second two terms separately and representing in fractional form, we get

$$v_1 = u_1 + u_2 = \frac{16}{11} + \frac{2}{13} = \frac{230}{143} \quad \text{and} \quad v_2 = u_3 + u_4 = \frac{4}{15} + \frac{2}{16} = \frac{94}{240}$$

The fractional representation for these is

$$\hat{v}_1 = 1.1001101111000000, \hat{v}_2 = 0.0110010001000101, \hat{X}_s$$
$$= 0.000000000000101$$

For the case corresponding to -1, we have on the other hand, $\hat{X}_s = 1.111111111100101$. Note that following Soderstrand *et al.* [78] suggestion of using 11 bits for the fractional part, the sum would be 0.0000000000 which does not give the correct sign.

6.5.2 Mixed Radix Conversion

In Mixed Radix conversion technique, the decoded number corresponding to a j moduli set is expressed as

$$B = x_1 + d_1 m_1 + d_2 m_1 m_2 + d_3 m_1 m_2 m_3 + \ldots + d_{j-1} m_1 m_2 m_3 \ldots m_{j-1} \quad (6.14)$$

In each step, one mixed radix digit d_i as well as several intermediate results $d'_{i+1}, d'_{i+2}, \ldots, d'_{j-1}$ are determined. At the end, the MRC digits are weighted following Eq. (6.14) to obtain the final decoded number. There is no need for final modulo reduction. The following example illustrates the technique.

m_1	m_2	m_3	7	5	3
x_1	x_2	x_3	3	2	1
$-x_3$	$-x_3$		-1	-1	
$(x_1 - x_3) \bmod m_1$	$(x_2 - x_3) \bmod m_2$		2	1	
$\times(1/m_3) \bmod m_1$	$\times(1/m_3) \bmod m_2$		$\times 5$	$\times 2$	
y_1	y_2		3	2	
$-y_2$			-2		
$(y_1 - y_2) \bmod m_1$			1		
$\times(1/m_2) \bmod m_1$			$\times 3$		
y_1			3		

The result is $X = y_1(m_2 m_3) + (y_2 m_3) + x_3 = 3(15) + 2(3) + 1 = 52$.

Note that in each step, the residue corresponding to one modulus is subtracted so that the result is exactly divisible by that modulus. The multiplication with multiplicative inverse accomplishes this division. The last step of evaluating Eq. (6.14) needs multiplications of bigger numbers, for example, $m_2 m_3$ in the three moduli example and addition of the resulting products using carry save adders followed by CPA. But, no final modulo reduction is needed in the case of MRC since the result is always less than $M = m_1 m_2 m_3$. It may be noted that MRC is sequential. However, the MRC algorithm can be pipelined. Note that the modulo subtractions can be carried out using logic and modulo multiplications can be carried out using logic or ROM. The conversion time is thus $(n-1)\Delta_{\text{modsub}} + (n-1)\Delta_{\text{modmul}} + \Delta_{\text{mult}} + \Delta_{\text{CSA}}(k\text{-}2) + \Delta_{\text{CPA}}$.

MRC is simpler in the case of some powers of two related moduli sets since some of the multiplicative inverses needed in successive steps are of the form 2^i or ± 1 so that the multiplication mod $2^n \pm 1$ can be realized using a data path of n bits by bit wise rotation of the operands and/or bit inversions and corrective additions as explained in previous section where n is the bit length of the moduli.

As an illustration for the case of the earlier considered moduli set $\{2^n - 1, 2^n, 2^n + 1\}$, the multiplicative inverses are as follows: $\left(\frac{1}{2^n+1}\right)_{2^n-1} = 2^{n-1}$, $\left(\frac{1}{2^n+1}\right)_{2^n} = -1$, $\left(\frac{1}{2^n}\right)_{2^n-1} = 1$. Thus, only modulo subtractions are needed and multiplication with 2^{n-1} mod $(2^n - 1)$ can be realized by rotation.

6.5.3 RNS to Binary conversion using New CRT

Variations of CRT have appeared in literature most important being New CRT I [80,81]. Using New CRT I, given the moduli set $\{m_1, m_2, m_3, \ldots m_n\}$, the weighted binary number corresponding to the residues $(x_1, x_2, x_3, \ldots x_n)$ can be found as

$$X = x_1 + m_1 \left(k_1(x_2 - x_1) + k_2 m_2(x_3 - x_2) + \ldots \right.$$
$$\left. + k_{n-1} m_2 m_3 \ldots m_{n-1}(x_n - x_{n-1}) \right)_{m_2 m_3 \ldots m_{n-1} m_n} \tag{6.15}$$

where $\left| k_1 m_1 \right|_{m_2 m_3 \ldots m_{n-1} m_n} = 1$, $\left| k_2 m_1 m_2 \right|_{m_3 \ldots m_{n-1} m_n} = 1$, $\ldots$, $\left| k_{n-1} m_1 m_2 m_3 \ldots m_{n-1} \right|_{m_n} = 1$

As an illustration, application to the moduli set $\{2^n, 2^n + 1, 2^n - 1\}$ yields

$$X = x_1 + 2^n \left((x_1 - x_2) + 2^{n-1}(2^n + 1)(x_3 - 2x_1 + x_2) \right)_{2^{2n}-1} \tag{6.16}$$

Thus, the value in the parantheses can be found and appending x_1 as LSBs, we can obtain X.

New CRT II [80] considers two moduli at a time and uses MRC to obtain the decoded number. Thus, for a k moduli system (considering k even) in the first level $k/2$ pairs of moduli are taken and decoded. This process continues in a tree of decoders using MRC in $\log_2 k$ levels to obtain the final binary number.

6.5.4 RNS to Binary conversion using Core Function

Some authors have investigated the use of core functionfor RNS to Binary conversion as well scaling. This will be briefly considered next. Consider the moduli set $\{m_1, m_2, m_3, m_4\}$. We need to choose first a value $C(M)$ known as core [82–85]. It can be the largest modulus in the RNS or product of two or more moduli. Then, we need to compute weights w_i which can be used to compute the core $C(n)$ of a given number n as

$$C(n) = n\frac{C(M)}{M} - \sum_{i=1}^{K} \frac{w_i r_i}{m_i} \tag{6.17}$$

The weights w_i are defined as

$w_i = ((\frac{1}{M_i})_{m_i} C(M))\bmod m_i$, where M and m_i are defined as in the case of CRT and $B_i = M_i(\frac{1}{M_i})_{m_i}$. Note that the weights are chosen to be small positive or negative numbers to reduce the noisiness of $C(n)$ and to ensure that $C(M) \ll M$. The weights also shall satisfy the condition that $\sum_{i=1}^{k} w_i M_i = C(M)$. For example, if $w_i = 4$ corresponding to modulus 5, it is written as -1. Note that $C(B_i)$ can be seen from Eq. (6.17) to be

$$C(B_i) = \frac{B_i C(M)}{M} - \frac{w_i}{m_i} \tag{6.18}$$

For residue to Binary conversion corresponding to residues $(r_1, r_2, r_3,..r_k)$, first $C(n)$ needs to be determined as

$$C(n) = \sum_{i=1}^{k} r_i C(B_i) \bmod C(M) \tag{6.19}$$

Next, n can be computed as

$$n = \left(\frac{M}{C(M)} \left(C(n) + \sum_{i=1}^{k} \frac{w_i}{m_i} r_i \right) \right) \bmod M \qquad (6.20)$$

The important property of $C(n)$ is that it monotonically increases with n. Thus, it is possible to find the location of "n" in the full dynamic range. However, since range of $C(n)$ is smaller than M, evidently, n is compressed from 0 to $(M-1)$ to 0 to $(C(M)-1)$.

Consider the moduli set $\{3, 5, 7, 11\}$. Let us choose $C(M) = 11$. Then $M = 1155$, $M_1 = 385$, $M_2 = 241$, $M_3 = 165$, and $M_4 = 105$. The values of $\left(\frac{1}{M_i} \right)_{m_i}$ are 1, 1, 2, 2. Thus, $B_i = M_i \left(\frac{1}{M_i} \right)_{m_i}$ are 385, 241, 330, 210. The w_i can be found as $-1, 1, 1, 0$. Next, we find $C(B_1) = 4$, $C(B_2) = 2$, $C(B_3) = 3$, and $C(B_4) = 2$. Consider the residues $(1, 2, 3, 8)$. Then we find $C(n) = (1 \times 4 + 2 \times 2 + 3 \times 3 + 8 \times 2)$ mod $11 = 0$. Next, n can be found as $n = 105 \left(0 + \frac{1 \times -1}{3} + \frac{2 \times 1}{5} + \frac{3 \times 1}{7} \right) = 52$. As another example, for the residues $(2, 4, 6, 10)$, $C(n)$ can found as $C(n) = (2 \times 4 + 4 \times 2 + 6 \times 3 + 10 \times 2)$ mod $11 = 10$. This means that the number is very close to the top of the dynamic range.

Since $C(M)$ can be chosen as a product of few moduli, scaling is possible using core extraction method. Several techniques for approximate scaling as well as accurate scaling have been described in literature [84,85]. Other RNS to Binary conversion techniques are based on quotient function [86] and Mixed Radix Chinese Remainder Theorem [87].

CRT has been applied to other moduli sets $\{2^{2n}, 2^{2n} - 1, 2^{2n} + 1\}$[88,89], $\{2^n, 2^n - 1, 2^{n+1} - 1\}$[90], $\{2^k, 2^k - 1, 2^{k-1} - 1\}$[91],$\{2^n + 1, 2^{n+k}, 2^n - 1\}$[92]. Reverse converters for the moduli set $\{2^k, 2^k - 1, 2^{k-1} - 1\}$ have been described by References [91,93–95]. Several authors have applied New CRTI to realize RNS to binary converters for various moduli sets. For higher moduli systems, usually combinations of MRC and CRT need to be used. Two and three level converters [96–100] have been suggested for moduli sets having four or more moduli. They use combination of CRT and MRC. Some reverse converters of four moduli sets are extensions of the converters for the three moduli sets. These use the optimum converters for the three moduli set and use MRC to get the final result to include the fourth modulus $2^n - 1$, $2^{n-1} + 1$, $2^{n-1} - 1$, $2^{n+1} + 1$, and so on, [96–99]. Reverse converters for the four moduli set $\{2^n - 1, 2^n + 1, 2^{n-3}, 2^n + 3\}$ have also been described, which use ROMs and combinational logic [88,100,101]. The four moduli set $\{2^n, 2^n - 1, 2^n + 1, 2^{2n} + 1\}$ [102] is attractive since New CRT-based reduction can be easily carried out. However, the bit length of one modulus is double that of the other three moduli.

The reverse converters for the moduli set $\{2^n - 1, 2^n + 1, 2^{2n+1}-1, 2^n\}$ and $\{2^n - 1, 2^n + 1, 2^{2n}, 2^{2n} + 1\}$ based on New CRT II and New CRT I respectively have been described in Reference [103]. The reverse converter for the five moduli set [104] $\{2^n - 1, 2^n, 2^n + 1, 2^{n+1}-1, 2^{n-1}-1\}$ for n even also is based on CRT. RNS to binary converters moduli sets using several conjugate moduli have been suggested [105,106]. An eight moduli RNS has also been investigated [107]. RNS with dynamic range upto $8n + 1$ bits also has been reported [108].

6.6 Scaling and Base Extension

It is often required to scale a number in DSP applications. Scaling by a power of two or by one modulus or product of few moduli will be desired. Division by arbitrary integer is exactly possible in RNS if the remainder of division is known to be zero.

As an example, consider the moduli set $\{3, 5, 7\}$. We wish to divide 39 by 13. This is possible by multiplication of residues of 39 with multiplicative inverse of 13. We know that $39 = (0, 4, 4)$. We can see that $(1/13) \bmod 3 = 1$, $(1/13) \bmod 5 = 2$, and $(1/13) \bmod 7 = 6$. Thus, multiplying $(0, 4, 4)$ with $(1, 2, 6)$, we obtain $(0, 3, 3)$ which corresponds to 3. The divisor shall be mutually prime to all moduli for division to be possible. On the other hand, if we wish to divide 40 by 13, it is not possible. If the residue mod 13 is first found and subtracted from 40, then only exact division is feasible, but the quotient is approximate.

The MRC technique described in Section 3.5.2 in fact performs scaling by first subtracting the residue corresponding to the modulus, and then multiplying with the multiplicative inverse. However, there will be a need for base extension which is explained next. Consider the previous MRC example. Consider that division by 3 of 52, that is, $\{1, 2, 3\}$ in the moduli set $\{3, 5, 7\}$ is desired. By subtracting the residue corresponding to modulus 3, that is, 1, we have 51 which if divided by 3 yields 17. Thus, division is accomplished. The residues corresponding to 17 in the moduli set $\{5, 7\}$ are now available. However, the result will be in complete RNS, only if residue of 17 mod 7 is also available. The computation of this residue is known as base extension.

Szabo and Tanaka describe an interesting technique for base extension [2]. But, it needs additional MRC. In this technique, we assume the desired residue to be found as x. Thus, we need to find binary number corresponding to $(x, 2, 3)$. But this time we start conversion from modulus 7 and decide the MRC digit corresponding to modulus 3. This should be zero, because our quotient 17 is less than 35. Thus, we can use this condition to find x as can be seen from the example.

3	5	7
x	2	3
$\underline{-3}$	$\underline{-3}$	
$(x{-}3)_3$	4	
$\underline{\times 1}$	$\underline{\times 3}$	
$(x{-}3)_3$	2	
$\underline{-2}$		
$(x{-}2)_3$		
$\underline{\times 2}$		
$(2x{-}1)_3$		

The condition $(2x{-}1)_3 = 0$ yields $x = 2$. Thus, the RNS number is (2, 2, 3). Alternative techniques based on CRT are available but they need the use of a redundant modulus and ROMs [109,110]. Shenoy and Kumaresan [109] have suggested base extension using CRT. This needs one extra (redundant) modulus. Consider the three moduli set $\{m_1, m_2, m_3\}$ and given residues (r_1, r_2, r_3). We wish to extend the base to modulus m_4. We need a redundant modulus m_r and we need to have the residue corresponding to m_r. In other words, all computations need to be done on moduli m_1, m_2, m_3, and m_r. Using CRT, we can obtain the binary number X corresponding to (r_1, r_2, r_3) as

$$X = \left(\sum_{i=1}^{3} M_i \left(\frac{1}{M_i} \right)_{m_i} r_i \right) - kM \tag{6.21a}$$

The residue $X \bmod m_4$ can be found from Eq. (6.21a), if k is known. Using a redundant modulus m_r, if $r_r = X \bmod m_r$ is known, k can be found from Eq. (6.21a) as

$$k = \left(\left(\left(\sum_{i=1}^{3} M_i \left(\frac{1}{M_i} \right)_{m_i} r_i \right) - r_r \right)_{m_r} \left(\frac{1}{M} \right)_{m_r} \right)_{m_r} \tag{6.21b}$$

After knowing k, $X \bmod m_4$ can be found from Eq. (6.21a).

Several scaling algorithms have been described recently [111–116] for powers of two related moduli set as well as general moduli sets. Jullien [111] has suggested scaling and base extension using LUTs based on Mixed Radix conversion technique. Scaling of a number in N moduli RNS by product S moduli, where $S < N$ is considered. After S steps (cycles) of MRC corresponding to division by S moduli, we obtain the residues of the quotient corresponding to

N–S moduli. Next base extension is needed to find the S residues corresponding to this quotient. This needs further MRC conversion and at the same time computing the residues corresponding to S moduli using Look up tables in N–S cycles. Note, however, that the LUT in the first level are addressed by $(S + 1)$ moduli and $(N$–$S)$ tables are required. In the second level S tables are needed that are addressed by $(N$–$S)$ inputs.

As an illustration, considering a four moduli RNS $\{m_4, m_3, m_2, m_1\}$, division by product of moduli m_1 and m_2 by performing MRC in two steps starting from r_1, yields the residues r'_4 and r'_3 corresponding moduli m_4 and m_3. From these, next we can get the residues r'_2 and r'_1 as

$$r'_2 = \left(r'_3 + (r'_4 - r'_3) \left(\frac{1}{m_3} \right)_{m_4} m_3 \right) \bmod m_2 \quad \text{and} \quad r'_1 = \left(r'_3 + (r'_4 - r'_3) \left(\frac{1}{m_3} \right)_{m_4} m_3 \right) \bmod m_1$$

Garcia and Lloris [112] suggested using two steps one for obtaining the scaled residues corresponding $(N$–$S)$ moduli and for obtaining scaled residues corresponding to the S moduli. The first step needs N-S LUTs addressed by S residues and another residue corresponding to m_j (each addressed by $S + 1$ residues). The second stage needs S LUTs each addressed by N–S residues.

CRT also can be used for scaling. In Julien's technique [111], the CRT expression is divided by the scaling factor and the fractions obtained are rounded by adding $1/2$. The residues of all the terms are added mod m_i for all the N–S moduli. Next, base extension is carried out to the S moduli using MRC as explained earlier.

Scaling without needing redundant moduli is also considered. Barsi and Pinotti [113] have suggested an exact scaling procedure by product of few moduli using look-up tables that does not need any redundant modulus. This needs two steps of base extension. Consider for illustration the six moduli set $\{m_1, m_2, m_3, m_4, m_5, m_6\}$ and residues $(r_1, r_2, r_3, r_4, r_5, r_6)$. We wish to scale the number by product of moduli $m_1, m_2,$ and m_3. The first step is to subtract the number corresponding to the three residues pertaining to these three moduli by performing a base extension to moduli $m_4, m_5,$ and m_6 to obtain the residues corresponding to these moduli. The result obtained is exactly divisible by $m_1 m_2 m_3$. Hence, the multiplication with multiplicative inverse of $(m_1 m_2 m_3)$ mod m_i $(i = 4, 5, 6)$ will yield the scaled result in the moduli set $\{m_4, m_5, m_6\}$. The next step is to perform base extension to the moduli set $m_1, m_2,$ and m_3 to get the scaled result. Note that exact scaling can be achieved.

Kong and Philips [114] have suggested scaling X by any factor K. We need to first base extend X to K and then the scaled residues can be computed by subtracting the residues of K from X and dividing the result by K (which means multiplication with the multiplicative inverse of K). Note, however, that the values of $(1/K)$ mod m_i shall be available *a priori*. Consider division by $m_1 m_2 m_3$

in a six moduli set $\{m_1, m_2, m_3, m_4, m_5, m_6\}$. We first base extend $\{x_1, x_2, x_3\}$ to the moduli m_4, m_5, and m_6. Then we subtract the result from the residues of m_4, m_5, and m_6. Note that the resulting quotient obtained by multiplying with $\left(\frac{1}{m_1 m_2 m_3}\right)_{m_4}$, $\left(\frac{1}{m_1 m_2 m_3}\right)_{m_5}$, $\left(\frac{1}{m_1 m_2 m_3}\right)_{m_6}$ is exactly represented in m_4, m_5, and m_6. We base extend this result to the moduli m_1, m_2, and m_3 to get the quotient in complete RNS.

Chang and Low [115] have described scaling by 2^n for the moduli set $\{2^n - 1, 2^n, 2^n + 1\}$. It uses MRC for finding the residues mod $(2^n - 1)$ and mod $(2^n + 1)$ in first step. Next, instead of base extension to obtain the residue corresponding to 2^n, they use Andraros and Ahmad technique [61] to obtain the $2n$-bit quotient $X/2^n$ and take the n LSBs as residue mod 2^n. Later, they have extended the technique for signed numbers also [116]. They observe that in this case, the residue mod 2^n must be incremented by 1. This, however, requires sign detection of $(X-r_2)/2^n$. The authors show that the sign can be negative if bit $(Y)_{2n-1}$ = 1 or if the $2n$-bit MSB Y of the decoded number is $2^{2n-1}-1$ and $x_{2,n-1} = 1$, where x_2 is the residue corresponding to modulus 2^n. However, detection of the condition for $Y = 2^{2n-1}-1$ needs a tree of AND gates. An alternative solution is possible in which the detection of the negative sign is possible under the three conditions: $(2n - 1)$th bit of Y is zero, $y_1 = 2^{n-1}-1$, and $y_{2,n-1} = 1$, where y_1, y_2 are residues of Y corresponding to moduli m_1 and m_2. Thus, a control signal generation block detecting the three conditions need to be added to the unsigned 2^n scaler architecture. The output of this block selects 0 or 1 to be added to $\tilde{y}_2$.

6.7 Magnitude Comparison and Sign Detection

Another operation that is often required is magnitude comparison. Unfortunately, unless both numbers that need to be compared are available in binary form, this is not possible. Solutions do exist but these are time consuming. For example, both RNS numbers can be converted into MRC form and by sequential comparison of Mixed radix digits starting from higher digit, comparison can be made [73,74]. Thus, the computation time involved is in the worst case n comparisons for a n moduli system preceded by MRC which can be done in parallel by having two hardware units.

Example 6.6. As an illustration, consider that comparison of 12 and 37 is needed in $\{3, 5, 7\}$. Mixed Radix conversion yields the mixed radix digits [0,1, 5] and [1,0, 2]. From most significant MRC digit, comparison can be made. Since $1 > 0$, 37 is greater than 12. $\qquad\square$

Diagonal function [117] also can be used for comparison of two numbers. For a n moduli set, sum of quotients is defined as SQ $= M_1 + M_2 + \ldots + M_n$, where $M_j = M/m_j$. Thus, we next compute $k_i = (-1/m_i) \bmod \text{SQ}$. Similar to CRT, we can compute a diagonal function $D(X)$ defined as

$$D(X) = (x_1 k_1 + x_2 k_2 + \ldots + x_n k_n) \bmod SQ. \tag{6.22}$$

This is a monotonic function. For comparing two numbers X and Y, we compute $D(X)$ and $D(Y)$. If $D(X) > D(Y)$, then $X > Y$ and if $D(X) < D(Y)$, then $X < Y$. In case $D(X) = D(Y)$, then we need to compare any two residues corresponding to any m_j and if $x_i < y_i$, $X < Y$ and if $x_i > y_i$, $X > Y$ and if $x_i = y_i$, then $X + Y$. This technique does not have any ambiguity. However, the computation of $D(X)$ is difficult due to the cumbersome mod SQ reduction. As an example consider the moduli set $\{3, 5, 7\}$. We have SQ $= 3 \times 5 + 5 \times 7 + 3 \times 7 = 71$ and $k_1 = 47$, $k_2 = 14$, and $k_3 = 10$. Note that $\sum_{i=1}^{n} k_i \bmod \text{SQ} = 0$. Consider two numbers $X = (1, 2, 3)$ and $Y = (2, 1, 5)$. We have $D(X) = (1 \times 47 + 2 \times 14 + 3 \times 10) \bmod 71 = 34$ and $D(Y) = (2 \times 47 + 1 \times 14 + 5 \times 10) \bmod 71 = 23$. Hence, $X > Y$. Note that $X = 52$ and $Y = 26$.

Sign detection is equally complicated since this involves comparison once again. Other techniques have been suggested in literature [118] but these are very involved. Tomczak [119] has derived the expression for the sign of a RNS number in the moduli set $\{2^n - 1, 2^n, 2^n + 1\}$ as

$$\text{sign}\{x_1, x_2, x_3\} = |t'|_{2^n} \geq 2^{n-1} \tag{6.23a}$$

In other words, it is the $(n-1)$th bit of t' where $t' = x_3 \text{-} x_2 + Y$ and

$$Y = \left(2^{n-1}(x_1 - x_3)\right)_{2^{n-1}} \tag{6.23b}$$

6.8 Error Correction and Detection

Error detection and correction in RNS can be carried out by using redundant moduli. Szabo and Tanaka [2] have suggested an exhaustive testing procedure to detect and correct the error, which needs two extra moduli. It needs $\sum_{i=1}^{K}(m_i - 1)K$ tests, where K is the number of moduli. This method is based on the observation that an error of "1" in any residue corresponding to modulus m_i cause a multiple of $\left(\frac{1}{P_i}\right)_{m_i} P_i$ to be added to the original error free number, where $P = \frac{m_{r1} m_{r2} M}{m_i}$. Hence, we need to find which multiple when added yields the correct number within the dynamic range. This needs to be carried out for all the moduli.

Jenkins [120–122] has suggested an error correction technique, which is based on projections. In this technique using two redundant moduli in addition to the original n moduli, the decoded words considering only $(n + 1)$ moduli at a time are computed using MRC. Only one of these will have most significant MRC digit as zero. As an illustration, consider the moduli set $\{3, 5, 7, 11, 13\}$ where 11 and 13 are the redundant moduli. Consider 52 changed as $\{1, 2, 4, 8, 0\}$ due to error in the residue corresponding to modulus 7. The various projections are as follows:

$\{3,5,7,11\}$ $\{1,2,4,8\}$ 382,	$\{3,5,7,13\}$ $\{1,2,4,0\}$ 1222	
$\{3,5,11,13\}$ $\{1,2,8,0\}$ 52,	$\{3,7,11,13\}$ $\{1,4,8,0\}$ 1768	
$\{5,7,11,13\}$ $\{2,4,8,0\}$ 767		

Evidently, 52 is the correct answer since the original dynamic range is $3 \times 5 \times 7 = 105$ only and in MRC form is $0 \times 165 + 3 \times 15 + 2 \times 3 + 1 = 52$ with the most significant MRC digit being zero.

Goh and Siddiqui [123] have described technique for multiple error correction using redundant moduli. This is based on CRT expansion as a first step to obtain the result. If this is within the dynamic range allowed by the nonredundant moduli, the answer is correctly decoded. Otherwise, it indicates wrong decoding. For double error correction as an illustration, for total number of moduli n, C_2^n possibilities exist. For each one of the possibilities, it can be seen that because of a double error in residues corresponding to moduli m_i and m_j, a multiple of the product of other moduli, that is, $M_{ij} = \frac{M}{m_i m_j}$ is added in the CRT expansion. Thus, by taking mod M_{ij} of the CRT expansion for all cases excluding two moduli at a time and picking among the results the smallest within the dynamic range due to the original moduli, we obtain the correct result.

An example will be illustrative. Consider the six moduli set $\{11,13, 17,19,23,29\}$, where 11,13 are nonredundant and 17,19,23,29 are the redundant moduli. The legitimate dynamic range is 0–142. Consider $X = 73$, which in RNS is (7, 8, 5, 16, 4, 15). Let it be changed to (7, 8, 11, 16, 4, 2) due to a double error. It can be seen that CRT gives the number 25121455, which obviously is wrong. Taking mod m_{ij} for all the 15 cases (excluding (1,2), (1,3),(1,4),(1,5),(1,6),(2,3), (2,4),(2,5),(2,6),(3,4), (3,5),(3,6), (4,5),(4,6),(5,6) since $C_2^6 = 15$, we have respectively, 130299, 79607,62265, 36629, 11435, 28915, 50926, 83464, 33722, 36252, 65281, 73, 23811, 16518, 40828. Evidently, 73 is the decoded word since rest of the values are outside the valid dynamic range.

6.9 Applications of RNS

Several applications of RNS for realizing FIR filters, Cryptographic hardware and digital communication systems have been described in literature. In this section, these will be briefly reviewed.

6.9.1 FIR Filters

FIR filters realized using RNS can use ROM-based multipliers [124]. The coefficients $h(i)$ and input samples $x(i)$ are in RNS form and the multiplications and accumulation needed for FIR filter operation are carried out in RNS. In order to avoid overflow, the dynamic range M of the RNS shall be chosen to be greater than the worst case weighted sum for the given coefficients of the FIR filter. Note that each modulus channel has a multiplier and accumulator mod m_j. Instead of weighting the input samples $x(i)$ by the coefficients $h(i)$ in RNS form, the values $h_j(i)(\frac{1}{M_j})_{m_j}$ can be stored so that CRT can be easily performed by

multiplying the final accumulated residues $r_{aj} = \sum_{i=0}^{l-1} \left(h_j(i)(\frac{1}{M_j})_{m_j} x_j(i) \right)_{m_j}$

with M_j and summing the results mod M, where l is the length of the FIR filter.

Note that instead of using multipliers, a bit-slice approach can be used. In this method, the MSBs of all the final accumulated residues r_{ai} address a ROM to get the decoded word. Next, this word is multiplied by 2 and the decoded word corresponding to the next bits of all the residues is obtained from ROM and accumulated. This process continues as many times as the word length of the residues.

Jenkins and Leon [124] have suggested the bit slice approach due to Peled and Liu [125] where the sum of the weighted residues for all possible l-bit words X for a l-tap filter are stored. Considering an illustration of a ten tap filter, 1024 locations of ROM contain weighted sum of the tap weights $\sum_{i=0}^{9} h(i)x_i$. Thus, as soon as new input sample is available, the MSBs of this sample and the previous nine samples are used to address the PROM to get the result corresponding to $\sum_{i=0}^{9} h(i)x_i$ and accumulated. The result is doubled and next the ROM is looked into corresponding to the next significant bit of all the samples and result accumulated. The advantage of the Peled and Liu method is that the architecture is independent of the number of taps in the filter but is dependent on the word length of the samples. This method is known as hybrid residue-combinatorial architecture.

Freking and Parhi [126] have derived condition for having the RNS basic hardware cost to be less than the binary implementations. Considering the use of a FIR unit comprising of adders needing an area of $n^2 + n$ full-adders assuming a binary multiplier with no rounding operation, a RNS with r mod-

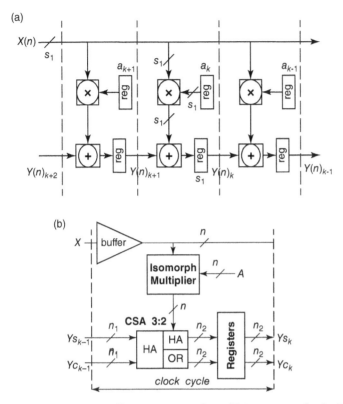

Figure 6.7 (a) RNS FIR filters in transpose form. (b) Tap structure for the RNS carry save architecture. (Adapted with permission from Reference [128]. Copyright 1989, IEEE.)

uli with a dynamic range of $2n$ bits, needs an area of $r(\frac{4n^2}{r^2} + \frac{4n}{r})$ full adders using the architecture of Figure 6.7a. Hence, if this needs to be less than $n^2 + n$, we obtain the condition that $r > \frac{4n}{n-3}$. Thus, a five moduli system may be optimal.

Several authors have suggested implementation of FIR filters using powers of two related moduli sets [127]. They have used multipliers using modified Booth's algorithm in transpose type FIR filters (see Figure 6.7a). Index calculus-based multipliers also have been used by many authors while using general moduli sets [128]. Instead of modulo reduction at each tap of the FIR filter, the accumulation corresponding to few taps is carried out and then reduction is carried out. The output of each tap is held in carry save form in order to reduce the delay due to addition (see Figure 6.7b) [128]. Note, however, that the register size to store the intermediate sum will be increased slightly due to this modification. Techniques for reducing power dissipation have been suggested for modulo adders [129]. The possibility of the sum exceeding the modulus is predicted thereby enabling computation of ether $A + B$ or $A + B - m_i$ only. This will be possible for most cases. In other cases both $A + B$ and $A + B - m_i$ need to be computed. QRNS implementations also have been described for realization of complex FIR filters [130], [131]. In these, multiplication of complex numbers is reduced to two real multiplications and few additions.

6.9.2 RNS in Cryptography

In cryptographic applications such as RSA encryption, Diffie-Hellman Key exchange, Elliptic curve cryptography, modulo multiplication (i.e., $(x \times y)$ mod n) and modulo exponentiation (i.e., x^α mod n) involving x, y, n, α of bit lengths varying between 160 bits to 2048 bits typically will be required [132,133]. The computation x^α mod n can be realized by successive squaring of x mod n, x^2 mod n, and so on, and multiplying these depending on the bits which are 1 in α. The fundamental operation is thus $(a \times b)$ mod n or $(a \times a)$ mod n, where a and b are intermediate values. Efficient computation is possible using Montgomery algorithm [134] wherein $(\frac{ab}{R})$mod N is computed where R is called Montgomery constant. The basic principle is to compute q such that $\frac{ab+qN}{R}$ is exact. We need to compute $\beta = (-\frac{1}{N})_R$. Next, we have $q = (ab\beta)$ mod R.

Consider $a = 10$, $b = 12$, and $R = 16$, $N = 23$. We find $\beta = (-\frac{1}{23})_{16} = 9$ and $q = (10 \times 12 \times 9)$ mod $16 = 8$. Thus, we compute $Y = (120 + 8 \times 23)/16 = 19$.

6.9.2.1 Montgomery Modular Multiplication

In Montgomery multiplication [135–141] using RNS (base B using k moduli) of dynamic range M, the Montgomery constant is chosen as M. While computation of integer q such that $(ab + qN)$ is exact, division by M in RNS is not possible since inverse of M does not exist modulo M. Hence, there is a need for a second RNS (extended base B' using k extra moduli). We perform multiplication by M^{-1} in the extended RNS B' with dynamic range M' and base extend the result back to the original RNS B.

The algorithm is presented in Figure 6.8 for computing $Z = \left(\frac{ab}{M}\right)$ mod N. First $t = ab$ is computed in both RNS in step 1. Next, in step 2, $q = t/N$ is computed in RNS1 (base B). In RNS2 (base B'), we compute $ab + \hat{q}N$ in step 4 (where $\hat{q}$ is obtained by base extension of q from RNS1 to RNS2 in step 3) and divide it by M. (Note that inverse of M exists in RNS2 and not in RNS1). This result is next base extended to RNS1 in step 5. Evidently, two base extensions are needed: one from RNS1 to RNS2 and another from RNS2 to RNS1. The base extension can be obtained by using CRT as discussed in Section 3.6 ([109,110])

Algorithm 1 : MM(a, b, N),   *RNS Montgomery Multiplication*

Input : Two RNS bases $\mathcal{B} = (m_1, \ldots, m_k)$, and $\mathcal{B}' = (m_{k+1}, \ldots, m_{2k})$, such that $M = \prod_{i=1}^{k} m_i$, $M' = \prod_{i=1}^{k} m_{k+i}$ and $\gcd(M, M') = 1$; a redundant modulus m_r, $\gcd(m_r, m_i) = 1$, $\forall i = 1 \ldots 2k$; a positive integer N represented in RNS in both bases and for m_r such that $0 < (k+2)^2 N < M$, M' and $\gcd(N, M) = 1$; (Note that M can be greater or less than M'.) two positive integers a, b represented in RNS in both bases and for m_r, with $ab < MN$.

Output : A positive integer $\hat{r}$ represented in RNS in both bases and m_r, such that $\hat{r} \equiv abM^{-1}$ (mod N), and $\hat{r} < (k+2)N$.

1: $t = ab$ in $\mathcal{B} \cup \mathcal{B}' \cup \{m_r\}$

2: $q = t(-n^{-1})$ in $\mathcal{B}$

3: $[q$ in $\mathcal{B}] \longrightarrow [\hat{q}$ in $\mathcal{B}' \cup \{m_r\}]$ *First base extension*

4: $\hat{r} = (t + \hat{q}N)M^{-1}$ in $\mathcal{B}' \cup \{m_r\}$

5: $[\hat{r}$ in $\mathcal{B}] \longleftarrow [\hat{r}$ in $\mathcal{B}']$ *Second base extension*

Figure 6.8 Montgomery multiplication using RNS. (Adapted with permission from Reference [137]. Copyright 2004, IEEE.)

but the multiple of M that needs to be subtracted needs to be known. The application of CRT in base B yields

$$q = \sum_{i=1}^{k} \sigma_i M_i - \alpha M \qquad (6.24a)$$

where M contains k moduli and $\sigma_i = \left(q_i \left| M_i^{-1} \right|_{m_i} \right) \bmod m_i$ and $\alpha < k$. We compute q in M' as

$$q = \left(\sum_{i=1}^{k} \sigma_i M_i - \alpha M \right)_{m_j} \qquad (6.24b)$$

for $j = k + 1, \ldots, 2k$. We compute next

$$\hat{r} = \frac{ab + \hat{q}N}{M} \qquad (6.24c)$$

where $\hat{q} = q + \alpha M = \sum_{i=1}^{k} \sigma_i M_i$. Note that the computed value $\hat{r} < M'$ and has a valid representation in base B'. From Equation (6.24a)–(6.24c), we get

$$\hat{r} = \left(\frac{ab + (q + \alpha M)N}{M} \right) \bmod N = (ab\, M^{-1}) \bmod N \qquad (6.24\text{d})$$

Thus, there is no need to compute α. Instead, $\hat{q}$ needs to be computed in B'. Once $\hat{r}$ is estimated, this needs to be extended to B, which can be done using Shenoy and Kumaresan [109,110] technique for which the redundant residue m_r is used. It may be noted that since in CRT, $\alpha < k$, $q < M$ and $ab < MN$, we have $\hat{q} < (k+1)M$ and hence $\hat{r} < (k+2)N < M'$. The condition $ab < MN$ implies that if we want to use $\hat{r}$ in the next step as needed in exponentiation algorithms, say squaring, we obtain the condition $(k+2)^2 N^2 < MN$ or $(k+2)^2 N < M$. Thus, if N is a 1024 bit number, and if we use 32 bit moduli, we need base B of size $k \geq 33$.

Note that lines 1, 2 and 4 in Figure 6.8 need $5k$ modular multiplications. The first and second base extensions (steps 3 and 5) each need $(k^2 + 2k)$ modular multiplications, respectively, thus needing over all $(2k^2 + 9k)$ modular multiplications. If the redundant modulus is chosen as a power of two, the second base extension needs $k^2 + k$ multiplications only [136].

Kawamura *et al.* [139] have described a Cox-Rower architecture for RNS Montgomery multiplication. They stipulate that $a, b < 2N$ so that

$$r = \frac{v}{M} = \frac{ab + qN}{M} < \frac{(2N)^2 + MN}{M} = N\left(\frac{4N}{M} + 1\right) \leq 2N \text{ if } 4N \leq M$$

For $r < M'$ and $v < MM'$, the condition $2N \leq M'$ is sufficient. In this, the base extension algorithm is executed in parallel by plural "Rower" units controlled by a "cox" unit. Each rower unit is a single precision modular multiplier accumulator whereas the Cox unit is typically a 7-bit adder. Referring to Eq. (6.24a), it is clear that the value of α (i.e., multiple of M) that needs to be subtracted needs to be determined so that base extension to carried out. Kawamura *et al.* [139] have suggested rewriting

$$q = \sum_{i=1}^{k} q_i \left(\frac{1}{M_i}\right)_{m_i} M_i - \alpha M = \sum_{i=1}^{nk} \sigma_i M_i - \alpha M \qquad (6.25\text{a})$$

as

$$\alpha + \frac{q}{M} = \sum_{i=1}^{n} \frac{\sigma_i}{m_i} \qquad (6.25\text{b})$$

Since $q < M$, $\sum_{i=1}^{k} \frac{\sigma_i}{m_i}$ is between α and $\alpha + 1$. Thus, $\alpha = \left\lfloor \sum_{i=1}^{k} \frac{\sigma_i}{m_i} \right\rfloor$ and $0 < \alpha < k$ holds. The value of α can be recursively estimated in the "cox" unit by approximating m_i in the denominator by 2^r where it is assumed that r is common to all moduli in spite of m_i being different and computing

$$\hat{\alpha} = \left\lfloor \sum_{i=1}^{n} \frac{trunc(\sigma_i)}{2^r} + \alpha \right\rfloor \tag{6.26}$$

where $trunc(\zeta_i) = \zeta_i \bigcap (1\ldots 10\ldots 0)_2$. (Note that the number of ones are q and number of zeroes are r-q and $\bigcap$ stands for bit-wise AND operation). The parameter α is an offset value to take into account the error caused due to the approximation. Note that Eq. (6.26) is computed as

$$\alpha_i = \alpha_{i-1} + \frac{trunc(\sigma_i)}{2^r}, k_i = \lfloor \alpha_i \rfloor, \alpha_i = \alpha_i - k_i \text{ for } i = 1, 2, \ldots n, \sigma_o = \alpha \tag{6.27}$$

Note that k_i is a bit and if it is 1, the rower unit subtracts M. Note that the error is transferred to the next step and only in the last step there is residual error. Kawamura *et al.* [139] have suggested the use of $\alpha = 0$ and $\alpha = 0.5$ for the first and second base extensions, respectively. Note that n clock cycles are needed to obtain the n number of k_i values. It can be seen that $n^2 + 2n$ operations are needed for each base extension and $5n$ other operations are needed for complete modulo multiplication.

Schinianakis and Stouraitis [141] have suggested the use of MRC for both the base extension operations in place of the approximate CRT-based techniques.

Note that Montgomery exponentiation [137–140] needs iteration of Montgomery modular multiplication Mont $(x,y) = (xyR^{-1})$ mod N, where R is the Montgomery constant. Given x and y, we need to first compute $\bar{x} = (xR)_N$ and $\bar{y} = (yR)_N$. In RNS Montgomery multiplication, M is used as the Montgomery constant. Hence, M mod N and M^2 mod N are precomputed so that *Mont* $(x,M^2 \text{mod} N)$ yields (xM)mod$N$ and similarly at the end, *Mont* $(zM,1)$ gives z mod N.

6.9.2.2 Elliptic Curve Cryptography using RNS

Schiniankis *et al.* [142] have realized an ECC processor using RNS. In order to reduce the division operations in affine representation, they have used projective coordinates. Consider the elliptic curve

$$y^2 = x^3 + ax + b \text{ over } F_p \tag{6.28a}$$

where $a, b \in F_p$ and $4a^3 + 27b^2 \neq 0 \bmod p$ together with a special point $\bigcirc$ called the point at infinity. The point at infinity is given by $\{0,0,0\}$. Substituting $x = \frac{X}{Z^2}, y = \frac{Y}{Z^3}$, using the Jacobian coordinate representation, Eq. (6.28a) changes as

$$E(F_p) : Y^2 = X^3 + aX\,Z^4 + b\,Z^6 \tag{6.28b}$$

The addition of two points $P_o = (X_o, Y_o, Z_o)$ and $P_1 = (X_1, Y_1, Z_1)$ thus will be given by

$$P_2 = P_0 + P_1 \rightarrow X_2 = R^2 - TW^2, 2Y_2 = VR - MW^3, Z_2 = Z_0Z_1W \tag{6.28c}$$

where

$$W = X_0Z_1^2 - X_1Z_0^2, R = Y_0Z_1^3 - Y_1Z_0^3, T = X_0Z_1^2 + X_1Z_0^2, M = Y_0Z_1^3 + Y_1Z_0^3,$$
$$V = TW^2 - 2X_2.$$

The doubling of point P_1 is given as

$$P_2 = 2P_1 \rightarrow X_2 = M^2 - 2S, Y_2 = M(S - X_2) - T, Z_2 = 2Z_1Y_1 \tag{6.29}$$

where $M = 3X_1^2 + aZ_1^4, S = 4X_1Y_1^2, T = 8Y_1^4$.

Note that the computation is intensive in multiplications and additions while division is avoided. The exponentiation follows the binary algorithm where successive squarings and multiplications depending on the position of 1 in the exponent are required. All the operations are mod p thus necessitating modulo adders and multipliers. If the field characteristic is 160 bits long, the equivalent RNS range to compute Eq. (6.28c) and (6.29) is about 660 bits. Hence, the authors use 20 moduli each of about 33 bit length. In the case of p being 192 bits, then the moduli size will be 42 bits. The authors used extended RNS using one redundant modulus to perform residue to binary conversion using CRT. The conversion from projective coordinates to affine is done using the relationships $x = \frac{X}{Z^2}, y = \frac{Y}{Z^3}$. Note that one modular inversion ($T_1 = 1/Z$) and four modular multiplications are required ($T_2 = T_{12}, x = XT_2, T_3 = T_1T_2$, and $y = YT_3$).

Schiniakis *et al.* [143] have further extended their work on ECC using RNS. They observe that for p of 192 bit length, the equivalent RNS dynamic range is 840 bits. As such, 20 moduli of 42 bits each have been suggested. The implementation of Eq. (6.28c) and (6.29) can take advantage of parallelism, between multiplication, addition or subtraction operations for both point addition as well as point doubling. They observe that 13 steps will be required for each.

Bajard *et al.* [144] suggest rewriting the formulae for addition/doubling to minimize the number of modular reductions. The authors consider RNS bases with moduli of the type $m_i = 2^k - c_i$ where c_i is small and sparse and $c_i \leq 2^{k/2}$. Several coprimes can be found, for example, for $m_i < 2^{32}$, $c_i = 2^{t_i} \pm 1$ with $t_i = 0,1,..,16$ for $c_i = 2^{t_i} - 1$ and with $t_i = 1,..,15$ for $c_i = 2^{t_i} + 1$. If more coprimes are needed, c_i of the form $2^{t_i} \pm 2^{S_i} \pm 1$ can be used. The reduction mod m_i in these cases needs few shift and add operations. Thus, the reduction part cost is 10% of the cost of multiplication and the cost of multiplication is negligible compared to the cost of reduction in RNS. Thus, considering each modulus to be of the size of a word, a RNS digit product is equivalent to a 1.1 word product (where word is k bits) and a RNS multiplication needs only $2n$ RNS digit products or $2.2n$ word products.

6.9.2.3 Pairing processors using RNS

Pairing has applications in three-way key exchange [145], identity-based encryption [146] and identity-based signatures [147] and noninteractive zero knowledge proofs [148]. PBC (pairing-based cryptography) relies on finite fields and is a function of two points on the elliptic curve groups to construct cryptographic systems. Pairings are a function of two points in the elliptic curves. Very few elliptic curves admit a usable pairing.

The pairing computation perhaps is most computationally intensive among all cryptographic algorithms. It can be broken down into multiplications and additions in the underlying prime fields. Around 10,000 modular multiplications are needed for a pairing processor for a 256-bit Ate Pairing. Efficient multiplication is thus crucial to the realization of fast pairing processors. Operations in extension field F_{p^k} are essential for pairing computation. Lazy reduction reduces the number of reductions in the multiplication over such extension fields.

The base moduli need to be chosen to reduce complexity of modulus reduction in RNS as mentioned in previous paragraph. Yao *et al.* [149] have suggested the selection of moduli in both the RNS used in base extension close so that $(b_k - c_j)$ values are small where b_k and c_j are the moduli in the two bases. This leads to the advantage that the bit lengths of the operands needed in base extension are small. They have suggested the use of eight-moduli sets for 256 bit dynamic range for 128 bit security as $B = 2^w - 1, 2^w - 9, 2^w + 3, 2^w + 11, 2^w + 5, 2^w + 9, 2^w - 31, 2^w + 15$, $C = 2^w, 2^w + 1, 2^w - 3, 2^w + 17, 2^w - 13, 2^w - 21, 2^w - 25, 2^w - 33$ so that the bit lengths of $\prod_{k=1, k \neq j}^{s} (b_i - c_j)$ are as small as possible (<25 bits).

Montgomery reduction in RNS has higher complexity than ordinary Montgomery reduction as explained earlier due to the time consuming base extensions. The multiplication is linear in word operations whereas the modular reduction is quadratic. However, this overhead of slow reduction can be partially removed by reducing the number of reductions also known as lazy reduction. In computing $ab + cd$ with a, b, c, d in F_p, where p is a n word prime number, we need $4n^2 + 2n$ word operations since two Montgomery modulo multiplications are needed. Recall that each multiplication, for example, ab needs n^2 word products and reduction takes $(n^2 + n)$ multiplications in conventional Montgomery technique. Two multiplications $a \times b$ and $c \times d$ need $2n^2$ word multiplications. Adding the results and having only one reduction thus overall needs only $3n^2 + n$ word products. This is called Lazy reduction and is possible for expressions like $AB \pm CD \pm EF$ in F_p, where the allowed RNS dynamic range shall be $3p^2$. The actual operating range of certain operations should be $22p^2$ for lazy reduction to be economical.

The architecture of pairing coprocessor due to Yao *et al.* [150] is presented in Figure 6.9. It uses eight PEs in the Cox-Rower architecture [139]. Each PE caters for one channel of RNS B and one channel of RNS C. Four DSP blocks can realize 35×35 multiplier whereas two DSPS can realize a 35×25 multiplier. A dual mode multiplier that can perform two element multiplications

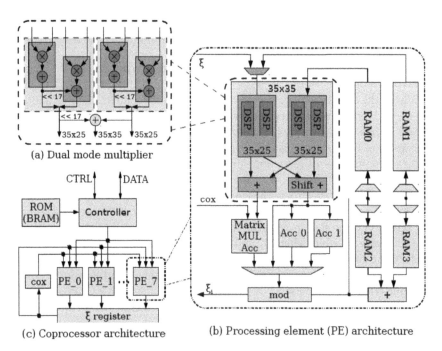

(a) Dual mode multiplier

(c) Coprocessor architecture

(b) Processing element (PE) architecture

Figure 6.9 Pairing coprocessor hardware architecture. (Adapted with permission from Reference [150]. Copyright 2011, IEEE.)

(a)

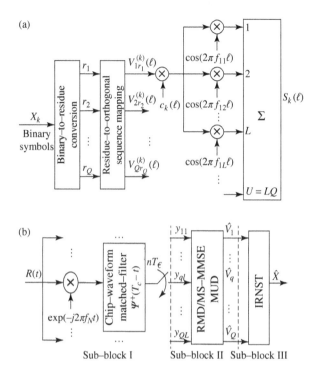

(b)

Figure 6.10 (a) Transmitter schematic for kth CRU (cognitive radio user) and (b) receiver block diagram in RRNS MC/DS-CCDMA system. (Adapted with permission from Reference [158]. Copyright 2012, IEEE.)

at a time is used. It uses two accumulators to perform the four multiplications in parallel and accumulate the results in parallel. Each PE performs two element multiplications at a time using the dual multiplier. The internal design of the PE is shown in Figure 6.9a comprising dual-mode multiplier, adder, 2 RAMS for adder inputs, 3 accumulators. The authors have implemented using Virtex-6 XC6VLX240T2 FPGA for implementing Ate and Optimal Ate pairing.

Duquesne and Guillermin [151] have implemented optimal pairing for Barreto–Naehrig (BN) curves [152] for 128-bit curves on Altera FPGAs. They have also employed lazy reduction and Montgomery reduction in RNS. Duquesne [153] has described pairing implementation using RNS for MNT curves [154] and BN curves for Tate, Ate, and R-Ate pairings.

6.9.3 RNS in Digital Communication Systems

RNS has been used for communication systems for protecting the information processed or transmitted [155–160]. This exploits the self-checking/error correction properties of redundant residue number systems.

Zhang *et al.* [159] have suggested use of RRNS (redundant residue number system) for multicarrier direct sequence CDMA multiple access for cognitive radio. The transmitter block diagram is presented in Figure 6.10a, wherein the input binary symbol that represents a message $0 \leq X \leq M$ is converted into residues using a front-end binary to RNS converter. These Q residues are mapped next into Q orthogonal sequences. Considering a Q moduli set, the number of orthogonal sequences are $\sum_{q=1}^{Q} m_q$. These are Hadamard–Walsh codes of length N_s. Next, each of these selected orthogonal sequences are spread using a user specific signature sequence. Next each of these Q spreading sequences are transmitted on L sub-carriers to achieve a Lth order frequency diversity. Thus, the RRNS MC/DS-CDDMA (code division dynamic multiple access multicarrier) system requires a total of $U = LQ$ number of sub-carriers. The receiver architecture is also shown in Figure 6.10b. After processing in the sub-block I, the results are sent to sub-block II where multiuser detection (MUD) is performed to suppress the multiuser interference. After the inverse RNS transform, the information transmitted by the k CRUs is recovered. The error and throughput performance is comparable to other techniques. Any erroneous residues can be discarded without affecting the data recovery provided that sufficient dynamic range remains in the reduced RRNS to unambiguously represent the nonredundant information. Madhukumar and Chin have suggested similar techniques with enhancements to reduce the number of carriers using PSK/QAM [160].

References

1 H. L. Garner, The residue number system, *IRE Trans. Electronic Comput.*, vol. 8, 1959, pp. 140–147.

2 N. S. Szabo and R. I. Tanaka, *Residue Arithmetic and Its Applications to Computer Technology*, Mc-Graw Hill, New York, 1967.

3 M. A. Soderstrand, W. K. Jenkins, G. A. Jullien, and F. J. Taylor, *Residue number system Arithmetic: Modern applications in Signal Processing*, IEEE Press, New York, 1986.

4 P. V. Ananda Mohan, *Residue Number Systems: Algorithms and Architectures*, Kluwer Academic Publishers, New York, 2002.

5 F. J. Taylor, Residue arithmetic: a tutorial with examples, *IEEE Comput.*, vol. 17, 1984, pp. 50–62.

6 A. Omondi and B. Premkumar, *Residue Number Systems: Theory and implementation*, Imperial College Press, London, 2007.

7 A. S. Navi, M. Molahosseini, and M. Esmaeildoust, How to teach residue number system to computer scientists and engineers, *IEEE Trans. Educ.*, vol. 54, 2011, pp. 156–163.

8 M. A. Bayoumi, G. A. Jullien and W. C. Miller, A VLSI implementation of residue adders, *IEEE Trans. Circuits Syst.*, vol. 34, 1987, pp. 284–288.

9 M. Dugdale, VLSI implementation of residue adders based on binary adders, *IEEE Trans. Circuits Syst., Part II*, vol. 39, 1992, pp. 325–329.

10 C. Efstathiou, D. Nikolos, and J. Kalanmatianos, Area-time efficient modulo $2^n - 1$ adder design, *IEEE Trans. Circuits Syst.*, vol. 41, 1994, pp. 463–467.

11 R. Zimmermann, Efficient VLSI implementation of modulo $(2^n \pm 1)$ addition and multiplication, *Proc. IEEE Symp. Comput. Arithmetic*, 1999, pp. 158–167.

12 L. Kalampoukas, D. Nikolos, C. Efstathiou, H. T. Vergos, and J. Kalamatianos, High speed parallel prefix modulo $(2^n - 1)$ adders, *IEEE Trans. Comput.*, vol. 49, 2000, pp. 673–680.

13 C. Efstathiou, H. T. Vergos, and D. Nikolos, Modulo $2^n \pm 1$ adder design using select-prefix blocks, *IEEE Trans. Comput.*, vol. 52, 2003, pp. 1399–1406.

14 R. A. Patel and S. Boussakta, Fast parallel-prefix architectures for modulo $2^n - 1$ addition with a single representation of zero, *IEEE Trans. Comput.*, vol. 56, 2007, pp. 1484–1492.

15 L. M. Liebowitz, A simplified binary arithmetic for the Fermat number transform, *IEEE Trans. ASSP*, vol. 24, 1976, pp. 356–359.

16 S. Efstathiou, H. T. Vergos, and D. Nikolos, Fast parallel prefix modulo $(2^n + 1)$ adders, *IEEE Trans. Comput.*, vol. 53, 2004, pp. 1211–1216.

17 H. T. Vergos, C. Efstathiou, A unifying approach for weighted and diminished-1 modulo $(2^n + 1)$ addition, *IEEE Trans. Circuits Syst. II: Express Briefs*, vol. 55, 2008, pp. 1041–1045.

18 H. T. Vergos, C. Efstathiou, and D. Nikolos, Diminished-1 modulo $2^n + 1$ adder design, *IEEE Trans. Comput.*, vol. 51, 2002, pp. 1389–1399.

19 Tso-Bing Juang, Chin-Chieh Chiu, and Ming-Yu Tsai, Improved area-efficient weighted modulo $2^n + 1$ adder design with simple correction schemes, *IEEE Trans. Circuits Syst. II: Express Briefs*, vol. 57, 2010, pp. 198–202.

20 E. Lu Lein Harn, J. Lee, and W. Hwang, A programmable VLSI architecture for computing multiplication and polynomial evaluation modulo a positive integer, *IEEE J. Solid-State Circuits*, vol. SC-23, 1988, pp. 204–207.

21 M. A. Soderstrand and C. Vernia, A high-speed low cost modulo p_i multiplier with RNS arithmetic applications, *Proc. IEEE*, vol. 68, 1980, pp. 529–532.

22 T. Stouraitis, S. W. Kim, and A. Skavantzos, Full adder based arithmetic units for finite integer rings, *IEEE Trans. Circuits Syst.*, vol. 40, 1993, pp. 740–744.

23 Z. Wang, G. A. Jullien, and W. C. Miller, An algorithm for multiplication modulo $(2^N - 1)$, *Proc. 39th Midwest Symp. Circuits Syst.*, Ames, IA, 1996, pp. 1301–1304.

24 C. Efstathiou, H. T. Vergos, and D. Nikolos, Modified booth modulo $2^n - 1$ multipliers, *IEEE Trans. Comput.*, vol. 53, 2004, pp. 370–374.

25 R. Muralidharan and C. H. Chang, Radix-8 booth encoded modulo $2^n - 1$ multipliers with adaptive delay for high dynamic range residue number system, *IEEE Trans. Circuits Syst. I: Regul. Pap.*, vol. 58, 2011, pp. 982–993.

26 G. W. Bevick, *Fast multiplication: algorithms and implementation*, Ph.D. dissertation, Stanford University, CA, 1994.

27 A. V. Curiger, H. Bonnennberg, and H. Keaslin, Regular architectures for multiplication modulo $(2^n + 1)$, *IEEE J. Solid-State Circuits*, vol. SC-26 1991, pp. 990–994.

28 X. Lai, *On the design and security of block ciphers*, Ph.D. dissertation, ETH Zurich, No.9752, 1992.

29 Z. Wang, G. A. Jullien, and W. C. Miller, An efficient tree architecture for modulo $(2^n + 1)$ multiplication, *J. VLSI Signal Process. Syst.*, 1996, pp. 241–248.

30 C. Efstathiou, H. T. Vergos, G. Dimitrakopoulos, D. Nikolos, Efficient diminished-1 modulo $2^n + 1$ multipliers, *IEEE Trans. Comput.*, vol. 54, 2005, pp. 491–496.

31 Y. Ma, A simplified architecture for modulo $(2^n + 1)$ multiplication, *IEEE Trans. Comput.*, vol. 47, 1998, pp. 333–337.

32 L. Sousa and R. Chaves, A universal architecture for designing efficient modulo $2^n + 1$ multipliers. *IEEE Trans. Circuits Syst.*, Part I, vol. 52, 2005, pp. 1166–1178.

33 J. W. Chen, R. H. Yao, and W. J. Wu, Efficient modulo $2^n + 1$ multipliers, *IEEE Trans. VLSI Syst.*, vol. 19, 2011, pp. 2149–2157.

34 R. Muralidharan and C. H. Chang, Area-power efficient modulo $2^n - 1$ and modulo $2^n + 1$ multipliers for $\{2^n - 1, 2^n, 2^n + 1\}$ based RNS, *IEEE Trans. Circuits Syst.*, vol. 59, 2012, pp. 2263–2274.

35 J. W. Chen and R. H. Yao, Efficient modulo $2^n + 1$ multipliers for diminished-1 representation, *IET Circuits Devices Syst.*, vol. 4, 2010, pp. 291–300.

36 R. Muralidharan and C. H. Chang, Radix-4 and Radix-8 Booth encoded multi-modulus multipliers, *IEEE Trans. Circuits Syst., Part-I*, vol. 60, 2013, pp. 2940–2952.

37 A. Wrzyszcz and D. Milford, A new modulo $2^\alpha + 1$ multiplier, *IEEE Int. Conf. Comput. Design: VLSI Comput. Processors*, 1993, pp. 614–617.

38 H. T. Vergos and C. Efstathiou, Design of efficient modulo $2^n + 1$ multipliers, *IET Comput. Digit. Tech*, vol. 1, 2007, pp. 49–57.

39 S. J. Piestrak, Design of squarers modulo A with low-level pipelining, *IEEE Trans. Circuits Syst. II: Analog Digit. Signal Process.*, vol. 49, 2002, pp. 31–41.

40 A. E. Cohen and K. K. Parhi, Architecture optimizations for the RSA public key cryptosystem: a tutorial, *IEEE Circuits Syst. Mag.*, vol. 11, 2011, pp. 24–34.

41 A. Spyrou, D. Bakalis, and H. T. Vergos, Efficient architectures for modulo $2^n - 1$ squarers, *Proc. IEEE Int. Conf. DSP*, 2009, pp. 1–6.

42 A. Strollo and D. Caro, Booth Folding encoding for high performance squarer circuits, *IEEE Trans. CAS, Part II*, vol. 50, 2003, pp. 250–254.

43 H. T. Vergos and C. Efstathiou, Diminished-1 modulo $2^n + 1$ squarer design, *Proc. IEE Comput. Digit. Tech.*, vol. 152, 2005, pp. 561–566.

44 H. T. Vergos and C. Efstathiou, Efficient modulo $2^n + 1$ squarers, *Proc. XXI Conf. Design Circuits Integrated Systems, DCIS*, 2006.

45 R. Muralidharan and C. H. Chang, Fixed and variable multi-modulus squarer architectures for triple moduli base of RNS, *Proc. ISCAS*, 2009, pp. 441–444.

46 D. Bakalis, H. T. Vergos, and A. Spyrou, Efficient modulo $2^n \pm 1$ squarers, *Integr. VLSI J.*, vol. 44, 2011, pp. 163–174.

47 D. Adamidis and H. T. Vergos, RNS multiplication/sum-of-squares units, *IET Comput. Digit. Tech.*, vol. 1, 2007, pp. 38–48.

48 G. Alia and E. Martinelli, A VLSI algorithm for direct and reverse conversion from weighted binary system to residue number system, *IEEE Trans. Circuits Syst.*, vol. 31, 1984, pp. 1033–1039.

49 S. J. Piestrak, Design of residue generators and multi-operand modulo adders using carry save adders, *Proc. 10th Symp. Comput. Arithmetic*, 1991, pp. 100–107.

50 P. V. Ananda Mohan, Efficient design of Binary to RNS converters, *JCSC*, vol. 9, 1999, pp. 145–154.

51 P. V. Ananda Mohan, Novel design for binary to RNS converters, *Proc. IEEE ISCAS*, London, U. K. 1994, pp. 357–360.

52 R. M. Capocelli and R. Giancarlo, Efficient VLSI networks for converting an integer from binary system to residue number system and vice versa. *IEEE Trans. Circuits Syst.*, vol CAS 35, 1988, pp. 1425–1430.

53 J. Y. S. Low and C. H. Chang, A new approach to the design of efficient residue generators for arbitrary moduli,. *IEEE Trans. Circuits Syst.-I: Regul. Pap.*, vol. 60, 2013, pp. 2366–2374.

54 A. B. Premkumar, A formal framework for conversion from binary to residue numbers, *IEEE Trans. Circuits Syst., Part II*, vol. 49, 2002, pp. 135–144.

55 A. B. Premkumar, E. L. Ang, and E. M. K. Lai, Improved memory-less RNS forward converter based on periodicity of residues, *IEEE Trans. Circuits Syst., Part II*, vol. 53, 2006, pp. 133–137.

56 P. K. Matutino, H. Pettenghi, R. Chave, and L. Sousa, Multiplier based binary to RNS converters modulo $(2^n \pm k)$, *Proc. 26th Conf. Des. Circuits Integr. Syst.*, 2011, pp. 125–130.

57 F. Pourbigharaz and H. M. Yassine, Simple binary to residue transformation with respect to $2^m + 1$ moduli, *Proc. IEE Circuits, Devices Syst.*, vol. 141, 1994, pp. 522–526.

58 S. J. Piestrak, Design of multi-residue generators using shared logic, *Proc. IEEE ISCAS*, 2011, pp. 1435–1438.

59 G. Bi, and E. V. Jones, Fast conversion between binary and residue numbers, *Electron. Lett.*, vol. 24 1988, pp. 1195–1197.

60 P. Bernardson, Fast memory-less over 64-bit residue to binary converter, *IEEE Trans. Circuits Syst.*, vol. 32, 1985, pp. 298–300.

61 S. Andraros and H. Ahmad, A new efficient memory-less residue to binary converter, *IEEE Trans. Circuits Syst.*, vol. 35, 1988, pp. 1441–1444.

62 K. M. Ibrahim and S. N. Saloum, An efficient residue to binary converter design, *IEEE Trans. Circuits Syst.*, vol. 35, pp. 1156–1158, 1988.

63 A. Dhurkadas, Comments on an efficient residue to binary converter design, *IEEE Trans. Circuits Syst.*, vol. 37, 1990, pp. 849–850.

64 P. V. Ananda Mohan and D. V. Poornaiah, Novel RNS to binary converters, *Proc. IEEE ISCAS*, 1991, pp. 1541–1544.

65 S. J. Piestrak, A high-speed realization of residue to binary system conversion. *IEEE Trans. Circuits Syst.*, Part II, vol. 42, 1995, pp. 661–663.

66 A. Dhurkadas, Comments on a high-speed realisation of a residue to binary number system converter, *IEEE Trans. Circuits Syst.*, *Part II*, vol. 45, 1998, pp. 446–447.

67 M. Bhardwaj, A. B. Premkumar, and T. Srikanthan, Breaking the $2n$-bit carry propagation barrier in residue to binary conversion for the $[2^n - 1, 2^n, 2^n + 1]$ moduli set, *IEEE Trans. Circuits Syst.*, *Part II*, vol. 45, 1998, pp. 998-1002.

68 R. Conway and J. Nelson, New CRT based RNS converter for restricted moduli set, *IEEE Trans. Comput.*, vol. 52, 2003, pp. 572–578.

69 Z. Wang, G. A. Jullien, and W. C. Miller, An improved residue to binary converter, *IEEE Trans. Circuits Syst.*, *Part I*, vol. 47, 2000, pp. 1437–1440.

70 P. V. Ananda Mohan, Comments on breaking the $2n$-bit carry propagation barrier in residue to binary conversion for the $[2^n - 1, 2^n, 2^n + 1]$ moduli set, *IEEE Trans. Circuits Syst.*, *Part II*, vol. 48, p. 1031, 2001.

71 Y. Wang, X. Song, M. Aboulhamid, and H. Shen, Adder based residue to binary number converters for $(2^n - 1, 2^n, 2^n + 1)$, *IEEE Trans. Signal Process.*, vol. 50, 2002, pp. 1772–1779.

72 W. Wang, M. N. S. Swamy, M. O. Ahmad, and Y. Wang, A study of the residue-to-binary converters for the three moduli sets, *IEEE Trans. Circuits Syst.*, *Part I*, vol. 50, 2003, pp. 235–243.

73 B. Vinnakota and V. V. B. Rao, Fast conversion techniques for binary to RNS. *IEEE Trans. Circuits Syst.*, *Part I*, vol. 41, 1994, pp. 927–929.

74 P. V. Ananda Mohan, Evaluation of fast conversion techniques for binary-residue number systems, *IEEE Trans. Circuits Syst.*, *Part I*, vol. 45, 1998, pp. 1107–1109.

75 D. Gallaher, F. E. Petry, and P. Srinivasan, The digit parallel method for Fast RNS to weighted number System conversion for specific moduli $(2^k - 1, 2^k, 2^k + 1)$, *IEEE Trans. Circuits Syst.*, *Part II*, vol. 44, 1997, pp. 53–57.

76 P. V. Ananda Mohan, On the digit parallel method for fast RNS to weighted number system conversion for specific moduli $(2^k - 1, 2^k, 2^k + 1)$, *IEEE Trans. Circuits Syst.*, *Part-II*, vol. 47, 2000, pp. 972–974.

77 A. S. Ashur, M. K. Ibrahim, and A. Aggoun, Novel RNS structures for the moduli set $\{2^n - 1, 2^n, 2^n + 1\}$ and their application to digital filter implementation, *Signal Process.*, vol. 46, 1995.

78 M. A. Soderstrand, C. Vernia, and J. H. Chang, An improved residue number system digital to analog converter, *IEEE Trans. Circuits Syst.*, vol. 30, 1983, pp. 903–907.

79 T. V. Vu, Efficient implementations of the Chinese remainder theorem for sign detection and residue decoding, *IEEE Trans. Comput.*, vol. 34, 1985, pp. 646–651.

80 Y. Wang, Residue to binary converters based on new Chinese remainder theorems, *IEEE Trans. Circuits Syst., Part II*, vol. 47, 2000, pp. 197–205.

81 P. V. Ananda Mohan, Comments on residue-to-binary converters based on new chinese remainder theorems, *IEEE Trans. Circuits Syst., Part-II, Analog Digit. Signal Process.*, vol. 47, 2000, p. 1541.

82 L. Akushskii, V. M. Burcev, and I. T. Pak, A new positional characteristic of non-positional codes and its application, in *Coding Theory and Optimization of Complex Systems*, (ed. F V. M. Amerbaev,) Nauka, Kazhakstan, 1977.

83 D. D. Miller, et al., Analysis of a residue class core function of Akushskii, Burcev and Pak. in *RNS Arithmetic: Modern Applications in DSP*, (ed. F G. A. Jullien), IEEE Press, 1986.

84 N. Burgess, Scaled and unscaled residue number systems to binary conversion techniques using the core function, *Proc.13th IEEE Symp. Comput. Arithmetic*, 1997, pp. 250–257.

85 N. Burgess, Scaling a RNS number using the core function, *Proc.16th IEEE Symp. Comput. Arithmetic*, 2003, pp. 262–269.

86 G. Dimauro, S. Impedevo, R. Modugno, G. Pirlo, and R. Stefanelli, Residue to binary conversion by the quotient function, *IEEE Trans. Circuits Syst., Part II*, vol. 50, 2003, pp. 488–493.

87 S. Bi and W. J. Gross, The mixed-radix Chinese remainder theorem and its applications to residue comparison, *IEEE Trans. Comput.*, vol. 57, 2008, pp. 1624–1632.

88 P. V. Ananda Mohan, Reverse converters for the moduli sets $\{2^{2n} - 1, 2^n, 2^{2n} + 1\}$ and $\{2^n - 3, 2^n + 1, 2^n - 1, 2^n + 3\}$, *SPCOM*, Bangalore, 2004, pp. 188–192.

89 P. V. Ananda Mohan, Reverse converters for a new moduli set $\{2^{2n} - 1, 2^n, 2^{2n} + 1\}$, *CSSP*, vol. 26, 2007, pp. 215–228.

90 P. V. Ananda Mohan, RNS to binary converter for a new three moduli set $\{2^{(n+1)} - 1, 2^n, 2^n - 1\}$, *IEEE Trans. Circ. Syst., Part II*, vol. 54, 2007, pp. 775–779.

91 P. V. Ananda Mohan, New residue to binary converters for the moduli set $\{2^k, 2^k-1, 2^{k-1}-1\}$, *IEEE TENCON* 2008, Digital Object Identifier: DOI 10.1109/TENCON. 2008.4766524

92 R. Chaves and L. Sousa, $\{2^n + 1, 2^{n+k}, 2^n - 1\}$: A new RNS moduli set extension, *Proc. Euro Micro Systems on Digital System Design*, 2004, pp. 210–217.

93 A. A. Hiasat, and H. S. Abdel-Aty-Zohdy, Residue to binary arithmetic converter for the moduli set $(2^k, 2^k-1, 2^{k-1}-1)$. *IEEE Trans. Circuits Syst., Part II,* vol. 45, 1998, pp. 204–209.

94 W. Wang, M. N. S. Swamy, M. O. Ahmad, and Y. Wang, A high-speed Residue-to-binary converter for thee moduli $\{2^k, 2^k-1, 2^{k-1}-1\}$RNS and a scheme for its VLSI implementation, *IEEE Trans. Circuits Syst., Part-II,* vol. 47, 2000, pp. 1576–1581.

95 W. Wang, M. N. S. Swamy, M. O. Ahmad, and Y. Wang, A note on a high-speed residue-to-binary converter for thee moduli $\{2^k, 2^k-1, 2^{k-1}-1\}$ RNS and a scheme for its VLSI implementation, *IEEE Trans. Circuits Syst., Part-II,* vol. 49, p. 230, 2002.

96 A. P. Vinod and A. B. Premkumar, A residue to binary converter for the 4-moduli superset $\{2^n - 1, 2^n, 2^n + 1, 2^{n+1}-1, JCSC,$ vol. 10, 2000, pp. 85–99.

97 M. Bhardwaj, T. Srikanthan, and C. T. Clarke, A reverse converter for the 4 moduli super set $\{2^n - 1, 2^n, 2^n + 1, 2^{n+1} + 1\}$, IEEE Conf. Comput. Arithmetic, pp. 1999.

98 B. Cao, T. Srikanthan, and C. H. Chang, Efficient reverse converters for the four-moduli sets $\{2^n - 1, 2^n, 2^n + 1, 2^{n+1}-1\}$ and $\{2^n - 1, 2^n, 2^n + 1, 2^{n-1}-1\}$, *IEE Proc. Comput. Digit. Tech.,* vol. 152, 2005, pp. 687–696.

99 P. V. Ananda Mohan and A. B. Premkumar, RNS to binary converters for two four moduli sets $\{2^n - 1, 2^n, 2^n + 1, 2^{n+1}-1\}$ and $\{2^n - 1, 2^n, 2^n + 1, 2^{n+1} + 1\}$, *IEEE Trans. Circuits Syst., Part I,* vol. 54, 2007, pp. 1245–1254.

100 M. H. Sheu, S. H. Lin, C. Chen, and S. W. Yang, An efficient VLSI design for a residue to binary converter for general balance moduli $(2^n-3, 2^n - 1, 2^n + 1, 2^n + 3)$, *IEEE Trans. Circuits Syst., Express Briefs,* vol. 51, 2004, pp. 52–55.

101 P. V. Ananda Mohan, New reverse converters for the moduli set $\{2^n-3, 2^n + 1, 2^n - 1, 2^n + 3\}$, *AEU,* vol. 62, 2008, pp. 643–658.

102 B. Cao, C. H. Chang, and T. Srikanthan, An efficient reverse converter for the 4-moduli set $\{2^n - 1, 2^n, 2^n + 1, 2^{2n} + 1\}$ based on the new Chinese remainder theorem, *IEEE Trans. Circuits Syst., Part-I,* vol. 50, 2003, pp. 1296–1303.

103 A. S. Molahosseini, K. Navi, C. Dadkhah, O. Kavehei, and S. Timarchi, Efficient reverse converter designs for the new 4-moduli sets $\{2^n - 1, 2^n, 2^n + 1, 2^{2n+1}-1\}$ and $\{2^n - 1, 2^n + 1, 2^{2n}, 2^{2n} + 1\}$ based on new CRTs, *IEEE Trans. Circuits Syst., Part I,* vol. 57, 2010, pp. 823–835.

104 B. Cao, C. H. Chang, and T. Srikanthan, A residue to binary converter for a new five-moduli set, *IEEE Trans. Circuits Syst., Part I,* vol. 54, 2007, pp. 1041–1049.

105 A. Skavantzos and T. Stouraitis, Grouped-moduli residue number systems for Fast signal processing. *Proc. IEEE ISCAS,* pp. 478–483, 1999.

106 A. Skavantzos and M. Abdallah, Implementation issues of the two-level residue number system with pairs of conjugate moduli, *IEEE Trans. Signal Process.,* vol. 47, 1999, pp. 826–838.

107 A. Skavantzos, M. Abdallah, T. Stouraitis, and D. Schinianakis, Design of a balanced 8-modulus RNS. *Proc. IEEE ISCAS,* pp. 61–64, 2009.

108 H. Pettenghi, R. Chaves, and L. Sousa, RNS reverse converters for moduli sets with dynamic ranges up to $(8n + 1)$ bits, *IEEE Trans. Circuits Syst.*, vol., pp., 2012.

109 A. P. Shenoy and R. Kumaresan, Fast base extension using a redundant modulus in RNS. *IEEE Transactions on Computers*, vol. 38, 1989, pp. 293–297.

110 A. P. Shenoy and R. Kumaresan, A fast and accurate scaling technique for high-speed signal processing, *IEEE Trans. Acoust., Speech Signal Process.*, vol. 37, 1989, pp. 929–937.

111 G. A. Jullien, Residue number scaling and other operations using ROM arrays, *IEEE Trans. Comput.*, vol. 27, no. 4, pp. 325–337, 1978.

112 A. Garcia and A. Lloris, A look up scheme for scaling in the RNS, *IEEE Trans. Comput.*, vol. 48, 1999, pp. 748–751.

113 F. Barsi and M. C. Pinotti, Fast base extension and precise scaling in RNS for look-up table implementation, *IEEE Trans. Signal Process.*, vol. 43, 1995, pp. 2427–2430.

114 Y. Kong and B. Phillips, Fast scaling in the residue number system, *IEEE Trans. VLSI Syst.*, vol. 17, 2009, pp. 443–447.

115 C. H. Chang and J. Y. S. Low, Simple, fast and exact RNS scaler for the three moduli set $\{2^n - 1, 2^n, 2^n + 1\}$, *IEEE Trans. Circuits Syst., Part I: Regul. Pap.*, vol. 58, 2011, pp. 2686–2697.

116 T. F. Tay, C. H. Chang, and J. Y. S. Low, Efficient VLSI implementation of 2^n scaling of signed integers in RNS $\{2^n - 1, 2^n, 2^n + 1\}$, *IEEE Trans. VLSI Syst.*, vol. 21, 2012, pp. 1936–1940.

117 G. Dimauro, S. Impedovo, and G. Pirlo, A new technique for fast number comparison in the residue number system, *IEEE Trans. Comput.*, vol. 42, 1993, pp. 608–612.

118 Z. D. Ulman, Sign detection and implicit explicit conversion of numbers in residue arithmetic, *IEEE Trans. Comput.*, vol. 32, 1983, pp. 5890–594.

119 T. Tomczak, Fast sign detection for RNS $(2^n - 1, 2^n, 2^n + 1)$, *IEEE Trans. Circuits Syst. I: Regul. Pap.*, vol. 55, 2008, pp. 1502–1511.

120 W. K. Jenkins, Residue number system error checking using expanded projections, *Electron. Lett.*, vol. 18, 1982, pp. 927–928.

121 W. K. Jenkins, The design of error checkers for self-checking residue number arithmetic, *IEEE Trans. Comput.*, vol. 32, 1983, pp. 388–396.

122 W. K. Jenkins and E. J. Altman, Self-checking properties of residue number error checkers based on mixed radix conversion. *IEEE Trans. Circuits Syst.*, vol. 35, 1988, pp. 159–167.

123 V. T. Goh and M. U. Siddiqui, Multiple error detection and correction based on redundant residue number systems, *IEEE Trans. Commun.*, vol. 56, 2008, pp. 325–330.

124 W. K. Jenkins and B. J. Leon, The use of residue number systems in the design of finite impulse response digital filters, *IEEE Trans. Circuits Syst.*, vol. CAS-24, 1977, pp. 191–201.

125 A. Peled and B. Liu, A new hardware realization of digital filters, *IEEE Trans. Acoust., Speech Signal Process.*, vol. ASSP-22, 1974, pp. 456–462.

126 W. L. Freking, and K. K. Parhi, Low-power FIR digital filters using residue arithmetic, *Conf. Record 31st Asil. Conf. Signals, Syst. and Comput. (ACSSC 1997)*, vol. 1, Pacific Grove, CA, USA 1997, pp. 739–743.

127 H. T. Vergos, A 200 MHz RNS core, *Proc. ECCTD*, vol. II, 2001, pp. 249–252.

128 D. Re, A. Nannarelli and M. Re, Implementation of digital filters in carry save Residue number system. *Conference Record 39th Asilomar Conf. Signals, Syst. Comput.*, 2001, pp. 1309–1313.

129 A. Nannarelli, G. C. Cardiralli, and M. Re, Power-delay trade-off in residue number system, *Proc. IEEE ISCAS*, vol. 5, 2003, pp. 413–416.

130 G. A. Jullien, R. Krishnan, and W. C. Miller, Complex digital signal processing over finite rings, *IEEE Trans. Circuits Syst.*, vol. 34, 1987, pp. 365–377.

131 W. K. Jenkins and J. V. Krogmeier, The design of dual-mode complex signal processors based on quadratic modular number codes, *IEEE Trans. Circuits Syst.*, vol. 34, 1987, pp. 354–364.

132 W. Stallings, *Cryptography and Network security*, Pearson Higher Education, 2013.

133 A. Menezes, P. van S Oorschot, and S. Vanstone, *Handbook of Applied Cryptography*, CRC Press, 1996.

134 P. L. Montgomery, Modular multiplication without trial division, *Math. Comput.*, vol. 44, 1985, pp. 519–521.

135 K. C. Posch and R. Posch, Modulo reduction in residue number systems. *IEEE Trans. Parallel Distrib. Syst.*, vol. 6, 1995, pp. 449–454.

136 C. Bajard, L. S. Didier, and P. Kornerup, An RNS Montgomery modular multiplication algorithm, *IEEE Trans. Comput.*, vol. 47, 1998, pp. 766–776.

137 J. C. Bajard, and L. Imbert, A full RNS implementation of RSA, *IEEE Trans. Comput.*, vol. 53, 2004, pp. 769–774.

138 H. Nozaki, M. Motoyama, A. Shimbo, and S. Kawamura, Implementation of RSA algorithm based on RNS Montgomery multiplication. in *Cryptographic Hardware and Embedded Systems–CHES*, (ed. F C. Paar), Springer-Verlag, Berlin, Germany, 2001, pp. 364–376.

139 S. Kawamura, M. Koike, F. Sano, and A. Shimbo, Cox-Rower architecture for fast parallel Montgomery multiplication, *Proc. of Int. Conf. Theory Appl. Crypt. Tech.: Adv. Cryptol.*, (EUROCRYPT), 2000, pp. 523–538.

140 F. Gandino, F. Lamberti, G. Paravati, J. C. Bajard and P. Montuschi, An algorithmic and architectural study on Montgomery exponentiation in RNS, *IEEE Trans. Comput.*, vol. 61, 2012, pp. 1071–1083.

141 D. Schinianakis and T. Stouraitis, A RNS Montgomery multiplication architecture, *Proc. IEEE ISCAS*, 2011, pp. 1167–1170.

142 D. M. Schinianakis, A. P. Kakarountas, and T. Stouraitis, A new approach to elliptic curve cryptography: an RNS architecture. *IEEE MELECON*, May 16–19, Benalmádena (Málaga), Spain, 2006, pp. 1241–1245.

143 D. M. Schinianakis, A. P. Fournaris, H. E. Michail, A. P. Kakarountas, and T. Stouraitis, An RNS implementation of an F_p elliptic curve point multiplier, *IEEE Trans. Circuits Syst. I: Regul. Pap.*, vol. 56, 2009, pp. 1202–1213.

144 J. C. Bajard, S. Duquesne, and M. Ercegovac, Combining leak resistant arithmetic for elliptic curves defined over F_p and RNS representation, *Cryptology Reprint archive* 311 2010.

145 A. Joux, A one round protocol for tri-partite Diffie-Hellman, *Algorithmic Number theory*, LNCS, 2000, pp. 385–394.

146 D. Boneh and M. K. Franklin, Identity based encryption from the Weil Pairing, *Crypto*, LNCS, vol. 2139, 2001, pp. 213–229.

147 D. Boneh, B. Lynn, and H. Shachm, Short signatures for the Weil pairing, *J. Cryptol.*, vol. 17, 2004, pp. 297–319.

148 J. Groth and A. Sahai, Efficient non-interactive proof systems for bilinear groups, *27th Ann. Int. Conf. Adv. Cryptol., Eurocrypt*, 2008, pp. 415–432.

149 G. X. Yao, J. Fn, R. C. C. Cheung, and I. Verbauwhede, Novel RNS parameter selection for fast modular multiplication, *IEEE Trans. Comput.*, vol. 63, 2014, pp. 2099–2105.

150 G. X. Yao, J. Fan, R. C. C. Cheung, and I. Verbauwhede, A high speed pairing Coprocessor using RNS and lazy reduction. *eprint.iacr.org*/2011/258.pdf. 2011.

151 S. Duquesne and N. Guillermin, A FPGA pairing implementation using the residue number System, *Cryptology ePrint Archive*, Report 2011/176, 2011, IR http://eprint.iacr.org/

152 P. Barreto and M. Naehrig, Pairing friendly elliptic curves of prime order, *SAC, 2005, LNCS, 3897, 2005*, pp. 319–331.

153 S. Duquesne, RNS arithmetic in F_{p^k} and application to fast pairing computation, *Cryptology ePrint Archive*, Report 2010/55. 2010, IR http://eprint.iacr.org

154 A. Miyaji, M. Nakabayashi, and S. Takano, New explicit conditions of elliptic curve traces for FR-reduction, *IEICE Trans. Fundamental.*, vol. 84, 2001, pp. 1234–1243.

155 L. L. Yang and L. Hanjo, Residue number system based multiple-code DS-CDMA system, *IEEE 49th Veh. Technol. Conf., Vol. 2*, 1999, pp. 1450–1454.

156 L. L. Yang and L. Hanjo, Ratio statistic assisted residue number system based parallel communication scheme, *IEEE 49th Veh. Technol. Conf.*, vol. 2, 1999, pp. 894–898.

157 L. L. Yang and L. Hanzo, A residue number system based parallel communication scheme using orthogonal signaling: part I - system outline, *IEEE Trans. Veh. Technol.*, vol. 51, 2002, pp. 1534-1546.

158 L. L. Yang and L. Hanzo, Residue number system assisted fast frequency-hopped synchronous ultra-wideband spread-spectrum multiple-access: a design alternative to impulse radio. *IEEE J. Select. Areas Commun.*, vol. 20, 2002, pp. 1652–1663.

159 S. Zhang, L. L. Yang, and Y. Zhang, Redundant residue number system assisted multicarrier direct-sequence code-division dynamic multiple access for cognitive radios, *IEEE Trans. Veh. Technol.*, vol. 61, 2012, pp. 1234–1250.

160 A. S. Madhukumar and F. Chin, Enhanced architecture for residue number system-based CDMA for high-rate data transmission, *IEEE Trans. Wirel. Commun.*, vol. 3, 2004, pp. 1363–1368.

7

Logarithmic Number System

Vassilis Paliouras[1] and Thanos Stouraitis[1,2]

[1] *Electrical and Computer Engineering Department, University of Patras, Greece*
[2] *Electrical and Computer Engineering Department, Khalifa University, UAE*

7.1 Introduction

7.1.1 The Logarithmic Number System

The representation of data as logarithms is long used as a means to simplify particular arithmetic operations and utilize improved numerical properties, even before the modern digital computer era. The logarithmic number system (LNS) is a formalization of the logarithmic representation of the data in a digital system and it can be conceived as a generalization of the floating-point representation. Swartzlander and Alexopoulos described the basics of the sign/logarithm arithmetic [1].

7.1.2 Organization of the Chapter

This chapter is organized as follows: The basics of the logarithmic representation are reviewed in Section 7.2. Section 7.3 describes the basic operations in logarithmic arithmetic. Forward and inverse conversion issues are studied in Section 7.4. Complex arithmetic with LNS is studied in Section 7.5. Processors based on the LNS are listed in Section 7.6. The impact of logarithmic arithmetic on the power dissipation is studied in Section 7.7, while several applications are discussed in Section 7.8. Finally, conclusions are discussed in Section 7.9.

Arithmetic Circuits for DSP Applications, First Edition. Edited by Pramod Kumar Meher and Thanos Stouraitis.
© 2017 by The Institute of Electrical and Electronics Engineers, Inc. Published 2017 by John Wiley & Sons, Inc.

7.2 Basics of LNS Representation

An LNS is defined as follows: A real number x is represented in a base-b logarithmic number system, as a triplet $\mathcal{X}$,

$$x \xrightarrow{\text{LNS}} \mathcal{X} \tag{7.1}$$

where $\mathcal{X}$ is defined as

$$\mathcal{X} = \{s_x, z_x, X\} \tag{7.2}$$

where s_x is a flag that denotes the sign of x, z_x is a zero flag that denotes whether x is a zero or not, and

$$X = \log_b |x| \tag{7.3}$$

While the choice $b = 2$ is common, base b of the logarithm has been proven to be a design parameter, the optimal choice of which can lead to hardware complexity minimization and power dissipation reduction [2].

7.2.1 LNS and Equivalence to Linear Representation

Traditionally, LNS has been considered as an alternative to floating-point representation [3]. Lewis discusses the equivalence of floating-point and LNS [4]. However, in this section, LNS is compared to an n-bit linear fixed-point representation and it is shown to provide substantial improvement in terms of power dissipation.

Two are the main issues in a finite word length number system, namely, the range of the numbers which can be represented and the precision of the representation [3]. The representational equivalence of an n-bit linear fixed-point system and of an LNS needs to be investigated, as the two representations differ in both range and precision behavior. In this section, two representations are considered equivalent, when they (i) cover equal data ranges and (ii) exhibit equal average representational error.

Let k and l denote the word length of the integer and fractional part of an LNS word, respectively. Let (k, l, b)-LNS denote an LNS of integer and fractional word length k and l, respectively, and of base b. The problem of equivalence between a (k, l, b)-LNS and an n-bit linear fixed-point system is to compute k and l in such a way that the two number representations satisfy a suitably defined criterion, for a particular base b.

The relative representational error, ε_{rel}, of a real number A is

$$\varepsilon_{\text{rel}} = \frac{|A - \widehat{A}|}{A} \tag{7.4}$$

where A is the actual value and $\widehat{A}$ is the value representable in the system. Notice that $A \neq \widehat{A}$ due to the finite length of the words. The relative representational error $\varepsilon_{\text{rel,LNS}}$, for an (k, l, b)-LNS is given by ([3], for $b = 2$)

$$\varepsilon_{\text{rel,LNS}} = b^{2^{-l}} - 1 \tag{7.5}$$

In the n-bit linear fixed-point case, Eq. (7.4) results in a relative error

$$\varepsilon_{\text{rel,FXP}} = 2^{-n}/A \tag{7.6}$$

where A denotes an n-bit fixed-point number. From Eqs (7.5) and (7.6), it can be noticed that $\varepsilon_{\text{rel,FXP}}$ depends on A, while $\varepsilon_{\text{rel,LNS}}$ does not.

The average representational error, ε_{ave}, is defined as

$$\varepsilon_{\text{ave}} = \frac{\sum_{A=A_{\min}}^{A_{\max}} \varepsilon_{\text{rel}}(A)}{A_{\max} - A_{\min} + 1} \tag{7.7}$$

where $A_{\min}$ and $A_{\max}$ define the range of representable numbers.

Due to definition (Eq. (7.7)), the average representational error for the fixed-point case is given by

$$\varepsilon_{\text{ave,FXP}} = \frac{1}{2^n - 1} \sum_{i=1}^{2^n - 1} \frac{1}{i} = \frac{\psi(2^n) + \gamma}{2^n - 1} \tag{7.8}$$

where γ is the Euler gamma constant and $\psi(x) = d(\ln \Gamma(x))/dx$, where $\Gamma(x)$ is the Euler gamma function. In LNS, as $\varepsilon_{\text{rel,LNS}}$ is constant, Eq. (7.5) dictates that

$$\varepsilon_{\text{ave,LNS}} = b^{2^{-l}} - 1 \tag{7.9}$$

The maximum number representable by an n-bit linear integer, $A_{\max}^{\text{FXP}}$, and by the (k, l, b)-LNS encoding of Eq. (7.1), $A_{\max}^{\text{LNS}}$, are

$$A_{\max}^{\text{FXP}} = 2^n - 1 \tag{7.10}$$

$$A_{\max}^{\text{LNS}} = b^{2^k + 1 - 2^{-l}} \tag{7.11}$$

Therefore, according to the equivalence criteria posed earlier, an LNS is equivalent to an n-bit linear fixed-point representation, when the following restrictions are simultaneously satisfied:

$$A_{\max}^{\text{LNS}} \geq A_{\max}^{\text{FXP}} \Leftrightarrow b^{2^k + 1 - 2^{-l}} \geq 2^n - 1 \tag{7.12}$$

$$\varepsilon_{\text{ave,LNS}} \leq \varepsilon_{\text{ave,FXP}} \Leftrightarrow b^{2^{-l}} - 1 \leq \frac{\psi(2^n) + \gamma}{2^n - 1} \tag{7.13}$$

obtained using Eqs. (7.8)–(7.11); when solved for k and l, Eqs. (7.12) and (7.13) give

$$l = \left\lceil -\log_2 \log_b (1 + \frac{\psi(2^n) + \gamma}{2^n - 1}) \right\rceil \tag{7.14}$$

$$k = \left\lceil \log_2 \left(\log_b (2^n - 1) + 2^{-l} - 1 \right) \right\rceil \tag{7.15}$$

The above analysis leads to introducing the following theorem.

Theorem 7.1 A (k, l, b)-LNS covers a range at least as long as an n-bit fixed-point system with an average representational error equal or smaller to that of the fixed-point system, when l and k are given by Eqs. (7.14) and (7.15), respectively.

While the word lengths k and l computed via Eqs. (7.14) and (7.15) meet the Eqs. (7.12) and (7.13), LNS covers a larger range than the equivalent fixed-point representation. Let n_{eq} denote the word length of a fixed-point system that can cover the range offered by an LNS defined through Eqs. (7.14) and (7.15), or, equivalently, let n_{eq} be the smallest integer which satisfies $2^{n_{eq}} - 1 \geq b^{2^k + 1 - 2^{-l}}$; thus,

$$n_{eq} = \left\lceil (2^k + 1 - 2^{-l}) \log_2 b \right\rceil \tag{7.16}$$

Notice that, when $n_{eq} \geq n$, the precision of the particular fixed-point system is better than that of the LNS derived by Eqs. (7.14) and (7.15).

7.3 Fundamental Arithmetic Operations

The use of a logarithmic representation of the data has an impact on the way that arithmetic operations are performed. In order to reduce conversion cost, operations need to be redefined, so that, given arguments in the logarithmic domain, the result is obtained in the logarithmic format. Some of the basic arithmetic operations are simplified, while others become more involved. In the following, the basic logarithmic arithmetic operations are discussed and a summary is offered in Table 7.1.

7.3.1 Multiplication, Division, Roots, and Powers

Assume that we seek the product $\mathcal{Z}$ of two LNS numbers $\mathcal{X}$ and $\mathcal{Y}$, which are the images of x and y. The zero flag z is asserted when at least one of the product

Table 7.1 Impact of LNS on arithmetic operations.

Linear Operation	Logarithmic
$Z = XY = b^x b^y = b^{x+y}$	$z = \log_b Z = x + y$
$Z = X/Y = b^x/b^y = b^{x-y}$	$z = x - y$
$Z = \sqrt[m]{X} = \sqrt[m]{b^x} = b^{\frac{x}{m}}$	$z = x/m, \quad m, \text{integer}$
$Z = X^m = (b^x)^m$	$z = mx, \quad m, \text{integer}$
$Z = X + Y = b^x + b^y = b^x(1 + b^{y-x})$	$z = x + \log_b(1 + b^{y-x})$
$Z = X - Y = b^x - b^y = b^x(1 - b^{y-x})$	$z = x + \log_b(1 - b^{y-x})$

factors is zero, therefore $z = z_x$ or z_y. The sign of the product is computed as $s = s_x$ xor s_y. By exploiting properties of the logarithm, we write

$$Z = \log_b |z| = \log_b |xy| \tag{7.17}$$
$$= \log_b |x| + \log_b |y| \tag{7.18}$$
$$= X + Y \tag{7.19}$$

Therefore, the computation of the product of two numbers in LNS is reduced to simple two-input logic operations to derive the sign and zero flags, as well as the addition $Z = X + Y$. Similarly, division in LNS can be performed as subtraction $D = X - Y$.

Suppose that we want to compute the nth power of x, denoted as P, in LNS,

$$P = \log_b |x^n| \tag{7.20}$$
$$= \log_b |x|^n \tag{7.21}$$
$$= n \log_b |x| = nX \tag{7.22}$$

Equation (7.22) reveals that exponentiation is reduced to multiplication. The special cases of square and square root can be computed as $P = 2X$ and $P = X/2$, respectively; these operations are implemented in digital hardware as simple shifts.

7.3.2 Addition and Subtraction

Assume that the logarithmic expression of the sum of two numbers x and y is sought and their logarithmic representation is available.

$$S = \log_b |x + y| \tag{7.23}$$
$$= \log_b \left| b^X + b^Y \right| \tag{7.24}$$

Assuming that $Y > X$, it follows that

$$S = \log_b |b^Y||b^{X-Y} + 1| \tag{7.25}$$
$$= Y + \log_b \left| b^{X-Y} + 1 \right| \tag{7.26}$$

Let d be defined as $d = -|X - Y|$. Then, Eq. (7.26) is written as

$$S = Y + \log_b |b^d + 1| = Y + f_a(d), \tag{7.27}$$

where $f_a(d) = \log_b |b^d + 1|$. Similarly, for the operation of subtraction, it follows that the difference of two numbers is given by

$$D = Y + \log_b |b^d - 1| = Y + f_s(d) \tag{7.28}$$

where $f_s(d) = \log_b |b^d - 1|$. While multiplication, division, roots, and powers can be implemented as binary operations, the operations of addition and subtraction require the computation of the nonlinear functions $f_a(d)$ and $f_s(d)$, respectively. Substantial amount of the LNS-related research focuses on the simplification of the evaluation of the nonlinear functions required for addition and subtraction. Complexity reduction is sought by optimizing the required wordlengths and the logarithmic base b and by using various types of approximation, as detailed in the next section. Notably, the implementation of subtraction is a great challenge due to the singularity of the function $f_s(d)$ at the point $d = 0$,

$$\lim_{d \to 0} f_s(d) = -\infty \tag{7.29}$$

7.3.2.1 Direct Techniques and Approximations

Several LNS algorithms and architectures have been proposed in the literature. Swartzlander *et al.* have presented an architecture for FFT using ROM lookup-based LNS operations [5]. Taylor *et al.* have presented a 20-bit LNS processor, featuring a memory compression scheme, which reduces the memory size by 75% over the direct lookup-table implementation. The processor contains 83.55 Kbits of ROM [6]. Henkel has suggested an improved algorithm for logarithmic addition, based on Chebyshev approximation with unequally spaced partition points [7]. This approach leads to significant memory reduction but holds for addition only. Lewis and Yu have developed a 30-bit LNS processor with a total memory size of 260 Kbits [8]. They use a linear interpolation-based method to reduce the required memory for addition and subtraction. Orginos *et al.* have presented a polynomial approximation-based architecture for multioperand addition/subtraction [9].

7.3.2.2 Cancellation of Singularities

Over the last years, LNS has attracted a lot of interest, because it can lead to VLSI architectures that demonstrate large dynamic range and precision, and simple and fast multiplication, division, roots, and powers [10]. However, LNS addition and subtraction are complicated, since they require the computation of the functions

$$s_b(x) = \log_b(1 + b^{-x}) \tag{7.30}$$

and

$$d_b(x) = \log_b(1 - b^{-x}) \tag{7.31}$$

respectively, while conversions to/from a logarithmic processor are needed and introduce inaccuracies as well as computational overhead [11]. This paragraph describes an algorithm for the computation of LNS subtraction. The main feature of the particular algorithm is that it circumvents the evaluation of the highly nonlinear function traditionally used in logarithmic subtraction. In some cases, this approach permits a reduction of more than 70% in memory requirements, thus making the realization of cost-effective LNS arithmetic units of very high accuracy feasible. The approach is extended to address certain algorithms that need to be implemented by an LNS processor. Moreover, alternative architectures that implement the computational approach and exhibit different time–area requirements are discussed. For the sake of completeness, upper bounds to the computational errors are offered and a memory optimization procedure is outlined. The computation of the function $\log_b(1 - b^{-x})$ is decomposed into the concurrent computation of a logarithm and the function $\log_b \left(\frac{1-b^{-x}}{x} \right)$. This approach is shown to greatly decrease the memory requirements of an LNS processor. While we focus on the case in which both the logarithm and $\log_b \left(\frac{1-b^{-x}}{x} \right)$ are computed by polynomial approximation, other approximation methods can also be applied depending on the specified accuracy.

Polynomial evaluation is of interest because of its fast convergence and because it leads to regular, modular, and pipelinable implementations with local interconnections. Moreover, piecewise polynomial approximation achieves sufficient accuracy with very low-degree polynomials. While speed improvements are achieved in this way, the area requirements increase with the number of intervals, since more coefficients need to be stored.

LNS Subtraction and Related Algorithms

In LNS, it is awkward to compute the difference of two almost equal numbers. A similar computational problem exists in the evaluation of the logarithm of any function with zeros at points within a range of interest. In this section, a technique for avoiding this problem is discussed. Let $f(x)$ be a function such that $f(0) = 0$ and $f'(0) \neq 0$, where $f'(x)$ is the first-order derivative of $f(x)$. Assume that the computation of $\log_2 f(x)$ is sought. Since it holds

$$\lim_{x \to 0} \log_2 |f(x)| = -\infty \tag{7.32}$$

the numerical evaluation of $\log_2 |f(x)|$ for x approaching zero can be difficult. However, by using the De l'Hospital rule, the following is obtained:

$$\lim_{x \to 0} \log_2 \left| \frac{f(x)}{x} \right| = \lim_{x \to 0} \log_2 \left| \frac{f'(x)}{1} \right| = \lim_{x \to 0} \log_2 |f'(x)| = \log_2 |f'(0)| \in \mathfrak{R},$$
$$\tag{7.33}$$

where $\mathfrak{R}$ denotes the set of the real numbers. Function $\log_2 f(x)$ can be computed as

$$\log_2 f(x) = \log_2(x \frac{f(x)}{x}) = \log_2 x + \log_2 \frac{f(x)}{x} \tag{7.34}$$

Both terms of the right side of Eq. (7.34) are easier to compute than the original function. Note that while $\log_2 x$ is singular at $x = 0$, its computation does not impose significant accuracy loss. Specifically, accurate computation of $\log_2 x$ for x close to zero can be done by representing x as a couple (m, I), such that

$$x = m2^I \tag{7.35}$$

where $m \in [0.5, 1)$ and I is an integer. Then it holds that

$$\log_2 x = I + \log_2 m \tag{7.36}$$

Common functions for which $f(0) = 0$ and $f'(0) \neq 0$ are $1 - 2^{-x}$, which is used for LNS subtraction, $\sin x$, and $\tan x$. In the following, the use of Eq. (7.34) in the computation of $\log_2(1 - 2^{-x})$ is demonstrated. The evaluation of other functions for which $f(x) = 0$ and $f'(0) \neq 0$ is similar.

Let A, B be two real nonzero numbers, which are represented in LNS as couples (s_A, a) and (s_B, b), where s_A and s_B are the signs of A and B, respectively, $a = \log_2 |A|$ and $b = \log_2 |B|$. The subtraction requires the computation of the function

$$f_s(x) = \log_2 (1 - 2^{-x}) \tag{7.37}$$

where $x = a - b$ [12]. Resembling the accuracy loss of the floating-point subtraction in the case of almost equal operands, it can be seen that Eq. (7.37) rapidly drops to $-\infty$ when $x \to 0$, or

$$\lim_{x \to 0} f_s(x) = -\infty \tag{7.38}$$

To alleviate the fast decay of $f_s(x)$, Eq. (7.33) can be utilized. In particular, Eq. (7.37) can be written as

$$f_s(x) = \log_2 (1 - 2^{-x}) = \tag{7.39}$$
$$= \log_2 \left(\frac{1 - 2^{-x}}{x} x \right) =$$
$$= \log_2 \left(\frac{1 - 2^{-x}}{x} \right) + \log_2 x$$

The first term $f_{s'}(x) = \log_2 \left(\frac{1 - 2^{-x}}{x} \right)$ of the sum has the property

$$\lim_{x \to 0} f_{s'}(x) = \lim_{x \to 0} \log_2 \left(\frac{1 - 2^{-x}}{x} \right) = \log_2(\ln 2)$$

The functions $f_s(x)$ and $f_{s'}(x)$ are plotted versus x in Figure 7.1. It can be seen that $f_{s'}(x)$ does not exhibit any singularity. By utilizing Eqs. (7.36)

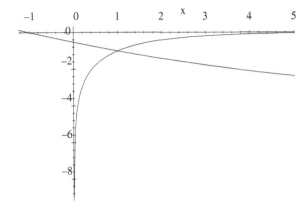

Figure 7.1 The functions $f_s(x)$ (solid) and $f_{s'}(x)$ (dashed).

and (7.39), an alternative approach to the computation of Eq. (7.37) can be derived

$$f_s(x) = \begin{cases} \log_2\left(\frac{1-2^{-x}}{x}\right) + \log_2 m + I, & x = m2^I, \, x \in R_1 \\ \log_2(1 - 2^{-x}), & x \in R_0 \end{cases} \tag{7.40}$$

The determination of the intervals R_1 and R_0 can be formulated as an optimization problem. Functions $\log_2(\frac{1-2^{-x}}{x})$, $\log_2 m$, and $\log_2(1 - 2^{-x})$ can be evaluated by using an Nth-order piecewise polynomial approximation.

Let $x_{\text{essential}}$ be the value, where the essential zero occurs, that is, for $x > x_{\text{essential}}$, it holds that $|f(x)| < 2^{-f}|$, where f is the number of fractional bits used in the representation and $f(x)$ is the function to be approximated. Assume that the sequence $\{d_i\}$ divides the interval of $(0, x_{\text{essential}})$ into ranges $[d_i, d_{i+1})$ of approximation intervals $[x_j, x_{j+1})$ with length h_i. In other words, all the approximation intervals $[x_j, x_{j+1})$, such that $d_i \leq x_j < x_{j+1} < d_{i+1}$, have the property $x_{j+1} - x_j = h_i$.

If piecewise second-order polynomial approximation is assumed for the computation of $\log_2(\frac{1-2^{-x}}{x})$, $\log_2 m$, and $\log_2(1 - 2^{-x})$, Eq. (7.40) can be written as

$$\widehat{f}_s(x) = \begin{cases} \displaystyle\sum_{j=0}^{2} \widehat{a}_j(x_i)x^j + \sum_{j=0}^{2} \widehat{b}_j(m_i)m^j + I, & x = m2^I, \, x \in R_1 \\ \displaystyle\sum_{j=0}^{2} \widehat{a}'_j(x_i)x^j, & x \in R_0 \end{cases} \tag{7.41}$$

where $\widehat{a}_j(x_i), \widehat{b}_j(m_i)$, and $\widehat{a}'_j(x_i)$ are the coefficients for the approximation of $\log_2\frac{1-2^{-x}}{x}$, $\log_2(1 - 2^{-x})$, and $\log_2 m$, respectively.

Table 7.2 A schedule for the two-MAC pipelined LNS subtractor.

Time\ Stage	0	1	2	3	4	5	6
1	Initialize						
2		$a_1 x$					
3		$a_2 x$	$a_1 x$				
4			$a_2 x$	$a_1 x + b_1$			
5				$a_2 x + b_2$	$A_1 x$		
6					$A_2 x$	$A_1 x$	
7						$A_2 x$	$A_1 x + C$
8							Sum

Architectures for LNS Subtraction

A variety of architectures can implement the algorithm. These architectures comprise a few basic building blocks, namely, multipliers, adders, look-up tables, address generators, and a normalizer. The normalizer returns two numbers, namely, the number of leading zeros of the input and the normalized input.

High throughput rates can be achieved by a pipelined implementation of the particular algorithm. In particular, a seven-stage implementation is described. At stage 0, the normalization, address generation, table look-up, and u-transform are performed. The operands are subsequently processed in a six-stage data path, which comprises a couple of two-stage pipelined multipliers and two adders. The scheduling of the pipelined data path is shown in Table 7.2. According to the schedule of Table 7.2, the approximation of $f_s(x)$ or $f_{s'}(x)$ is "multiplexed" with the approximation of $\log_2 x$ during the six stages of the data path. When stage i computes a partial result of the approximation of $f_{s'}(x)$ or $f_s(x)$, then stage $i - 1$ computes a partial result of the approximation of $\log_2 x$. Finally, in stage 6 at the eighth clock, the two partial results are added. The clock period is determined by the delay of a stage of the pipelined multiplier. New results can be computed at a rate of two clock periods. The latency of the pipelined LNS subtractor is eight clock periods. The pipelined two-MAC architecture is depicted in Figure 7.2.

Memory Requirements and Comparisons

The main advantage of the particular algorithm is that it can lead to architectures with reduced memory requirements. The benefits are demonstrated by comparing the actual requirements to those of the second-order approximation of $\log_2(1 - 2^{-x})$. Different computational techniques, like the interleaved-memory function interpolator [12] that reduces the memory size needed for the direct approach, can also be applied in this approach. The number of intervals

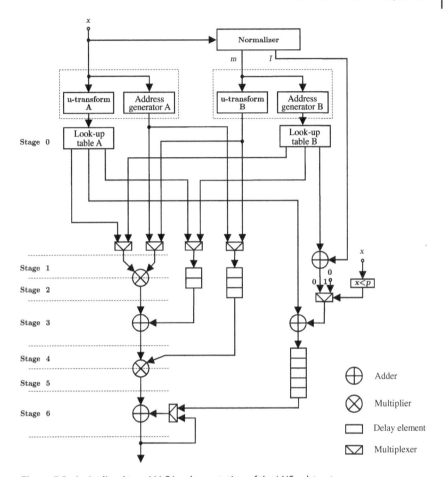

Figure 7.2 A pipelined two-MAC implementation of the LNS subtractor.

required for this approach and for the direct one are shown in Table 7.3, while the memory requirements are presented in Table 7.4. In the direct approximation intervals close to zero, coefficients with large integer part are needed, that is, in excess of 25 bits. In this case, the overall memory can be partitioned into several small ones to avoid wasted memory space. Another approach could be to sacrifice accuracy in the intervals close to zero. The memory sizes for the approximation of $\log_2(1 - 2^{-x})$ shown in Table 7.4 have been obtained by following the latter approach. Hence, the direct approach requires more memory space and achieves lower accuracy than the one discussed here.

The computation of certain nonlinear functions such as $\log_2(1 - 2^{-x})$, $\log_2 \sin x$, and $\log_2 \tan x$, by an LNS processor has been presented. The particular approach reduces the memory requirements of the LNS subtraction

Table 7.3 Number of intervals required by the two approaches.

Accuracy	Number of intervals			
(Fractional bits)	This work			$\log_2(1 - 2^{-x})$
	Total	Function	$\log_2 x$	
8	24	22	2	23
10	24	21	3	35
12	34	30	4	48
14	41	34	7	88
16	54	44	10	126
18	72	56	16	182
20	93	66	27	291
22	131	94	37	553
24	221	157	64	626
26	336	233	103	1021
28	500	354	146	1837
30	1079	823	256	2464
32	1273	862	411	4044

Table 7.4 Memory requirements of the discussed architecture.

Accuracy	Memory size (bits)		Savings
(Fractional bits)	This work	$\log_2(1 - 2^{-x})$	(%)
8	798	828	4
10	990	1470	33
12	1576	2304	32
14	2187	4752	54
16	3250	7560	57
18	4768	12012	60
20	6684	20952	68
22	10198	43134	76
24	18535	52584	65
26	30446	91890	67
28	47938	176352	73
30	110570	251328	56
32	138306	436752	68

more than 70%, allowing the low-cost implementation of LNS processors of increased accuracy, even beyond the 32 fractional bits. In this way, LNS may provide an interesting alternative to floating-point arithmetic.

Transformation Techniques

As discussed above, the singularity in the subtraction function increases the complexity of the corresponding hardware. The technique above removed this difficulty by reducing it to the computation of an easy-to-approximate function and a logarithm, for which efficient techniques exist. Arnold *et al.* have suggested the use of a dual redundant logarithmic number system (DRLNS) [13]

$$x = b^{\hat{X}_P} - b^{\hat{X}_N} \tag{7.42}$$

where x is the real value represented in DRLNS as $(\hat{X}_P, \hat{X}_N)$. The particular redundant representation allows subtraction without the use of the subtraction function [13].

Arnold *et al.* have suggested a cotransformation-based technique that reduces the problem of computing values close to the singularity to the computation of values of the well-behaved addition function [14].

Coleman has addressed the problem using a different type of cotransformation that maps the original problem to the computation of the same function in a region far from the singularity [15]. In this procedure, the subtraction function is required to be computed more than once.

7.4 Forward and Inverse Conversion

The conversion from the linear to logarithmic domain and vice versa is necessarily the computation of a logarithm and an exponential, or antilogarithm, respectively. Several authors have disclosed efficient circuits for such computations, optimized for a variety of implementation platforms. The main constraint driving the implementation is the required accuracy. Low-precision applications may find very simple approaches sufficient [18], for example,

$$\log_2(1 + m) \approx m, \qquad 0 \le m < 1 \tag{7.43}$$

Further improvements in the accuracy of Mitchell's approximation have been introduced by Mahalingam and Ranganathan [19]. However, high-precision applications require more elaborated processing, involving interpolations [20,21].

7.5 Complex Arithmetic in LNS

Complex numbers are conceived as points in the complex plane, represented as pairs (x, y) of reals. There exists a so-called rectangular representation of a complex number z, that is,

$$z = x + j \cdot y \tag{7.44}$$

where x and y are real and j denotes a root of $x^2 + 1 = 0$. Swartzlander *et al.* have used the rectangular representation of complex numbers in an LNS implementation of an FFT processor [5]. The polar representation of a complex number is a product,

$$z = r \exp(j\theta) \tag{7.45}$$

where $r = \sqrt{x^2 + y^2}$ and

$$\theta = \arg(z) = \begin{cases} \arctan\left(\frac{y}{x}\right), & x > 0 \\ \arctan\left(\frac{y}{x}\right) + \pi, & x < 0 \quad \text{and} \quad y \geq 0 \\ \arctan\left(\frac{y}{x}\right) - \pi, & x < 0 \quad \text{and} \quad y < 0 \\ \frac{\pi}{2}, & x = 0 \quad \text{and} \quad y > 0 \\ -\frac{\pi}{2}, & x = 0 \quad \text{and} \quad y < 0 \end{cases} \tag{7.46}$$

The polar representation (Eq. (7.45)) can be combined with the LNS, forming the complex LNS (CLNS) representation [14]:

$$z_{\text{CLNS}} = \log_b r + j\theta \log_b e \tag{7.47}$$

When the logarithm base $b = e$ is used, the representation is reduced to $z_{\text{CLNS}} = \log r + j\theta$. The adoption of the combined logarithmic-polar representation extends the simplicity of the real logarithmic multiplication and division to the corresponding complex operations, avoiding the need for the computation of cross-product terms required, for example, in the case of complex multiplication using the rectangular representation.

Let two CLNS numbers be expressed as $X = X_L + jX_\theta$ and $Y = Y_L + jY_\theta$. The CLNS expression of the product $Z = Z_L + jZ_\theta$ can be obtained as

$$Z_L = X_L + Y_L \tag{7.48}$$
$$Z_\theta = \left(X_\theta + Y_\theta\right) \bmod 2\pi \tag{7.49}$$

By appropriately scaling the angle of the logarithmic representation by $4/\pi$, it becomes possible to reduce the mod 2π operation in Eq. (7.49) to a modulo-by-a-power-of-two operation, which is trivial to implement in hardware [14].

As in the real domain, the operations of addition and subtraction remain complicated in the complex domain as well. The computation of the sum S requires the computation of $S = X + S_b(Y - X)$ where function S_b is the complex counterpart of the logarithmic addition function in the real domain and it is defined as

$$S_b(Z) = \log_b(1 + b^Z) \tag{7.50}$$
$$= \log_b(1 + \cos(Z_\theta)b^{Z_L} + j\sin(Z_\theta)b^{Z_L}) \tag{7.51}$$

The sum $S = S_L + jS_\theta$ is obtained as

$$S_L = Y_L + \Re[S_b(X - Y)] \tag{7.52}$$
$$S_\theta = \left(Y_\theta + \Im[S_b(X - Y)]\right) \bmod 2\pi \tag{7.53}$$

Subtraction can be performed in this sense, by $\hat{Y}$ instead of Y, with $\hat{Y} = Y + j\pi$. Clearly, the complexity is in the computation of $S_b(X - Y)$. Arnold *et al.* introduce the use of cotransformations to address the problem of implementation complexity [14]. Recently, Arnold and Collange have proposed a formulation of the complex logarithmic addition function leading to hardware-efficient architectures for performing complex addition in the log-polar domain that can reuse real LNS ALUs [22].

7.6 LNS Processors

Several researches have developed processor architectures based on the LNS, targeting mostly DSP applications. The particular architectures appear as alternative solutions to single-precision floating-point processors. Therefore, the focus of these early contributions is on achieving the required precision in a hardware-efficient manner. Huang *et al.* have presented two techniques that allow the hardware computation of $\log_2(1.x)$ and $2^{0.x}$, namely, digit-partition (DP) and iterative difference by linear approximation (IDLA) [23]. In the former case, the concept is to partition the input word and employ small look-up tables, the results from which are subsequently combined with simple operations. In the latter case, an iterative algorithm is used to tune the slope of a linear approximation, until a precision constraint is met. Several authors have sought high-precision operations, resorting to piecewise linear and higher order interpolations ([12]. Paliouras *et al.* have presented a VLIW processor architecture that utilizes multiple LNS ALUs [24] capable of addition and multiplication. This processor is discussed in section 7.6.1. Later, Arnold [25] discussed a very long instruction word (VLIW) architecture equiped with a single-cycle multiplier and a pipelined adder. Recently, Coleman *et al.* have presented the European logarithmic processor (ELM) [26]. The main characteristic of the ELM is that it supports four single-cycle ALUs that provide multiplication, division, or square root operations and two multicycle ALUs that provide addition and subtraction.

7.6.1 A VLIW LNS Processor

The organization and the VLSI implementation of a very long instruction word (VLIW) digital signal processor are discussed in this paragraph [24]. The processor operates in the logarithmic number system (LNS) and it features two LNS execution units. Each execution unit contains a single-clock multiply/divide unit and a pipelined adder/subtractor of dynamic range equivalent to the single-precision IEEE 754 standard requirements. The architecture is optimized for the execution of FIR and IIR filters, as well as any sum-of-products-based digital signal processing algorithm. The full exploitation of the independent resources is facilitated by porting the complexity of resolving the dependencies among the instructions from special control hardware to the organization of the application software, an off-line task. The main characteristics of the introduced processor include the ability to issue concurrently up to 30 elementary instructions, the two LNS execution units including division, and the dynamic range offered by the 33-bit data word. The chip has been designed and simulated in a commercial $0.7\,\mu$m CMOS technology.

Recently, the VLIW paradigm in processor architecture has attracted a lot of interest, as it provides a means for fully exploiting independent hardware resources by avoiding the burden of instruction-level dependency analysis performed by the processor [27]. The VLIW paradigm naturally suits applications, which exhibit inherent parallelism, such as signal and image processing algorithms, dominated by sum-of-products operations. The LNS is especially suited for signal processing applications. Benefits from using the LNS stem from compressing the dynamic range of data, a feature commonly exploited in communication systems (a-law, m-law). The particular property leads to reduced LNS data word lengths, when compared to equivalent conventional representations. Furthermore, operations such as multiplication, division, roots, and powers are simplified. However, the actual performance improvement depends largely on the algorithms to be implemented and on the imposed accuracy requirements. Due to its amenability for DSP applications, the LNS has attracted a lot of attention. This section describes the architecture of a programmable digital signal processor. The architecture of the introduced processor reconciles VLIW and super-scalar techniques with RISC and DSP elements, as well as the processing capabilities of the LNS. The objective in the presented endeavour has been to evaluate the synergism of state-of-the-art processor architecture trends and of nonconventional number system techniques. It is found that the LNS is expedient for the new generation of digital signal processors, as it facilitates the integration of several execution units in a chip by reducing the hardware complexity. Hardware complexity reduction is achieved by both compressing the dynamic range of the data and by replacing the traditional operations hardware with low-complexity LNS counterparts. The discussed processor is found to outperform several commercial processors in particular applications, while avoiding a high-system clock frequency and the consequent design difficulties.

7.6.1.1 The Architecture

The processor is composed of the instruction-set hardware, a set of FSMs which supervise the boot procedure, the instruction execution and general I/O, the separate data and program memories, and the LNS execution units. The organization of the introduced processor is depicted in Figure 7.3. A characteristic of the particular organization is that operations on data stored in memory is not supported. Instead, data should be initially loaded in a register and subsequently participate in an operation. The particular architecture style (register–register architecture) is common in RISC processors [28]. The processor is build around a pair of LNS execution units, which communicate through a network of six data busses and a shared data memory of six ports. The instruction-set hardware consists of two independent submodules, each with a program memory, program counter, and instruction and immediate data registers. An instruction is decoded by the instruction decoder to generate the control signals for the execution units. The particular organization permits the independent operation of each execution unit, thus ensuring high resource utilization in a variety of applications. The control of multiple registers, busses, and memory banks can be efficiently implemented in application assembly code, due to the VLIW paradigm. Each instruction word consists of 58 bits, organized into two 29-bit subwords. The organization of a subword is depicted in Table 7.5 and it controls one of the available execution units. It should be noted that the architecture is scalable, in that it can incorporate more than two execution units. In this case the instruction word is augmented with the appropriate number of sub words.

Prominent is the role of the instruction execution FSM, which supervises the fetch, decode, and execute phases in instruction processing, as well as the flow of program execution. The three phases in processing successive instructions are partially overlapped, as shown in Figure 7.4, to accelerate program execution.

7.6.1.2 The Arithmetic Logic Units

Each Erithmetic logic unit is based on the LNS and it is depicted in Figure 7.5. It is composed of an LNS adder/subtractor and an LNS multiplier/divider. The LNS multiplier/divider is implemented as a simple fixed-point adder/subtractor. The use of the LNS greatly reduces the complexity of implementing division. Therefore, the cost of including a dedicated divider per ALU becomes practically

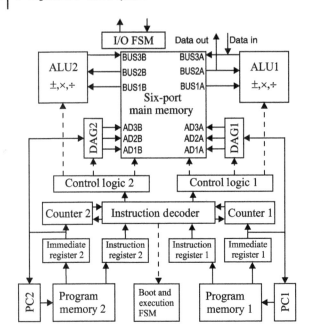

Figure 7.3 The block diagram of the introduced processor.

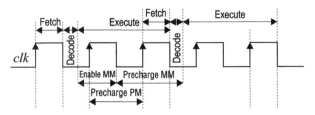

Figure 7.4 Timing diagram of instruction execution pipeline.

negligible. Hence, the processor is particularly efficient for multiplication/division-intensive applications. The LNS addition/subtraction is based on a second-order polynomial interpolation. The adder/subtractor is implemented as a four-stage pipeline plus a stage for ROM lookup table accessing. Each function point is stored in a ROM lookup table as a 28-bit word. A total of 2210 words are stored. To accelerate LNS addition and subtraction, an interleaved memory organization is adopted for the required ROM lookup tables. The particular memory architecture permits the simultaneous fetching of three-stored words, only at the cost of a 50% increase in total memory space. A new operation can be issued at each clock on both subunits of an ALU, leading to a peak performance of four operations per cycle. The composite operations of

Table 7.5 Instruction subword bit allocation. In the current implementation, an instruction comprises two subwords.

Set/increase pointers p1, p2, p3	bits 28–25
Set/increase pointer p4	bits 24–23
Set/increase counters cntr1, cntr2	bits 22–20
Load registers AX, AY, MX, MY	bits 19–17
Load registers AR, MR	bits 16–15
Add/sub	bit 14
Mul/div	bit 13
Composite operation	bit 12
Select address port B	bit 11
en_rd, en_wr for memory	
en1, en2 for bus 3	bits 10–9
en1, en2, en3 for bus 1	bits 8–7
en1, en2, en3 for bus 2	bits 6–5
en1, en2 for interim bus 1	bit 4
en1, en2 for interim bus 2	bit 3
en1, en2 for interim bus 3	bit 2
Jump, jump CE, jump NCE	bits 1–0

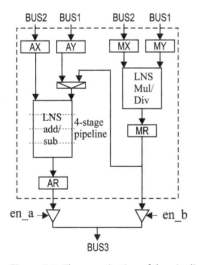

Figure 7.5 The organization of the pipelined LNS ALU.

multiply–add/subtract and divide–add/subtract are supported, with no need to access the main memory for intermediate result storage. The composite operations are particularly useful in sum-of-products intensive DSP algorithms.

7.6.1.3 The Register Set

Traditionally, the organization of the assortment of registers in a processor is a dominant issue from a programmers viewpoint, as it influences the structure of application assembly code and compiler performance. The registers in the introduced architecture can be distinguished in ALU registers, memory pointers, counters, and program sequencer registers. Each ALU contains six 33-bit data registers, namely, AX, AY, MX, MY, AR, and MR. The registers AX, AY, MX, and MY are located at the inputs of the add/subtract and multiply/divide sub-units of the ALU, while AR and MR are located at their outputs (Figure 7.5). The registers AX and MX can be loaded via BUS1, while BUS2 corresponds to the registers AY and MY. A particular data bus driving a pair of input registers cannot read data from the memory port corresponding to the other input register pair. Finally, BUS3 is exploited for transferring the contents of the output registers to the main memory. The three different busses per ALU allow the parallel load of ALU registers, thus facilitating high-rate data transfer. The memory pointers are four nine-bit counters per DAG unit, required to address the main memory. All pointers can be loaded from the corresponding immediate data register (IMD_REG) detailed below, while they are controlled by the instruction decoder or the I/O FSMs. Two nine-bit counters are included and they can be exploited to count up to 512 steps in a subroutine. When the contents of a counter reach zero (the counter expires), the related subroutine concludes. Two flags are related to each counter and can be used for program flow control, namely, counter expires-CE and not counter expires-NCE, denoting if a counter is zeroed or not, respectively. Finally, the fourth class of registers and counters are directly related to program flow control. Each of the two submodules of the instruction-set hardware contains two program counters (PC_A and PC_B), a 29-bit instruction register, and an immediate loads register (IMD_REG). The PCs increase their contents to facilitate sequential program execution, while PC_B can be directly loaded from IMD_REG to implement jumps in program execution.

7.6.1.4 Memory Organization

The memory organization in the introduced processor adopts the principle of separate program and data memory, resembling a modified Harvard architecture. However, the conventional organization is modified to adapt into the VLIW paradigm. In particular, multiple data memories are implemented as a six-port register file, called main memory, while the program memory is partitioned into two independent two-port RAM units. Furthermore, immediate data are stored in program memory to reduce communication burden within the processor. Each execution unit accesses three of the main memory ports, one write and two read ports. The addresses to the main memory are provided by two data address generators (DAGs) producing three address words each. A DAG comprises a set of counters that can be loaded from several sources to allow flexibility in addressing the main memory. The program memory is composed of two dual-port RAM blocks, which contain the stored program and the immediate load data. The units are physically independent to simplify the concurrent operation of the two execution units. Each memory block has two ports, denoted as Port A and Port B. Port B is a read-only port, while Port A is a read/write port. The program memory banks are addressed by the seven-bit wide program counters. Instruction words and immediate load data are stored in consecutive memory locations. Memory positions of odd addresses contain instruction data, while even positions contain immediate load data.

7.6.1.5 Applications and the Proposed Architecture

The architecture is fine tuned for digital filter implementation by providing composite operations, automated generation of data address sequences by the DAGs, as well as separated data/program memories. The main addressing mode provided is the register indirect one and it is implemented on the DAG counters (memory pointers). Immediate addressing can be implemented by means of the immediate register. In an FIR filter software implementation developed for the introduced VLIW processor, it has been found that a rate of 30 elementary instructions per clock can be sustained out of a peak performance of 30 instructions. Notice that the task of identifying data dependencies and resource conflicts is ported out of the processor into the software application development. To simplify the task of assembly code writing, simple parsing software has been developed to aid in resource conflict identification.

7.6.1.6 Performance and Comparisons

The processor can operate at a clock frequency of 12 MHz. However, its performance can reach a top of 360 million instructions per second (MIPS), while the sustained rate of logarithmic operations per second is 48 millions. The total hardware complexity is equivalent to approximately 310 K two-input NOR gates, placing the processor well within the limits of contemporary VLSI technology. A performance comparison to commercial DSPs is offered in Table 7.6.

7.7 LNS for Low-Power Dissipation

In the recent years, the amount of power dissipated in an electronic system has evolved into an important design issue, mainly due to the growing need for

Table 7.6 Performance comparison to commercial devices.

Processor	MHz	MIPS	MOPS
This work	12	360	48 (LNS)
ADSP-2105	13.84	13.84	27
VIPER	25	100	—

portable electronics equipment, as well as the requirement for very high-speed processors [29].

In this section, we discuss how the logarithmic number system (LNS) may be exploited to save power [2]. The equivalence of an LNS to a linear fixed-point system is initially explored and a related theorem is introduced. Then, activity is studied for both uniform and gaussian correlated input distributions. LNS is shown to reduce the average bit assertion probability by more than 60%, in certain cases, over an equivalent linear representation. The impact of LNS to power dissipation through architecture simplification is also discussed.

Power dissipation minimization is sought at all levels of design abstraction, ranging from system-level software/hardware partitioning down to technology-related issues. The average power dissipation in a circuit is given as

$$P_{\text{ave}} = a f_{\text{clk}} C_L V_{\text{dd}}^2 \tag{7.54}$$

where f_{clk} is the clock frequency, C_L is the total switching capacitance, V_{dd} is the supply voltage, and a is the average activity in a clock period. A wide variety of design techniques have been proposed, to reduce the factors of product (7.54) [29]. Among them, the selection of the number system and the optimization of corresponding arithmetic circuits has been considered as a power dissipation minimization technique [30], affecting all factors of Eq. (7.54).

In this section, it is shown that the adoption of LNS [31] can lead to substantial power dissipation savings, since it both reduces the average bit activity and it simplifies certain arithmetic operations. Building on the equivalence of LNS and linear representations discussed in Section 7.2.1, wordlength reductions become possible. Furthermore, Lewis discusses the equivalence of floating-point and LNS [4].

7.7.1 Impact of LNS Encoding on Signal Activity

Let $p_{0 \to 1}(i)$ be the bit assertion probabilities, that is, the probability of the ith bit transition from 0 to 1. Assuming that data are temporaly independent, it holds that

$$p_{0 \to 1}(i) = p_0(i) p_1(i) = \left(1 - p_1(i)\right) p_1(i) \tag{7.55}$$

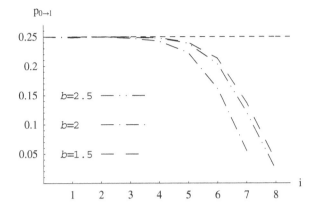

Figure 7.6 Activities against bit significance i, in an LNS word, for $n = 8$ and various values of the base b. The horizontal dashed line is the activity of the corresponding n-bit fixed-point system.

where $p_0(i)$ and $p_1(i)$ are the probability of the ith bit being 0 and 1, respectively. Assuming uniform data distribution, it is $p_0(i) = p_1(i) = 1/2$, which, due to Eq. (7.55), gives $p_{0\rightarrow1}(i) = 1/4$, for $i = 0, 1, \ldots, n-1$.

The $p_{0\rightarrow1}^{\text{LNS}}$ probabilities of bit assertions in LNS words are not constant; they depend on the significance of the ith bit. To evaluate the probabilities $p_{0\rightarrow1}^{\text{LNS}}(i)$, the following experiment is performed. For all possible values of X in a n-bit system, the corresponding $\lfloor \log_b X \rfloor$ values in a (k, l, b)-LNS format are derived and probabilities $p_1(i)$ for each bit are computed. Then, $p_{0\rightarrow1}^{\text{LNS}}(i)$ are obtained by Eq. (7.55) and are depicted in Figure 7.6. It can be seen that $p_{0\rightarrow1}(i)^{\text{LNS}}$ for the more significant bits is substantially lower than $p_{0\rightarrow1}^{\text{LNS}}(i)$ for the less significant bits. Also, it can be seen that $p_{0\rightarrow1}^{\text{LNS}}(i)$ depends on b. This behavior, which is due to the inherent data compression property of the logarithm function, leads to a reduction of the average activity in the entire word. Average activity savings percentage, S_{ave} is

$$S_{\text{ave}} = \left(1 - \frac{\sum_{i=0}^{k+l-1} p_{0\rightarrow1}^{\text{LNS}}(i)}{0.25n}\right) 100\% \tag{7.56}$$

where it has been used that $p_{0\rightarrow1}^{\text{FXP}}(i) = 1/4$, for $i = 0, 1, \ldots, n-1$, n is the fixed-point word length, and k and l are computed via Theorem 1. Savings percentage S_{ave} is demonstrated in Figure 7.7a for various values of n and b, and it is found to be more than 15% in certain cases.

As implied by the definition (7.16) of n_{eq}, the linear system which provides an equivalent range to that of a (k, l, b)-LNS, requires n_{eq} bits. If the reduced precision of (k, l, b)-LNS compared to n_{eq}-bit fixed-point system is acceptable

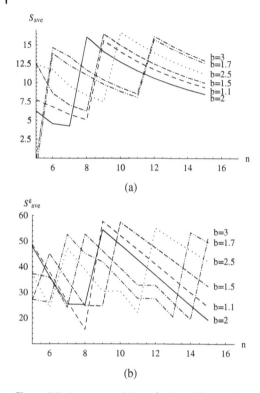

Figure 7.7 Average activity reduction (%) over (a) an n-bit and (b) an n_{eq}-bit linear fixed-point system, for several b.

for a particular application, the savings demonstrated in Figure 7.7b for various values of n and b are achieved. Savings are found to exceed 50% in some cases.

Figure 7.7 reveals that, for a particular word length n, the proper selection of logarithm base b can significantly affect the average activity. Therefore, the choice of b is important in designing a low-power LNS-based system.

In the following, activity is studied assuming correlated gaussian input. Assume that two successive inputs to an LNS-based system are the LNS representations of two random variables, X and Y, which follow a bivariate joint gaussian probability density function with zero mean, $\mu_X = \mu_Y = 0$, equal standard deviation, $\sigma_x = \sigma_y = \sigma$, and correlation factor ρ. The particular distribution is exploited in the derivation of the dual bit type method for modeling activity [32].

Activity measurements are performed by generating a random sample that follows the particular distribution and subsequently applying it to an LNS encoder.

The bit assertion probability per bit position $p_{0 \to 1}$, for two's complement encoding and for base-b LNS encoding are depicted in Figure 7.8. Figure 7.8

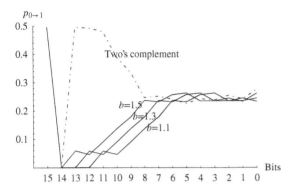

Figure 7.8 Probability $p_{0\rightarrow1}$ per bit position for two's complement and LNS encoding for $\rho = -0.99$.

shows the signal activity per bit position, for a zero-mean gaussian anticorrelated ($\rho = -0.99$) input signal, with $\sigma = 1500$, and word length $n = 14$. The applicability of the dual bit type method [32] to model the activity depicted in Figure 7.8 is apparent. The bit assertion activity, assuming the same signal is encoded in LNS for various bases b shown in Figure 7.8, reveals that significant activity reduction is possible, despite the fact that the sign bit activity is essentially identical in both cases, and the word length of the LNS representation is slightly increased. In addition, the dominating role of b on bit activity is clearly demonstrated.

Activity savings for $n = 16$, and various values of correlation ρ and $b = 1.1$, versus the standard deviation of the input distribution are depicted in Figure 7.9. Figure 7.9 reveals that for anticorrelated signals, significant activity savings are achieved. However, the situation is reversed for correlated signals. Finally, by relaxing the error specification and comparing the activity of an LNS signal to the activity of an n_{eq}-bit two's complement signal, the performance depicted in Figure 7.10 is achieved.

7.7.2 Power Dissipation and LNS Architecture

LNS exploits properties of the logarithm to reduce the strength of several arithmetic operations. By reducing the area complexity, the switching capacitance C_L of Eq. (7.54) can be reduced. Furthermore, reduction in latency allows for further reduction in supply voltage, which also reduces power dissipation [29].

Multiplication is also substantially simplified. Assume that a n-bit carry-save array multiplier, which has a complexity of $n^2 - n$ one-bit full adders (FAs) is replaced by an n-bit adder, which, assuming $k + l = n$, has a complexity of n FAs, for a ripple-carry implementation [3]. Therefore, the adoption of LNS reduces area complexity by a factor $r_{C_L} = (n^2 - n)/n = n - 1$.

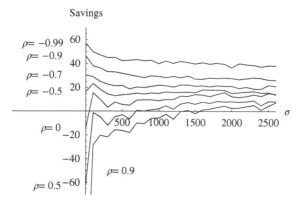

Figure 7.9 Savings percentage (%) for gaussian input versus σ for $n = 16$, several correlation factors ρ, and $b = 1.1$.

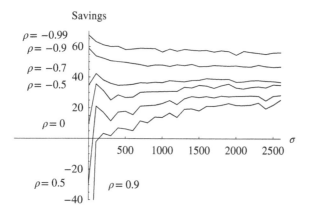

Figure 7.10 Activity savings percentage (%) over an n_{eq}-bit system ($n = 14, b = 1.1$) for correlated gaussian input.

However, addition and subtraction are complicated in LNS, since they require a table look-up operation for the evaluation of the functions $\log_b(1 \pm b^{y-x})$, shown in Table 7.1, although different approaches have been proposed in the literature [11]. A table look-up operation requires a ROM of $n \times 2^n$ bits, a size which can inhibit LNS utilization for large values of n. In an attempt to alleviate the particular problem, efficient table reduction techniques have been proposed [6]. Hence, in applications the computational load of which is dominated by operations of simple LNS implementation, power dissipation reduction due to the LNS impact on reducing the architecture complexity can be expected. As multiplication–additions are important in DSP applications, the power requirements of an LNS and an equivalent fixed-point mutiplier-adder are compared.

Table 7.7 Power dissipated (mW/MHz) in LNS subtraction/addition and $n \times n$-bit fixed-point multiplication, for several n and b.

n (bits)	Base b	LNS Add	LNS Subtract	LNS Average	Fixed-point multiply
8	1.5	0.99	0.99	0.99	2.02
	2	0.90	0.89	0.89	
	2.5	0.98	0.98	0.98	
10	1.5	1.14	1.16	1.15	2.64
	2	1.18	1.18	1.18	
	2.5	1.09	1.11	1.10	
12	1.5	1.49	1.52	1.50	3.35
	2	1.58	1.58	1.58	
	2.5	1.46	1.49	1.48	
14	1.5	2.38	2.46	2.42	4.13
	2	2.62	2.61	2.62	
	2.5	2.26	2.36	2.31	

From a hardware architecture viewpoint, the LNS multiplier is functionally and structurally identical to a fixed-point adder. Hence, the problem of comparing the amount of power dissipated by implementing a multiplication–addition operation, is transformed to comparing the power dissipation figures of the LNS adder/subtractor with a corresponding fixed-point multiplier. A quantitative power dissipation comparison is offered in Table 7.7, where it is shown that, assuming identical input activity, the LNS adder/subtractor dissipates half the power of the equivalent fixed-point multiplier. The activity savings achieved by LNS are not considered, and can improve the overall savings. The power dissipation data in Table 7.7 refer to a $0.7\,\mu$m CMOS technology, operating at a supply voltage of 5 V. The LNS adder/subtractor is not implemented as a single look-up table (LUT) but instead as an assortment of small LUTs that operate in parallel, as depicted in Figure 7.11. The breakdown of the LUT required to store the addition/subtraction function values is motivated by the form of the functions $\log_b(1 \pm b^d)$, $d = y - x$. As the value of d increases, both functions approach zero; hence, fewer bits are required to represent the stored value. This observation is particularly important for LNS subtraction, which is very steep near zero, thus requiring a longer word length. At the cost of introducing hardware complexity overhead for the generation of the addresses to the small LUTs, the architecture of Figure 7.11 is derived, in which each LUT corresponds to a region of the approximation interval, where the function values are

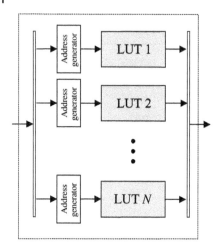

Figure 7.11 LUT architecture for storing $\log_b(1 \pm b^{y-x})$.

of equal word length. To further reduce memory size, it is noted that the most significant bit can be omitted, since it is always asserted. The adopted approach resembles the one taken by Taylor *et al.* [6].

The power P_{tot} dissipated in an LNS addition/subtraction

$$z = x + \log_b(1 \pm b^{y-x}), \quad x \geq y \tag{7.57}$$

is given by

$$P_{\text{tot}} = P_{\text{comp}} + P_{\text{address}} + P_{\text{LUT}} + P_{\text{add}} \tag{7.58}$$

where P_{comp} is dissipated to determine the larger of x and y, P_{address} is dissipated for address generation, P_{LUT} is dissipated for LUT access, and P_{add} is dissipated for the final addition.

Complexity overhead is imposed by linear-to-logarithmic and logarithmic-to-linear conversion. However, as the number of operations grows, the conversion overhead remains constant, thus its contribution to the overall budget becomes negligible.

LNS impact onto power dissipation has been investigated. The discussion is based on proposed LNS/fixed-point equivalence conditions. Subsequently, it has been shown that LNS can lead to significant average bit activity reduction, for uniformly distributed and anticorrelated gaussian signals. It has been found that the efficiency of the LNS representation is dominated by the choice of word lengths k and l, and—the often underestimated design parameter—logarithm base, b. Savings are more important when the dynamic range of the signal is smaller than the range provided by the word length. Furthermore, the impact of LNS onto the architecture has been briefly discussed. It has been shown that savings are possible for multiply–add operations, as well as for several more complex in the linear domain operations, such as division, roots, and powers. Therefore, the LNS, due to savings in signal activity and architectural simplification, can be a successful candidate for the implementation of low-power computationally intensive systems.

7.8 Applications

LNS and its variants provide simplification of certain operations combined with a reduction of the word length required for data representation without causing a degradation of perceived quality of the result. Furthermore, the roundoff error behavior of the LNS is found to outperform floating-point and linear representations, in certain cases.

Compressed data word lengths, operation simplification, and superior roundoff-error performance offered by LNS have motivated its use in several applications in signal processing and communications, video, graphics, and control. Logarithmic representations have been used in cabin air-pressure control in commercial aircraft [14]. LaMaire and Lang have studied the use of LNS in the implementation of digital regulators [33].

7.8.1 Signal Processing and Communications

Several authors have studied the effects of roundoff noise on digital filters implemented using LNS [34,35]. The motivation for using LNS in digital filtering comes mainly from its roundoff behavior [10] and potential for word length reduction, as the computational load for a digital filter comprises comparable amounts of multiplications and additions/subtractions. Chandra has provided expressions for the ratio of the output variance in the presence of roundoff errors over the roundoff error-free output variance for logarithmic FIR filters and has compared it to floating-point equivalents [34].

Kurokawa *et al.* have studied the impact of roundoff noise on recursive filters [35] in comparison to floating-point implementations. They provided theoretical and experimental results that show that for a particular word length, logarithmic filters accumulate less roundoff error than floating-point filters for the case of stochastic inputs. Kwa *et al.* have studied the quantization noise of logarithmic encoders when excited by sinusoidal and speech signals [36]. Logarithmic digital filters have found application in digital hearing aids. Morley *et al.* have implemented a multichannel digital processor that implements frequency selective filtering and limiting [37]. An interesting feature of the particular architecture is the use of logarithmic A/D and D/A converters to avoid the corresponding digital conversion circuitry.

Early LNS works have studied the logarithmic implementation of discrete transforms. Specifically, Swartzlander *et al.* [38] and Hongyuan and Lee [39] worked on the Fast Fourier Transform and provided LNS butterfly architectures and error analysis, assuming rectangular LNS representation of the complex numbers. The use of the log-polar CLNS and its impact on FFT implementation is discussed by Arnold *et al.* [40]. They report substantial word length reduction, specifically 9–12 bits in word size for 256–1024-point FFTs, compared to fixed-point realizations, while producing comparable accuracy. Brokalakis and Paliouras [41] have used an optimized LNS representation, in terms of word length, to implement FFT hardware for wireless modems. A hybrid approach used as LNS was employed to facilitate multiplications by the twiddle factors.

LNS research has produced several solutions for the simplification of addition and subtraction. Interestingly, these results are applicable in certain important communications-related algorithms, used in iterative decoders. Kang *et al.* have applied LNS techniques in the implementation of a turbo decoder and report a 27.6% reduction in the power consumption with negligible impact on the signal-to-noise ratio [42]. Turbo decoders contain implementation of the log-maximum aposteriori probability (log-MAP) algorithm, which requires the max* operation, also known as the Jacobian logarithm, defined as

$$\max{}^{\star}(a, b) = \ln(\exp(a) + \exp(b)) = \max(a, b) + \ln(1 + \exp(-|a - b|))$$
$$(7.59)$$

Due to the similarity of the max* operations with the logarithmic addition function, techniques developed for the simplification of LNS adder implementations can be used in the implementation of log-MAP decoders, employed in the implementation of turbo decoders. Wang *et al.* also discuss approximation techniques for the multivariate Jacobian logarithm, aiming at the implementation of log-SPA [43]. They build on the recursive computation of the Jacobian logarithm [44]. Peng and Chen discuss the use of LNS in decoders for nonbinary LDPC codes [45]. Using an FFT-based log-sum product decoding algorithm (log-SPA) algorithm, they follow a hybrid approach to avoid LNS addition and subtraction by transforming data to the linear domain, in order to execute the particular operations.

General signal processing applications such as recursive least squares QR filtering have been implemented on the European logarithmic microprocessor (ELM), a programmable processor based on logarithmic arithmetic [26]. Performance comparisons reveal that for 32-bit applications the particular device outperforms conventional superscalar pipelined floating-point processors.

7.8.2 Video Processing

Several authors have investigated the use of LNS in video and computer graphics applications [46–48]. Arnold has demonstrated that it is possible to reduce the word length required during MPEG decoding by using LNS, without compromising the quality of the visual result; in particular, he has shown that LNS can lead to simple inverse discrete cosine transform (IDCT) implementations, which in turn can lead to reduced power consumption [46]. IDCT has a key role in MPEG. Going beyond word length reduction, further arithmetic transformation techniques have been investigated that allow the efficient handling of the sum-of-products nature of the IDCT. As additions and particularly subtractions are relatively involved operations, the use of a dual redundant LNS (DRLNS) has been studied as a means to avoid LNS subtraction complexity by resorting to redundancy, in the context of an MPEG decoder application and in comparison to a signed LNS [49]. A different approach to constructing a DCT in LNS has been followed by Lee [50]. Instead of using redundancy, Lee employs a hybrid approach. The awkward LNS operations of addition and subtraction are performed in the linear domain, while multiplications are performed in the logarithmic domain. Combined benefits become available at an increased cost of linear-to-logarithmic conversion and vice versa. Although conversions introduce both complexity and computational errors, the overall hybrid LNS-linear architecture is found to be quite efficient.

Research efforts have been extended to other components of a video compression system, seeking techniques that outperfom current solutions and can exploit the simplicity of LNS. To achieve this goal, Ruan and Arnold have introduced cost functions that simplify the procedure of motion estimation in an MPEG encoder. In particular, in order to avoid the commonly-used mean absolute difference (MAD) measure, defined as

$$\text{MAD} = \sum_{i=0}^{n-1} \sum_{j=0}^{n-1} \left| X_{ij} - Y_{ij} \right| \tag{7.60}$$

they proposed a pixel value ratio, the mean larger ratio (MLR), defined as

$$\text{MLR} = \prod_{i=0}^{n-1} \prod_{j=0}^{n-1} \max\left(\frac{X_{ij}}{Y_{ij}}, \frac{X_{ij}}{Y_{ij}} \right) = \prod_{i=0}^{n-1} \prod_{j=0}^{n-1} \frac{\max(X_{ij}, Y_{ij})}{\min(X_{ij}, Y_{ij})} \tag{7.61}$$

where X_{ij} and Y_{ij} denote pixel values of the blocks to be compared. Notable, the MLR cost function when implemented in LNS has a hardware complexity similar to MAD operating in a fixed-point system. This is due to the simple LNS multiplication and division, which easily provide the products and quotients required by MLR, and in contrast to MAD, which has a rather cumbersome

LNS implementation, being a sum of absolute differences. Other LNS-friendly motion-estimation cost functions include the threshold MLR (TMLR) [51].

The derivation of efficient low-power hardware solutions for video compression has motivated research on the integration issues of hybrid processing units in existing systems. Specifically, potential mismatch errors, caused by the use of different IDCTs in the encoding and decoding parts of a video system have been studied. In particular, Lee [52] has investigated the behavior of a system composed of a video encoder equiped with an IEEE-1180-compliant IDCT and a corresponding decoder with a hybrid linear-LNS IDCT. He reports that error may accumulate due to mismatch, rendering the transmission of intraframes more frequent to maintain overall quality.

Clearly, LNS and hybrid-LNS solutions have been proven efficient for applications such as MPEG, which supports frequent intraframes, or JPEG, which does not use a feedback path with an IDCT, as they integrate seamlessly and support low-cost and low-power hardware architectures.

7.8.3 Graphics

Graphics applications have long benefit from the LNS, due to the rich variety of complicated arithmetic involved, which contains multiplications and divisions, roots, and powers and other nonlinear operations, susceptible to efficient logarithmic implementation. Hybrid LNS solutions have also found application in graphics processors. Lai and Wu [47] have investigated hybrid floating-point and LNS systems for graphics applications. Nam and Yoo [53] and Nam *et al.*[54] discussed a hybrid fixed-point and LNS implementation of a 3D vector processor for handheld graphics systems, capable of complicated arithmetic operations, such as power, logarithm, trigonometric functions, vector-SIMD multiplication, division, square root, and vector dot product. The arithmetic employed is the so-called FXP-HNS, a hybrid number system which executes addition/subtraction in a variable-precision linear fixed-point format, while the remainder of complicated arithmetic operations are performed in the logarithmic domain. Again, as in all hybrid systems, the accuracy of linear-to-logarithmic conversion and vice versa has a significant role, due to the introduction of roundoff errors and complexity.

Graphics applications, such as ray tracing have also been executed on the ELM [26], demonstrating the efficiency of the particular device and the underlying logarithmic arithmetic in comparison to a conventional superscalar processor.

7.9   Conclusions

The LNS is a promising data representation for application-specific processing. Its applicability has been proven in the literature for several application domains, from computer graphics systems to modern wireless telecommunication systems.

Several arithmetic operations are substantially simplified in LNS. The inherent data compression property allows for the use of relatively short word lengths, thus allowing for hardware complexity reductions. Roundoff error behavior is found superior over several conventional representations. Operations awkward in LNS when suitably optimized, are shown to impose moderate complexity on the overall system, while efficient forward and reverse converters have been developed.

Clearly, LNS is a promising candidate for application-specific hardware implementations.

References

1 E. Swartzlander and A. Jr. Alexopoulos, The sign/logarithm number system, *IEEE Trans. Comput.*, vol. C-24, no. 12, 1975, pp. 1238–1242.

2 V. Paliouras and T. Stouraitis, Signal activity and power consumption reduction using the logarithmic number system, in *Proc. 2001 IEEE Int. Symp. Circuits Syst. (ISCAS)*, vol. 2, Sidney, 2001, pp. 653–656.

3 I. Koren, *Computer Arithmetic Algorithms*. Prentice-Hall, New Jersey, 1993.

4 D. M. Lewis, 114 MFLOPS logarithmic number system arithmetic unit for DSP applications, *IEEE J. Solid State Circuits*, vol. 30, no. 12, 1995, pp. 1547–1553.

5 E. Swartzlander, Jr., D. Chandra, H.T. Nagle, Jr., and S. Starks, Sign/Logarithm arithmetic for FFT implementation, *IEEE Trans. Comput.*, vol. 32, no. 6, 1983, pp. 526–534.

6 F. Taylor, R. Gill, J. Joseph, and J. Radke, A 20 bit logarithmic number system processor, *IEEE Trans. Comput.*, vol. 37, no. 5, 1988, pp. 190–199.

7 H. Henkel, Improved addition for the logarithmic number system, *IEEE Trans. Acoust., Speech, and Signal Process.*, vol. 37, no. 2, 1989, pp. 301–303.

8 D. Lewis and L. Yu, Algorithm design for a 30 bit integrated logarithmic processor, in *Proc. 9th Symp. on Comput. Arithmetic*, 1989, pp. 192–199.

9 I. Orginos, V. Paliouras, and T. Stouraitis, A novel algorithm for multi-operand logarithmic number system addition and subtraction using polynomial approximation, in *Proc. 1995 IEEE Int. Symp. Circuits Syst. (ISCAS'95)*, 1995, pp. 1992–1995.

10 T. Stouraitis and F. Taylor, Analysis of logarithmic number system processors, *IEEE Trans. Circuits Syst.*, vol. 35, no. 5, 1988, pp. 519–527.

11 V. Paliouras and T. Stouraitis, A novel algorithm for accurate logarithmic number system subtraction, in *Proc. of the 1996 IEEE Symp. Circuits Syst. (ISCAS'96)*, vol. 4, 1996, pp. 268–271.

12 D. M. Lewis, Interleaved memory function interpolators with application to an accurate LNS arithmetic unit, *IEEE Trans. Comput.*, vol. 43, no. 8, 1994, pp. 974–982.

13 M. Arnold, T. Bailey, J. Cowles, and J. Cupal, Redundant logarithmic number systems, in *Proc. 9th Symp. Comput. Arithmetic*, 1989, pp. 144–151.

14 M. G. Arnold, T. A. Bailey, J. R. Cowles, and M. D. Winkel, Arithmetic co-transformations in the real and complex logarithmc number system, in *Proc. IEEE Symp. Comput. Arithmetic*, 1997, pp. 190–199.

15 J. N. Coleman, Simplification of table structure in logarithmic arithmetic, *Electron. Lett.*, vol. 31, no. 22, 1995, pp. 1905–1906.

16 S. Collange, J. Detrey, and F. de Dinechin, Floating-point or LNS: choosing the right arithmetic on an application basis, in *Proc. 9th Euromicro Conf. Digit. Syst. Desi. (DSD'06)*, 2006, pp. 197–203.

17 K. Johansson, O. Gustafsson, and L. Wanhammar, Implementation of elementary functions for logarithmic number systems, *IET Comput. Digit. Tech.*, vol. 2, no. 4, 2008, pp. 295–304. Available at: http://link.aip.org/link/?CDT/2/295/1

18 J. N. Mitchell, Computer multiplication and division using binary logarithms, IRE Trans. Electron. Comput. vol. 11, 1962, p. 512–517.

19 V. Mahalingam and N. Ranganathan, Improving accuracy in Mitchell's logarithmic multiplication using operand decomposition, *IEEE Trans. Comput.*, vol. 55, no. 12, 2006, pp. 1523–1535.

20 S. Paul, N. Jayakumar, and S. Khatri, A fast hardware approach for approximate, efficient logarithm and antilogarithm computations, *IEEE Trans. VLSI Syst.*, vol. 17, no. 2, 2009, pp. 269–277.

21 R. Muscedere, V. Dimitrov, G. Jullien, and W. Miller, Efficient techniques for binary-to-multidigit multidimensional logarithmic number system conversion using range-addressable look-up tables, *IEEE Trans. Comput.*, vol. 54, no. 3, 2005, pp. 257–271.

22 M. G. Arnold and S. Collange, A dual-purpose real/complex logarithmic number sysetm ALU, in *Proc. 19th IEEE Int. Symp. Comput. Arithmetic*, 2009, pp. 15–24.

23 S.-C. Huang, L.-G. Chen, and T.-H. Chen, The chip design of a 32-bit logarithmic number system, in *Proc. ISCAS'94*, 1994, pp. 167–170.

24 V. Paliouras, J. Karagiannis, G. Aggouras, and T. Stouraitis, A very-long instruction word digital signal processor based on the logarithmic number system, in *Proc. 5th IEEE Int. Conf. Electron., Circuits Syst. (ICECS'98)*, vol. 3, Lisbon, Portugal, 1998, pp. 59–62.

25 M. G. Arnold, A VLIW architecture for logarithmic arithmetic, in *Proc. Euromicro DSD*, 2003.

26 J. Coleman, C. Softley, J. Kadlec, R. Matousek, M. Tichy, Z. Pohl, A. Hermanek, and N. Benschop, The European logarithmic microprocesor, *IEEE Trans. Comput.*, vol. 57, no. 4, 2008, pp. 532–546.

27 J. Gray, A. Naylor, A. Abnus, and N. Bagherzadeh, VIPER: A VLIW integer microprocessor, *IEEE J. Solid State Circuits*, vol. 28, no. 12, 1993, pp. 1377-1382.

28 J. Hennessy and D. Patterson, *Computer Architecture: A Quantitative Approach*, 2nd ed. Morgan-Kaufmann, 1996.

29 A. P. Chandrakasan and R. W. Brodersen, *Low Power Digital CMOS Design*. Kluwer Academic Publishers, Boston, 1995.

30 B. Parhami, *Computer Arithmetic: Algorithms and Hardware Designs*. Oxford University Press, New York, 2000.

31 E. Swartzlander and A. Alexopoulos, The sign/logarithm number system, *IEEE Trans. Comput.*, vol. 24, no. 12, 1975, pp. 1238–1242.

32 P. E. Landman and J. M. Rabaey, Architectural power analysis: the dual bit type method, *IEEE Trans. VLSI Syst.*, vol. 3, no. 2, 1995, pp. 173–187.

33 R. LaMaire and J. Lang, Performance of digital linear regulators which use logarithmic arithmetic, *IEEE Trans. Automatic Control*, vol. 31, no. 5, 1986, pp. 394–400.

34 D. Chandra, Error analysis of FIR filters implemented using logarithmic arithmetic, *IEEE Trans. Circuits Syst. II Analog Digit. Signal Process.*, vol. 45, no. 6, 1998, pp. 744–747.

35 J. Kurokawa, T. Payne, and S. Lee, Error analysis of recursive digital filters implemented with logarithmic number systems, *IEEE Trans. Acoust., Speech Signal Process.*, vol. 28, no. 6, 1980, pp. 706–715.

36 S. W. Kwa, G. L. Engel, and R. E. Morley, Quantization noise analysis of sign/logarithm data encoders when excited by speech or sinusoidal inputs, *IEEE Trans. Signal Process.*, vol. 48, no. 12, 2000, pp. 3578–3581.

37 R. E. Morley, Jr., G. L. Engel, T. J. Sullivan, and S. M. Natarajan, VLSI based design of a battery-operated digital hearing aid, in *Proc. IEEE Int. Conf. Acoust., Speech Signal Process.*, 1988, pp. 2512–2515.

38 E. Swartzlander, D. Chandra, H. Nagle, and S. Starks, Sign/logarithm arithmetic for FFT implementation, *IEEE Trans. Comput.*, vol. C-32, no. 6, 1983, pp. 526–534.

39 W. Hongyuan and S. Lee, Comments on sign/logarithm arithmetic for FFT implementation, *IEEE Trans. Comput.*, vol. C-35, no. 5, 1986, pp. 482–484.

40 M. Arnold, T. Bailey, J. Cowles, and C. Walter, Analysis of complex lns ffts, in *2001 IEEE Workshop Signal Process. Syst.*, 2001, pp. 58–69.

41 A. Brokalakis and V. Paliouras, Using the arithmetic representation properties of data to reduce the area and power consumption of FFT circuits for wireless OFDM systems, in *2011 IEEE Workshop Signal Process. Syst. (SiPS)*, 2011, pp. 7–12.

42 B. Kang, N. Vijaykrishnan, M. J. Irwin, and T. Theocharides, Power-efficient implementation of turbo decoder in SDR system, in *Proc. IEEE Int. SOC Conf.*, 2004, pp. 119 – 122.

43 H. Wang, H. Yang, and D. Yang, Improved log-MAP decoding algorithm for turbo-like codes, *Commun. Lett., IEEE*, vol. 10, no. 3, 2006, pp. 186–188.

44 P. Robertson, E. Villebrun, and P. Hoeher, A comparison of optimal and suboptimal MAP decoding algorithms operating in the log domain, in *Proc. IEEE Int. Conf. Commun.*, Seattle, Washington, 1995, pp. 1009–1013.

45 R. Peng and R.-R. Chen, Application of nonbinary LDPC codes for communication over fading channels using higher order modulations, *Global Telecommun. Conf., 2006. (GLOBECOM '06. IEEE)*, 2006, pp. 1–5.

46 M. G. Arnold, Reduced power consumption for MPEG decoding with LNS, in *Proc. IEEE Int. Conf. Appl. Specific Syst., Archit. Processors, (ASAP '02)*, 2002.

47 F. Lai and C. Wu, A hybrid number system processor with geometric and complex arithmetic capabilities, *IEEE Trans. Comput.*, vol. 40, no. 8, 1991, pp. 952–962.

48 T. Kurokawa and T. Mizukoshi, Computer graphics using logarithmic number systems, *IEICE Trans.*, vol. E74, no. 2, 1991, pp. 447–451.

49 M. G. Arnold, Redundant logarithmic arithmetic for MPEG decoding, in *Proc. Int. Symp. Opt. Sci. SPIE Annu. Meeting*, 2004, pp. 112–122.

50 P. Lee, An evaluation of a hybrid-logarithmic number system DCT/IDCT algorithm, in *IEEE Int. Symp. Circuits Syst., (ISCAS 2005)*, vol. 5, 2005, pp. 4863–4866.

51 J. Ruan and M. Arnold, Threshold Mean Larger Ratio motion estimation in MPEG encoding using LNS, in *Integrated Circuit and System Design. Power and Timing Modeling, Optimization and Simulation*, Lecture Notes in Computer Science (eds. E. Macii, V. Paliouras, and O. Koufopavlou), Springer Berlin Heidelberg, vol. 3254, 2004, pp. 208–217.

52 P. Lee, An estimation of mismatch error in IDCT decoders implemented with hybrid-LNS arithmetic, in *Proc. 15th Eur. Signal Process. Conf. (EUSIPCO 2007)*, pp. 683–687.

53 B.-G. Nam and H.-J. Yoo, A low-power vector processor using logarithmic arithmetic for handheld 3D graphics systems, in *Proc. 33rd Eur. Solid State Circuits Conf., (ESSCIRC 2007)*, 2007, pp. 232–235.

54 B.-G. Nam, H. Kim, and H.-J. Yoo, A low-power unified arithmetic unit for programmable handheld 3-D graphics systems, *IEEE J. Solid State Circuits*, vol. 42, no. 8, 2007, pp. 1767–1778.

8

Redundant Number System-Based Arithmetic Circuits

G. Jaberipur

Computer Science and Engineering Department, Shahid Beheshti University,Tehran, Iran

8.1 Introduction

Carry propagation in radix-based number systems, like the conventional binary or decimal number systems, is the main issue that slows down the arithmetic operations. The latency of simple ripple-carry adders, where the carry signal in the radix position i depends on the one in position $i - 1$, increases linearly with the bit-width of the operands. This is often proclaimed as O(n), for n-bit operands. However, following are some well-known solutions to handle the carry propagation problem:

- *Carry Look-Ahead Addition.* The carry look-ahead adder (CLA) [1], improves the addition speed to logarithmic order (i.e., O($\log n$)), via employing extra logic to predict all the carry signals independently.
- *Residue Number Systems.* In a k-modulus residue number system (RNS) [2], the latency of addition is reduced to O $\left(\lceil\frac{n}{k}\rceil\right)$ or even to O $\left(\log\lceil\frac{n}{k}\rceil\right)$, if combined with CLA. However, conversion operations from conventional to RNS and the reverse are normally slow and more complex (and slower for larger k), which can be compensated only in applications [3] with several addition operations that take place between the required conversions.
- *Redundant Number Systems.* Constant-time (i.e., O(1)) addition is possible by redundant number systems (RDNS), such that a generated carry in a weighted position can propagate at most up to some predefined more significant position, where the attendant digit set is redundant (see Definition 8.7).

Arithmetic Circuits for DSP Applications, First Edition. Edited by Pramod Kumar Meher and Thanos Stouraitis.
© 2017 by The Institute of Electrical and Electronics Engineers, Inc. Published 2017 by John Wiley & Sons, Inc.

In this chapter, we aim to study a variety of redundant number systems, representations, and encodings, as well as their applications and impact on the performance of digital arithmetic operations. Contents of the rest of this chapter are as follows. Fundamentals of redundant number systems are discussed in Section 8.2. and 8.3. Section 8.4 deals with circuit realization of redundant arithmetic operations and number encodings that facilitate design process and enhance performance. The conversion from binary to redundant number systems and the reverse are discussed in Section 8.5. Special arithmetic circuits such as arithmetic shifters are provided in Section 8.6. The examples of some applications are discussed in Section 8.7. Finally, conclusions and a list of further readings are provided at the end of this chapter.

8.1.1 Introductory Definitions and Examples

In the rest of this section, you can find some useful definitions and a table of symbols, abbreviations, and acronyms (Table 8.3) that are used in this chapter.

Definition 8.1: *Weighted Positional Number System* A k-digit radix-r number X in a weighted positional number system is represented as $X = \left(x_{k-1}x_{k-2}\ldots x_1 x_0\right)_r$, where $x_i \in [\alpha, \beta]$, for $(0 \le i \le k-2)$, $x_{k-1} \in \left[\alpha', \beta'\right]$, $(\alpha, \alpha' \le 0$, and $\beta, \beta' \ge 0)$ (to allow representation of 0), $\beta \ge \alpha - 1 + r$ (to allow contiguity), and $X = \sum_{i=0}^{k-1} r^i x_i$.

Note that digit set of the most significant position can be smaller, as in Example 8.1.

Example 8.1: *Signed Decimal Number System* Consider a decimal number system that contains numbers in $[-100, 99]$. This interval can be represented by three radix-10 digits, from digit sets $\{-1, 0\}$, $[0, 9]$, and $[0, 9]$, respectively. For instance, -84 and $+76$ are represented as $(-1, 1, 6)_{10}$ and $(0, 7, 6)_{10}$, respectively.

Definition 8.2: *Binary Encoding of a Digit Set* A binary encoding for a digit set $D = [\alpha, \beta]$ is defined as a weighted bit set (WBS) [4] that can collectively represent values of all digits in D.

Example 8.2: *Digit Sets* $[0, 15]$, $[0, 18]$, $[0, 9]$, *and* $[-8, 7]$
Table 8.1 contains four digit sets, their binary encodings that are described via bit weights, and their WBS dot notations, where • represents one bit.

Table 8.1 Binary encodings.

Digit set	Encoding weights	Dot notation	ξ	$\varepsilon = \xi/2^b$
[0, 15]	8, 4, 2, 1	• • • •	16	1
[0, 18]	8, 4, 2, 2, 1, 1	• • • • • •	19	19/64
[0, 9]	4, 2, 2, 1	• • • •	10	10/16
[0, 9]	8, 4, 2, 1	• • • •	10	10/16
[−8, 7]	−8, 4, 2, 1	• • • •	16	1

Definition 8.3: *Faithful and Unfaithful Digit Set Encodings* An encoding for digit set D, per Definition 8.2, is faithful, if and only if its represented values exactly coincide with those in the digit set, otherwise (i.e., representing at least one value out of D) it is unfaithful.

All encodings of Table 8.1 are faithful, except for (8, 4, 2, 1) encoding of [0, 9] that can represent six other values in [10, 15]. Contents of the two right columns will be explained later.

8.1.1.1 Posibits and Negabits

Definition 8.4: *Posibits, Negabits, and Their Arithmetic Values* We name a normal bit x as *posibit* (short for positive bit), whose arithmetic value x^p is the same as the bit's logical state 0 or 1, thus $x^p = x$. In contrast, we define a negative bit y as *negabit*[1], again with logical state 0 or 1, whose arithmetic value y^n is −1 or 0, respectively (i.e., $y^n = y - 1$). In dot notation a negabit is represented as ○.

Note that the most significant bit of a k-bit two's complement (TC) number is conventionally considered as a $\left(-2^{k-1}\right)$-weighted posibit. However, it can be interpreted as a 2^{k-1}-weighted inverted negabit.

Example 8.3: *TC Encoding of Digit Set* [−4, 3] Table 8.2 represents conventional and negabit TC (NTC) encodings of digit set [−4, 3], where NTC codes are nicely sorted in accordance to their arithmetic values. One application of this encoding is in exponent representation that obviates the need for exponent biasing in floating point arithmetic [1].

---

1 This term was originally defined in Reference [9].

Table 8.2 Two encodings for 3three-bit TC numbers.

Digit	TC ●●●	NTC ○●●
−4	100	000
−3	101	001
−2	110	010
−1	111	011
0	000	100
1	001	101
2	010	110
3	011	111

Three handy arithmetic cells that are widely used in arithmetic circuits, including those based on redundant number systems, are full adder (FA), half adder (HA), and (4; 2) compressor ($^4_2$C). A very useful property of negabits is that any three-tuple of posibits and negabits can be manipulated by standard FAs [5]. For example, Eqs (8.1) and (8.2) describe such FA functionality for two posibits [negabits] and one negabit [posibit], where $x + y + z = 2c + s$, represents the conventional FA functionality that produces sum s and carry c on inputs x, y, and z.

$$x^p + y^p + z^n = x + y + z - 1 = 2c + s - 1 = 2c^p + s^n \quad (8.1)$$

$$x^n + y^n + z^p = x - 1 + y - 1 + z = 2c + s - 2 = 2(c - 1) + s = 2c^n + s^p \quad (8.2)$$

Figure 8.1 depicts four different functionalities of FAs and three of HAs on different posibit–negabit tuples [5].

Any other arithmetic cell that is functionally equivalent to a circuit with only FAs and HAs, as building blocks, can exhibit the same functionality on the corresponding input mix of posibits and negabits. For example, (4; 2) compressors,

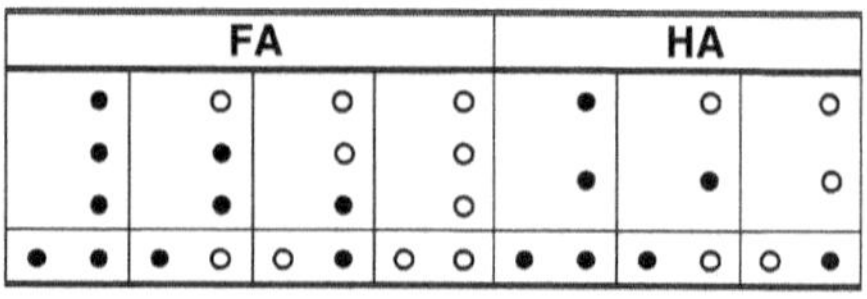

	FA				HA	
●	○	○	○	●	○	○
●	●	○	○	●	●	○
●	●	●	○			
● ●	● ○	○ ●	○ ○	● ●	● ○	○ ●

Figure 8.1 Functionality of FA and HA on different posibit—negabit mixes.

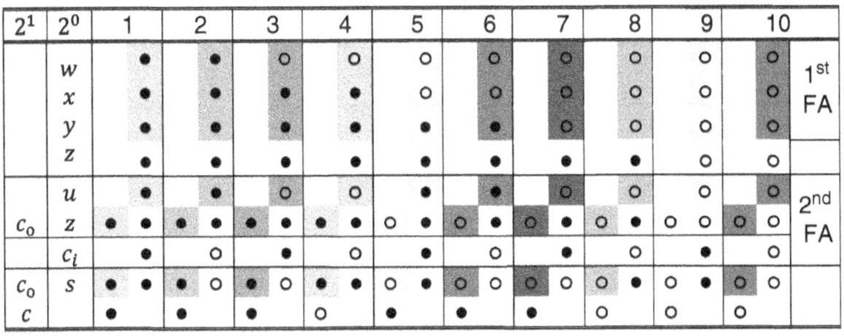

Figure 8.2 $^4_2$C functionality on 10 different posibit—negabit input mixes.

which are functionally equivalent to two FAs, can accept and produce correct output on any five-tuple of posibits and negabits [5]. Figure 8.2 depicts how $^4_2$C operates on 10 different posibit/negabit mixes. In each case, there are four main inputs that are represented by w, x, y, z, and a fifth input c_i. The equally shaded bits represent inputs and outputs of the 1st FAs, whose sum and carry-out bits are denoted as $u = w \oplus x \oplus y$ and $c_o = wx \vee (w \vee x) y$.

Recalling Definition 8.4, more flexibility can be introduced in assigning different arithmetic values to logical states of a bit, which leads to other two-valued digits besides posibits and negabits. For example, the quotient bits in non-restoring binary division [1] can assume one of the two unit values -1 and 1, corresponding to logical states 0 and 1, respectively. Such two-valued digits are called unibits (short for unit bit) and are represented in dot notation as **o** [4]. However, a two-valued digit [4] can be in general defined and designated as flexible bit (fit for short), where posibit, negabit, and unibit are special fits.

Definition 8.5 : *Flexible Bit or Fit* The logical states 0 and 1 of a fit can assume arbitrary arithmetic values λ and $\lambda + \gamma$, respectively, where λ is the lower arithmetic value and $\gamma > 0$ represents the gap between two arithmetic values of the fit $\{\lambda, \lambda + \gamma\}$. For example, $\lambda = 0, -1, -1$, and $\gamma = 1, 1, 2$, for posibits, negabits, and unibits, respectively.

Definition 8.6: *Encoding Efficiency* Let ξ denote cardinality of digit set D and assume that digits of D are represented by b bits. Then encoding efficiency of D is defined as $\varepsilon = \xi / 2^b$.

Example 8.4: *Encoding Efficiency* See the rightmost column of Table 8.1.

We provide some useful definitions and a table of symbols, abbreviations, and acronyms (Table 8.3), in order to enhance the readability of the entire chapter.

Table 8.3 Description of symbols, abbreviations, and acronyms.

Symbol, abbreviation, acronyms	Description
●	Posibit
○	Negabit
◉	Unibit
■	End marker
^4_2C	(4; 2) compressor
BSD	Binary SD
CS	Carry save
FA	Full adder
Fit	Flexible bit
FMA	Fused multiply add
GSD	Generalized SD
HA	Half adder
HA^{+1}	Enforced carry FA
MRSD	Maximally redundant SD
NTC	Negabit TC
O	Algorithmic order
RDNS	Redundant number system
RNS	Residue number system
SD	Signed digit
TC	Two's complement
WBS	Weighted bit set

8.2 Fundamentals of Redundant Number Systems

We begin by defining the redundant digit sets, a history of using them in digital arithmetic, and proceed in Section 8.3 with formal definition of redundant number systems and carry-free addition.

8.2.1 Redundant Digit Sets

Definition 8.7: *Redundant Digit Set and Redundancy Index* Let cardinality of a radix-r integer digit set $D = [\alpha, \beta]$ (i.e., number of digits in D) be denoted by $\xi_D = \beta - \alpha + 1$. Then D is redundant if and only if $\xi_D > r$. Redundancy index [6] of D is defined as $\rho = \xi_D - r$.

Example 8.5: *Redundancy Index* A digit set D with $\rho < 0$ (i.e., $\xi_D < r$) is not practical since it does not cover an interval of integers, while $\rho = 0$ corresponds to the conventional nonredundant digit sets such as binary $\{0, 1\}$ or decimal $[0, 9]$. However, in the case of $\rho > 0$ some digit values can be represented as two-digit numbers on the same digit set. For instance, consider, radix-10 digit set $\{0 \ldots 9, A\}$ with $\rho = 1$, where $A = 9 + 1$ can also be represented as two-digit decimal number 10. As another instance, observe radix-3 digit set $T = \{2', 1', 0, 1, 2\}$ with $\rho = 2$, where primed figures represent negative values (e.g., $2' = -2$). Therefore, all nonzero digits in T can be equivalently represented as two-digit radix-3 numbers, namely, $(2 = 1\,1' = 3 - 1)$, $(1 = 1\,2' = 3 - 2)$, $(-1 = 1'2 = -3 + 2)$, and $(-2 = 1'1 = -3 + 1)$. Note that more the redundancy index more the number of digits with multiple representations.

First instance of using a redundant digit set in digital arithmetic occurred in the presentation of the well known carry-save (CS) or stored-carry addition [7], where the employed binary digit set is $\{0, 1, 2\}$ and the corresponding faithful weighted bit-set (WBS) encoding (see Definitions 8.2 and 8.3) uses two equally weighted bits.

The radix-r symmetric signed digit (SD) sets of the form $[-\alpha, \alpha]$, with $\alpha \geq \lceil \frac{r+1}{2} \rceil$, were introduced by Avizienis [8], where cardinality $\xi_o = 1 + 2\alpha \geq 1 + (r + 1) \Rightarrow \rho \geq 2$, for odd r, and $\xi_e = 1 + 2\alpha \geq 1 + (r + 2) \Rightarrow \rho \geq 3$, for even r; hence, a redundant digit set per Definition 8.7. Three decades later, SD was extended by Parhami [6] as generalized signed digit (GSD) set of the form $[-\alpha, \beta]$, with $\alpha, \beta \geq 0$. GSD includes both CS and SD and a number of other useful redundant digit sets not used before (e.g., stored carry-or-borrow $[-1, 2]$).

Over the years, it was noticed that certain encodings of redundant digit sets led to more compact circuits for arithmetic with redundant operands. For example, numerous implementation efforts established the equally weighted negabit–posibit pair (vis--vis $\langle n, p \rangle$ [6]) encoding to be superior for representing binary signed digits (BSD). This encoding is a special case of WBS encoding, whose general form was introduced in Reference [9], developed in Reference [4] and was further generalized in Reference [10] for binary encoding of digit sets of the form $[\alpha, \beta]$, where $\alpha < \beta$ with the possibility of $\alpha > 0$.

Definition 8.8 *Minimally Redundant Digit Set* A radix-r redundant digit set D_{min} is denoted as minimally redundant if its redundancy index ρ_{min} is 1. Binary representation of such digits requires at least $b_{min}\left(= \lceil \log \xi_{D_{min}} \rceil = \lceil \log (r + 1) \rceil \right)$ bits. $b_{min} = h + 1$, for $r = 2^h$.

Definition 8.9: *Maximally Redundant Digit Set* A redundant digit set D_{max} is denoted as maximally redundant if $\xi_{D_{max}} = 2^{b_{min}}$, that is, D_{max} has maximum cardinality while minimum number of bits are used to represent its digits. $\xi_{D_{max}} = 2^{b_{min}} - 1$, for symmetric digit sets.

Example 8.6: *Minimally and Maximally Redundant Digit Sets* Recall radix-10 digit set $D_{min} = \{0 \dots 9, A\}$ from Example 8.5, which is minimally redundant since $\xi_{D_{min}} = 11 \Rightarrow \rho_{min} = 11 - 10 = 1$, where $b_{min} = \lceil \log 11 \rceil = 4$. Therefore, decimal digit set $D_{max} = [0, 15]$ (of practical use in Reference [11–13]) is maximally redundant since $\xi_{D_{max}} = 16 = 2^4$. Also $[-7, 7]$ (again practically used in Reference [14,15]) is maximally redundant and symmetric, with $\xi = 2^4 - 1$. Similarly, radix-16 digit set $[-8, 8]$ is minimally redundant with $b_{min} = 5$, and hexadecimal symmetric digit set $[-15,15]$, with cardinality $31 = 2^5 - 1$ is maximally redundant, also of practical use [16,17]. Note that with $(b_{min} - 1)$ bits we get at nonredundant digit sets (e.g., four bits with $\rho = 0$, in case of hexadecimal digit sets).

8.3 Redundant Number Systems

Definition 8.10: *Redundant Number Systems (RDNS)* A weighted positional number system (see Definition 8.1) is redundant, if and only if, the attendant digit set in at least one weighted position is redundant (vis--vis a redundant position).

Definition 8.11: *Homogeneous RDNS* In almost all practical cases, the employed RDNS is homogenous in that all positions hold the same redundant digit set. Therefore, in the rest of this chapter, we assume homogeneity of all RDNSs, unless otherwise specified.

The main property of redundant number systems (RDNS) is that there may be multiple representations for the same number. One reason is that, given a redundant digit set $D = [\alpha, \beta]$, $\rho = \beta - \alpha + 1 - r > 0 \Rightarrow \beta - r \geq \alpha \Rightarrow \beta - r \in D$. Therefore, a number with consecutive digits $x_{i+1} (< \beta)$ and $x_i (= \beta)$ can be equivalently represented with digits $X_{i+1} (= x_{i+1} + 1)$ and $X_i = (\beta - r)$, in the same positions, respectively (see also Example 8.5).

Example 8.7: *Homogeneous Decimal RDNS* In a four-digit RDNS, with $\{0, \dots 9, A\}$ as the digit set in all four positions, decimal number 2000 can also be represented as $1A00$, $19A0$, and $199A$.

Multiple representation property can lead to the most important benefit of RDNS, that is, possibility of constant-time addition. Unlike nonredundant number systems, worst-case time for adding two k-digit redundant numbers

Table 8.4 Classification of k-digit positional number systems.

α	β	r	ρ	l	Classification CL/CF	NR
≤ 0	≥ 0	≥ 2	0	$k-1$	NA	N
≤ 0	≥ 0	2	≥ 1	1	CL	R
≤ 0	$\alpha + r$	> 2	1	1	CL	R
-1	r	> 2	2	1	CL	R
$-r$	1	> 2	2	1	CL	R
0	$\geq r+1$	> 2	≥ 2	0	CF	R
$\leq -(r+1)$	0	> 2	≥ 2	0	CF	R
≤ -2	≥ 2	> 2	≥ 2	0	CF	R

CL: Carry-limited, CF: carry-free, N: nonredundant, R: redundant

$X = x_{k-1} \ldots x_0$ and $Y = y_{k-1} \ldots y_0$, from the same RDNS, does not depend on k. In other words, worst-case carry propagation chain can occur within a limited number l of consecutive digits. The special case of $l = 0$ is called carry-free.

8.3.1 Constant Time Addition

Classification of k-digit positional number systems, into redundant and nonredundant, and conditions for constant-time addition, with thorough details and proofs, can be found elsewhere [6,10]. However, some more useful definitions and examples are given below, followed by a summary of aforementioned classification in Table 8.4.

Definition 8.12: *Digit Addition* Digit addition in position $0 \leq i \leq k - 1$ of two k-digit radix-r operands $X = x_{k-1} \ldots x_0$ and $Y = y_{k-1} \ldots y_0$, is conventionally defined as $rc_{i+1} + s_i = p_i + c_i$, for $0 \leq i \leq k - 1$, $c_0 = 0$, where $p_i = x_i + y_i$ is called position sum, with c_i and c_{i+1} being carry digits entering into and outgoing from position i, respectively, and $s_i, x_i, y_i \in [\alpha, \beta]$.

Definition 8.13: *Carry-Sink Position* A redundant position i (per Definition 8.10) in adding X and Y (per Definition 8.12) is called carry-sink position if c_{i+1} does not depend on c_i for any $x_i, y_i, x_{i-1}, y_{i-1} \in [\alpha, \beta]$. We postulate that position 0 is carry-sink.

Example 8.8: *Carry-Sink* Consider a radix-10 RDNS with maximally redundant decimal digit set D = $[0, 15]$. Since position sums (see Definition 8.12) $p \in [0, 30]$ the carries of digit additions (per Definition 8.12) satisfy $c \in [0, 3]$.

Equation (8.3) shows that carry-out of position i (i.e., c_{i+1}) does not depend on carry-in c_i; hence, position i with D = [0, 15] is carry-sink.

$$c_{i+1} = \begin{cases} 0, & \text{if } 0 \le p_i \le 9 \\ 1, & \text{if } 10 \le p_i \le 19 \\ 2, & \text{if } 20 \le p_i \le 29 \\ 3, & \text{if } p_i = 30 \end{cases} \tag{8.3}$$

Note that $0 \le s_i \le 12$, since given $0 \le c_i \le 3$ the following holds.

$$10c_{i+1} \le p_i < 10\,(c_{i+1} + 1) \Rightarrow 0 \le s_i = p_i + c_i - 10c_{i+1} < 10 + 3$$

Definition 8.14: *Carry-Free Addition* If all radix positions of an RDNS are carry-sink, addition of two numbers X and Y (per Definition 8.12) is carry-free, in that a carry c_{i+1}, which is generated in position i, and its weight is r^{i+1}, will sink in (i.e., not propagate beyond) position $i + 1$.

The maximally redundant number system of Example 8.8 has the carry-free property. However, it is interesting to note that the carry derivation scheme, described therein by Eq. (8.3), actually shrinks the input digit set D = [0, 15] to output digit set [0, 12], since $0 \le s_i \le 12$. To take advantage of the full capacity of maximally redundant digit set [0, 15], we can introduce some imprecision (or overlapping) in p_i intervals that lead to different assignments for c_{i+1}. This is shown by Eq. (8.4), which leads to $s_i \in [0, 15]$.

$$c_{i+1} = \begin{cases} 0, & \text{if } 0 \le p_i \le 13 \\ 1, & \text{if } 10 \le p_i \le 23 \\ 2, & \text{if } 20 \le p_i \le 30 \end{cases} \tag{8.4}$$

Definition 8.15: *Carry-Limited Addition* In the process of adding two numbers X and Y (per Definition 8.12), if any s_i ($2 \le i \le k - 1$), as well as on x_i, y_i, x_{i-1}, and y_{i-1}, depends on limited number of other less significant digits (e.g., x_{i-2} and y_{i-2}), the addition operation is called carry-limited.

Example 8.9: *RDNS with Carry-Limited Property* Recall the radix-10 digit set $\{0, \ldots 9, A\}$ from Example 8.7 and assume that in a decimal position $i - 2$ there are two A-valued digits to be added, where $p_{i-2} = A + A = 20$. This has to be decomposed either to $c_{i-1} = 2$, and $w_{i-2} = 0$, or to $c_{i-1} = 1$, and $w_{i-2} = A$. In the former case a possible $w_{i-1} = 9$, and in the latter $w_{i-1} = A$, both lead to $s_{i-1} = 11$, which is out of the valid range. However, this invalid sum digit can be further decomposed to $c_i' = 1$ $w_{i-1}' = 1$, which leads to $s_i' \le 10$, since in the second decomposition $w'_{i-1} \leq 9$. This is actually a double decomposition scheme, where s'_i depends on w'_i (and thus to x_i and y_i, the input digits in position i), and c'_i, which depends on s_{i-1}, which in turn depends on w_{i-1} (and thus to x_{i-1} and y_{i-1}) and c_{i-1}, where finally the latter is a function of x_{i-2} and y_{i-2} due to its dependency to p_{i-2}.

Algorithm 8.1: *General Carry-Free Addition*
Inputs: Radix-r k-digit numbers $X = x_{k-1} \dots x_0$ and $Y = y_{k-1} \dots y_0$, with x_i, $y_i \in [\alpha, \beta]$ ($0 \leq i \leq k - 1$), such that α and β satisfy the conditions in Table 8.4 for $l = 0$.
Output: k-digit sum $S = s_{k-1} \dots s_0$ and overall carry digit c_k, such that $r^k c_k + S = X + Y$.

(1) Compute, in parallel, all positional sums $p_i = x_i + y_i$, for $0 \leq i \leq k - 1$.
(2) Extract, in parallel, all carry digits c_{i+1} and interim sum digits w_i, for $0 \leq i \leq k - 1$, such that $p_i = w_i + rc_{i+1}$ and $\alpha - c_i^{\min} \leq w_i \leq \beta - c_i^{\max}$, where $c_i^{\min}$ and $c_i^{\max}$, as functions of α and β, are minimum and maximum c_i-values that satisfy the conditions required for position i to be a sink per Definition 8.13.
(3) Compute, in parallel, all sum digits $s_i = w_i + c_i$, for $0 \leq i \leq k - 1$.

Regarding Step (2), recall Example 8.8, where $c_i^{\min} = 0$ and $c_i^{\max} = 3$ and in the modified version shown by Eq. (8.4), $c_i^{\max} = 2$.

Some of many ways for circuit realization of this algorithm, will be discussed in Section 8.4, where in order to speed up carry-free addition, we describe few techniques to effectively combine the above three steps into one with minimal or no hardware overhead.

8.3.2 Carry-Save Addition

Definition 8.16 : *Carry-Save Representation* Carry-save representation of a radix-r minimally redundant (Definition 8.8) digit set $D_{\min} = [\alpha, \alpha + r]$ (with $\rho = 1$) consists of the following two parts:

1) Main part contains a $b_{\min}$-bit unsigned (for $\alpha = 0$), or NTC (for $\alpha < 0$) number that represent $[\alpha, \alpha + r)$.
2) An extra least significant posibit, as the saved carry.

We remind the reader that $b_{\min} = \lceil \log \xi_{D_{\min}} \rceil = \lceil \log (r + 1) \rceil$. The rationale for "CS" designation will be explained within Definition 8.17.

Example 8.10 : *Binary, Decimal, and hexadecimal CS Digit Sets* Binary CS representation is well known and used as intermediate redundant encoding for binary partial products. Binary CS digit set is $[0, 2]$ and is normally represented

	CS	BSD	DCS	HCS	HSD
Main	•	○	• • • •	• • • •	○ • • •
Stored	•	•	•		•

Figure 8.3 Dot notation representation of CS digits.

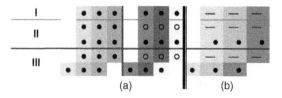

Figure 8.4 (a): Binary CS and (b): decimal CS addition.

as two equally weighted posibits. Binary-signed digits, when encoded as an equally weighted negabit–posibit pair, can also be regarded as a CS encoding of digit set [−1, 1], where the main part is a single negabit. The minimally redundant radix-10 digit set [0, 10], radix-16 digit set [0, 17], and radix-16 digit set [−8, 8] can also be represented by CS encodings, where the main parts are unsigned, unsigned, and four-bit NTC numbers, respectively. Figure 8.3 depicts CS, BSD, decimal CS (DCS), hexadecimal CS (HCS), and minimally redundant hexadecimal SD (HSD) in dot notation.

Definition 8.17: *Carry-Save Addition* One of the operands and result of carry-save addition (CSA) are CS numbers in $[\alpha, \alpha + r]$, and the other operand is an unsigned number in $[0, r − 1]$, such that positional sum digits belong to $[\alpha, \alpha + 2r − 1]$, which are decomposed to a main part in $[\alpha, \alpha + r − 1]$ and a carry to be saved in the next higher position. Note that the operation is carry-free, per Definition 8.14, since all weighted positions are carry-sink, that is, the saved carry c_{i+1} that is generated in a radix-r position i depends only on the same-position operands.

Example 8.11: *Binary and Decimal CSA* Figure 8.4a [b] depicts binary [decimal] CSA in dot-notation representation, where equally shaded bits represent inputs and outputs of the same adder cells (i.e., FAs [decimal FAs]), — denotes a BCD digit, Part I contains nonredundant three-digit numbers, and Part II [III] represents three [four]-digit CS and DCS redundant numbers.

Although CSA is carry-free, limited carry propagation occurs in adding two CS numbers. The reason, as indicated in Table 8.4, is that $\rho = 1$. Nevertheless, such fully redundant CS addition is of practical use in design and implementation of binary multipliers, where the main partial product reduction cell is a

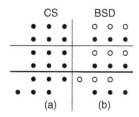

Figure 8.5 Application of (4;2) compressors in reducing two CS (a) and BSD (b) numbers.

Weights	4	2	1	4	2	1
WBS	○	○	●	‾●	‾●	‾○
	●				‾○	
Digit set	[-6,5]			[-5,6]		

Figure 8.6 WBS negation

(4; 2) compressor ($_2^4C$). An array of k $_2^4C$s is capable of adding two k-digit CS numbers with one-digit carry propagation. The same structure is capable of reducing two BSD numbers to one, which is of practical use when partial products are represented in BSD [8]. Figure 8.5a [b] depicts reduction of three-digit CS [BSD] numbers via an array of three $_2^4C$s.

8.3.3 Borrow Free Subtraction

Subtraction can be done borrow-freely in redundant number systems. However, as in nonredundant number systems, it is normally performed via adding the minuend and the negated subtrahend. One way of negating a number is to negate all digits in parallel. If the digits are WBS encoded, the simplest way is to invert the logical states of all bits and regard every negabit [posibit] as a posibit [negabit]. This is justified by Eq. (8.5), where the arithmetic values of x and y (per Definition 8.4), when representing posibits [negabits] are denoted as x^p [x^n] and y^p [y^n].

$$-x^p = -x = -(1 - \bar{x}) = \bar{x} - 1 = \bar{x}^n, \quad -y^n = -(y - 1) = 1 - y = \bar{y} = \bar{y}^p \quad (8.5)$$

Example 8.12: *WBS Negation* Figure 8.6 depicts a WBS encoding of digit set [−6, 5], and that of its negated set [−5, 6]. For example, assume logical state 0 for both negabits and 1 for both posibits, which leads to $-6 + 5 = -1$, for the original digit in the left. If we invert logical states of all bits to get at 1 [0] for posibits [negabits] in the right digit, its value will be $6 - 5 = 1$.

Example 8.13 : *Two-Deep WBS Encoding* Encoding of digit sets in Figure 8.6 are faithful and two-deep; properties that are both desired in realization of arithmetic operations. To derive two-deep faithful WBS encoding

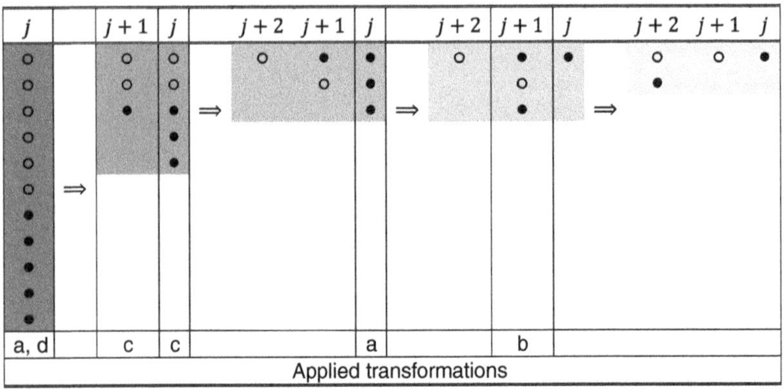

j		$j+1$	j	$j+2$	$j+1$	j	$j+2$	$j+1$	j	$j+2$	$j+1$	j
a, d		c	c			a			b			

Applied transformations

Figure 8.7 Evolution of 2-deep WBS encoding of digit set $[-6, 5]$.

of an arbitrary digit set $[-\alpha, \beta]$, one can start with a 2^j-weighted single-column WBS, consisting of α negabits and β posibits, and apply appropriate transformation of Table 8.5, until no column with depth more than two exists [9]. For instance, Figure 8.7 depicts such transformation for $\alpha = 6$, and $\beta = 5$ (i.e., the same digit set as in Example 8.12).

Table 8.5 Transformations for derivation of 2-deep WBS encoding.

		Position			WBS range
		j	$j+1$	j	
Transformation	**a**	• • •	•	•	$[0, 3]$
	b	• •	•	○	$[-1, 2]$
	c	○ •	○	•	$[-2, 1]$
	d	○ ○	○	○	$[-3, 0]$

8.4 Basic Arithmetic Circuits for Redundant Number Systems

Logical circuits that implement arithmetic operations are normally composed of basic arithmetic cells such as full adders (FA), half adders (HA), pseudo half adders (HA^{+1}; i.e., enforced carry FA), (4; 2) compressors ($_2^4C$), parallel prefix computing nodes, carry look-ahead blocks, and other less common cells (e.g., (7; 2) compressors). There are more specialized cells in high-radix division (e.g., quotient digit selection block), and decimal arithmetic (e.g., decimal full adder, precomputed multiples logic). However, due to widespread use of FAs and $_2^4C$s, considerable design effort has been put toward optimized transistor-level realization of such cells (e.g., [18–24]) that encourages the design engineers to use them in place of customized gate-level or transistor-level designs. For example, the redundant binary adder of Reference [25], which is actually a BSD adder, can be replaced by a $_2^4C$ (see Figure 8.5), with far less area consumption [5].

The key clue to capability of standard FAs and $_2^4C$s in handling arbitrary mix of posibits and negabits is the inverse encoding of negabits versus the conventional interpretation of most significant bits of two's complement numbers [4]. For example, conventional $_2^4C$s implement arithmetic Eq. (8.6), corresponding to Figure 8.5a, where w, x, y, and z are the main posibit inputs, c_i and c_o are posibit carry-in and carry-out, respectively, and s and c are posibit sum and saved carry. However, in case of adding two BSD numbers, each $_2^4C$ of the corresponding $_2^4C$ realization implements either Eq. (8.7) (corresponding to Figure 8.5b), or Eq. (8.8), depending on the polarity of carry-in. Figure 8.8 depicts FA realizations of such $_2^4C$s.

$$w + x + y + z + c_i = 2\left(c_o + c\right) + s \tag{8.6}$$

$$w + x^n + y + z^n + c_i = w + x - 1 + y + z - 1 + c_i = w + x + y + z + c_i - 2$$
$$= 2\left(c_o + c\right) + s - 2 = 2\left(c_o + c - 1\right) + s = 2\left(c_o + c^n\right) + s \tag{8.7}$$

$$w^n + x + y^n + z + c_i^n = w - 1 + x + y - 1 + z + c_i - 1$$
$$= w + x + y + z + c_i - 3 \tag{8.8}$$
$$= 2\left(c_o - 1 + c\right) + s - 1 = 2\left(c_o^n + c\right) + s^n$$

8.4.1 Circuit Realization of Carry-Free Adders

Figure 8.9 depicts an implementation of Algorithm 8.1 for three consecutive positions $i - 1$, i, and $i + 1$. The comparator in position i checks p_i against α and β, in order to decide on value of carry-digit c_{i+1}. In the practical case of $r = 2^h$, the adders in Steps (1) and (3) can be any $(h + 1)$-bit adder. However,

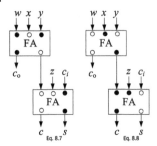

Figure 8.8 $\frac{4}{2}C$ realizations for Eqs (8.7) and (8.8).

a variety of digit set encodings are employed in practice, in order to speed up carry-free addition, so that there is only one $(h + 1)$-bit operation.

8.4.2 Fast Maximally Redundant Carry-Free Adders

Maximally redundant radix-2^h digits in $\left[-2^h, 2^h - 1\right]$ (e.g., $[-16, 15]$, for $h = 4$) can be represented by $(h + 1)$-bit NTC numbers. A fast addition scheme, with only one h-bit addition is described by Figure 8.10, for $h = 4$. The bits that belong to consecutive digits are distinguished by different shadings. The constant negabits (o_1) are inserted, in the primary WBS sum, to help in getting at the right bit pattern for final sum digits, where index 1 denotes logical state of a negabit with arithmetic value 0. This cost-free four-deep WBS sum is reduced to the next one via a carry-save adder. This is followed by one h-bit adder per radix-2^h position (distinguished via shading strength) to reach at the required sum digits, where gray-shaded bits belong to preceding and succeeding positions.

Figure 8.11 depicts a subtraction scheme similar to that of addition of Figure 8.10, where the subtrahend digits are two's complemented, as all the relevant inverted bits have over bars. However, the required +1 per digit is

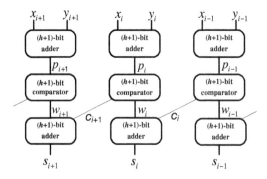

Figure 8.9 Carry-free addition architecture based on Algorithm 8.1.

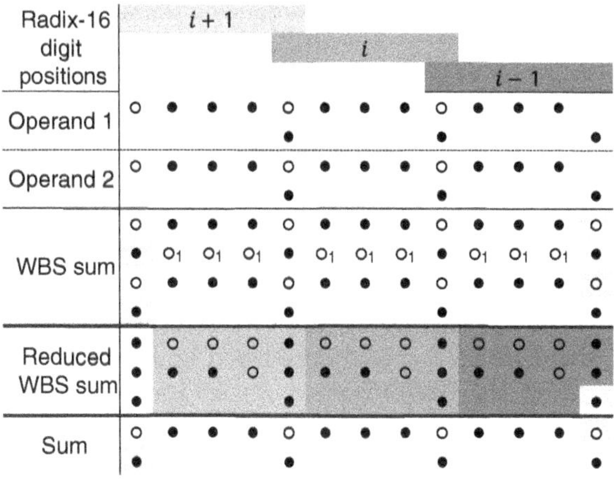

Figure 8.10 Carry-free addition of maximally redundant numbers.

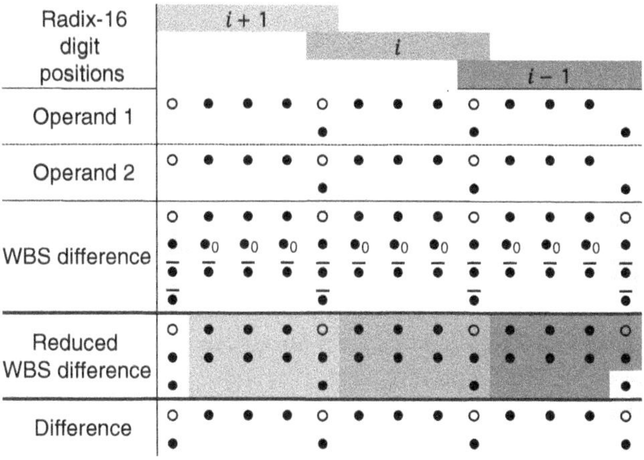

Figure 8.11 Carry-free subtraction of maximally redundant numbers.

not enforced directly (except for the least significant digit). Instead, the already inverted negabit of the less significant subtrahend digit is replaced by a posibit with the same logical state, since arithmetic value of a negabit x is $x - 1$. An advantage of this subtraction scheme is that a unified add/subtract circuit can be devised as in Figure 8.12, where $S = 0\,[1]$ controls add [sub] function of the circuit.

Logical state of indexed negabits of Figure 8.10 and indexed posibits of Figure 8.11 both are equal to $\bar{S}$. Note that the overhead of this unification is one XOR gate per bit of subtrahend and replacement of HA^{+1}s (HAs) of carry-save stage of Figure 8.10 (Figure 8.11) with FAs (see Figure 8.12).

8.4.3 Carry-Free Addition of Symmetric Maximally Redundant Numbers

Carry-free add/subtract scheme of Section 8.4.2 can be applied to symmetric maximally redundant numbers with digit set $\left[-2^h + 1,\ 2^h - 1\right]$ (e.g., $[-15, 15]$, for $h = 4$), which is also known as maximally redundant signed digit (MRSD) set [17]. However, in case of subtraction, particular input digits may lead to a difference digit as low as -2^h, which is off the valid digit set. For example, as a numerical instance, consider Figure 8.13, where index-free bits can assume either logical state. Note that the value of middle difference digit is -16. This might not cause any problem for some applications. However, for some others, it may be desirable that the digit set is exactly preserved under arithmetic operations. Therefore, in this section, we provide an MRSD addition scheme that satisfies this property of digit-set preservation.

Recalling Algorithm 8.1 (carry-free addition), Eq. (8.9) describes carry extraction from p_i, where maximum possible overlapping in p_i intervals is allowed.

$$c_{i+1} = \begin{cases} -1, \text{if} - 2^{h+1} + 2 \le p_i \le -2 \Rightarrow -2^h + 2 \le w_i \le 2^h - 2 \\ 0, \quad \text{if} - 2^h + 2 \le p_i \le 2^h - 2 \Rightarrow -2^h + 2 \le w_i \le 2^h - 2 \\ 1, \quad \text{if} \qquad 2 \le p_i \le 2^{h+1} - 2 \Rightarrow -2^h + 2 \le w_i \le 2^h - 2 \end{cases} \quad (8.9)$$

If c_{i+1} and w_i can be directly extracted, in constant time (i.e., not depending on h), from radix-2^h operand digits x_i and y_i, then the only O(h) operation would be $s_i = w_i + c_i$. Let x_i and y_i belong to MRSD set $\left[-2^h + 1,\ 2^h - 1\right]$ and be represented by $(h + 1)$-bit NTC numbers. Then, we let p_i to be represented as a two-deep WBS number that is obtained at no cost, as is illustrated in Figure 8.14 for $h = 4$. Now decompose p_i, again at no cost, to u_i and v_i such that $p_i = 2^{h-1}u_i + v_i$, where the WBS representation of u_i and v_i are distinguished by light and strong shading.

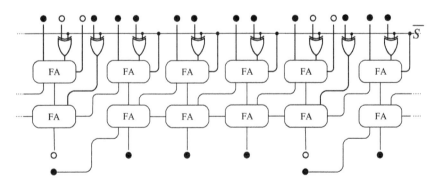

Figure 8.12 Unified add/subtract architecture for radix-16 maximally redundant numbers.

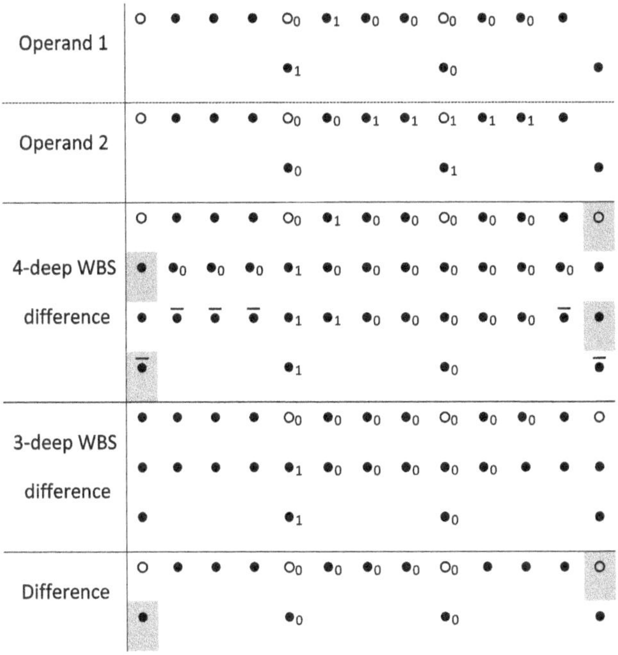

Operand 1

Operand 2

4-deep WBS

difference

3-deep WBS

difference

Difference

Figure 8.13 Special subtraction case.

Equation (8.10) shows that the key operation in finding a WBS encoding of sum digit s_i is to decompose u_i, as $u_i = 2c_{i+1} + z_i$, after which NTC version of s_i can be easily obtained via reducing the WBS sum to a two-deep one, followed by an h-bit addition. Table 8.6 shows the required decomposition, such that w_i and s_i fall within valid ranges, and the digit set $\left[-2^h + 1, 2^h - 1\right]$ is preserved (see s_i in top and bottom rows). Note that in case of $u_i = -4$ [-3], both [one of] x_i and y_i are negative and greater than -2^h, which leads to $v_i \geq 2$ [1].

$$p_i = 2^{h-1}u_i + v_i = 2^h c_{i+1} + 2^{h-1}z_i + v_i \Rightarrow w_i = 2^{h-1}z_i + v_i \Rightarrow s_i$$
$$= 2^{h-1}z_i + v_i + c_i \tag{8.10}$$

The ranges of c_{i+1} and z_i, in Table 8.6, allow for representing the former as an equally weighted posibit/negabit pair (c'_{i+1}, c''_{i+1}) and the latter as two equally weighted negabits (z'_i, z''_i). The corresponding logical expressions can be easily

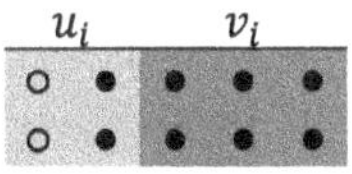

u_i v_i **Figure 8.14** Decomposition of p_i.

Table 8.6 Decomposition of u_i leading to valid ranges for w_i and s_i.

u_i	c_{i+1}	z_i	v_i	$w_i = 2^{h-1}z_i + v_i$	$s_i = w_i + c_i$
-4	-1	-2	$[2, 2^h - 2]$	$[-2^h + 2, -2]$	$[-2^h + 1, -1]$
-3	-1	-1	$[1, 2^h - 2]$	$[-2^{(h-1)} + 1, 2^{(h-1)} - 2]$	$[-2^{h-1}, 2^{h-1} - 1]$
-2	-1	0	$[0, 2^h - 2]$	$[0, 2^h - 2]$	$[-1, 2^h - 1]$
-1	0	-1	$[0, 2^h - 2]$	$[-2^{h-1}, 2^{h-1} - 2]$	$[-2^{h-1} - 1, 2^{h-1} - 1]$
0	0	0	$[0, 2^h - 2]$	$[0, 2^h - 2]$	$[-1, 2^h - 1]$
1	1	-1	$[0, 2^h - 2]$	$[-2^{h-1}, 2^{h-1} - 2]$	$[-2^{h-1} - 1, 2^{h-1} - 1]$
2	1	0	$[0, 2^h - 2]$	$[0, 2^h - 2]$	$[-1, 2^h - 1]$

driven via a 16-entry truth table in terms of bits of u_i, as in Eq. (8.11), where u_i is represented by two negabits (x_i^h, y_i^h) and two posibits (x_i^{h-1}, y_i^{h-1}).

$$c'_{i+1} = \left(x_i^h \vee y_i^h\right)\left(x_i^{h-1} \vee y_i^{h-1}\right), \; c''_{i+1} = x_i^h y_i^h,$$

$$z'_i = \overline{x_i^{h-1}}\left(x_i^h \vee y_i^h\right) \vee y_i^{h-1}, \qquad z''_i = \overline{y_i^{h-1}}\left(x_i^h \vee y_i^h\right) \vee x_i^{h-1}$$

(8.11)

Figure 8.15 depicts WBS representation of s_i that is composed of four bit sets; (1) v_i (six-posibit dark shaded), which is directly wired from input operands, (2) z_i and (3) c_i that are obtained via decomposition of u_i and u_{i-1} (light shaded), respectively, and (4) the one-indexed negabits (with arithmetic worth of 0) that are inserted, in order to help in getting at the desired NTC output. The first three rows are reduced, via a CSA, to dark-shaded bits within the next WBS s_i, where the carry bits are kept intact. To get at the final sum, this WBS s_i is manipulated by an h-bit adder, where arithmetic value of the negabit carry-out together with the leftmost dark shaded negabit cannot be less than -1 according to s_i values in Table 8.6. Therefore, the negabit of final s_i is derived via ANDing the just mentioned two negabits. The overall delay is $2(h + 2)\Delta G$, as is shown in Figure 8.16, which depicts the required circuitry for addition scheme of Figure 8.15. Several other designs for addition of maximally redundant signed digit operands can be found in Reference [12]. Note that the latency of this addition scheme is the same as that of Figure 8.10.

8.4.4 Addition and Subtraction of Stored-Carry Encoded Redundant Operands

Another way of optimized realization of Algorithm 8.1 is the removal of addition in Step (3) via storing c_i along the interim sum w_i. However, this saving is advantageous, only if the algorithm is modified and efficiently implemented for stored-carry inputs. Note that such stored-carry representation is different from CS, in that here the stored-carry is normally a three-valued digit with at

	z_i	v_i	$\leftarrow c_i$
	○	● ● ●	
s_i	○	● ● ●	
	○$_1$ ○$_1$	○$_1$	○$_1$
			○ $\leftarrow c_i'$
			● $\leftarrow c_i''$
	○ ○ ○	○	
s_i	○ ● ● ●	○	
			●
s_i	○ ● ● ●	●	

Figure 8.15 Reduction of WBS s_i to NTC via CSA reduction and h-bit addition.

least two-bit encoding. The extra bit, compared to when both operands are CS, is actually the cost of carry-free addition versus limited-carry propagation in CSA. For example, in case of signed digit addition, the stored-carry in $[-1, 1]$ is faithfully represented by an equally weighted posibit/negabit pair. Table 8.7 describes this situation for radix-16 maximally redundant signed digit set $[-15, 15]$, where the five-bit two's complement main part is normally used unfaithfully (i.e., it can also represent -16), as was the case in Figure 8.14. However, the weighted bit set that is composed of the main part and stored-carry actually represents the digit set $[-17, 16]$. The wider digit set leads to wider representation range that might be indeed desirable, since it hides possible intermediate overflows in multioperand additions. One drawback, though, is that the encoding efficiency (see Definition 8.6) is dropped by 73%. Another one is that the required carry-digit set, by the carry-free addition Algorithm 8.1, widens to $[-2, 2]$, thus requiring at least three bits for carry digits.

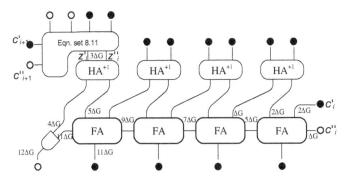

Figure 8.16 Circuit realization of Figure 8.15.

Table 8.7 Evolution of stored unibit representation.

Dot notation	Digit set	Encoding efficiency
○ ● ● ● ●	$[-15,15]$	$31/32$
○ ● ● ● ●		
●	$[-17,16]$	$34/128 = 17/64$
○		
○ ● ● ● ●		
● ○	$[-17,16]$	$34/64 = 17/32$

To increase the drastically lowered encoding efficiency, observe that the three least significant bits of the stored-carry representation form the digit set $[-1, 2]$. This digit set can be equivalently represented by a posibit/unibit pair (see last row of Table 8.7, where the unibit is denoted by ○). The rationale for this conversion is explained via Table 8.8. Furthermore, contents of this table exactly coincide with an FA truth table; that is converting the source three-bit collection to posibit/unibit pair can be done via an FA, where carry [sum] output represents the unibit [posibit]. This conversion can be equivalently described via Eq. (8.12), where $c^u = 2c - 1$ represents arithmetic value of unibit c (i.e., $c^u = -1$ [1], for $c = 0$ [1]).

$$x + y + z^n = (x + y + z) - 1 = (2c + s) - 1 = (2c - 1) + s = c^u + s \qquad (8.12)$$

8.4.4.1 Stored Unibit Carry Free Adders

Figure 8.17 depicts three consecutive digit-slices of carry-free addition process for two radix-16 stored unibit numbers. The two equally weighted unibits are replaced by a doubly weighted posibit/negabit pair, which is justified by truth

Table 8.8 3:2 fit reduction.

●	●	○	Sum	○	●
0	0	0	-1	0	0
0	0	1	0	0	1
0	1	0	0	0	1
0	1	1	1	1	0
1	0	0	0	0	1
1	0	1	1	1	0
1	1	0	1	1	0
1	1	1	2	1	1

Table 8.9 Transformation of two unibits.

⦿	⦿	Sum	●	○
0	0	−2	0	0
0	1	0	0	1
1	0	0	1	0
1	1	2	1	1

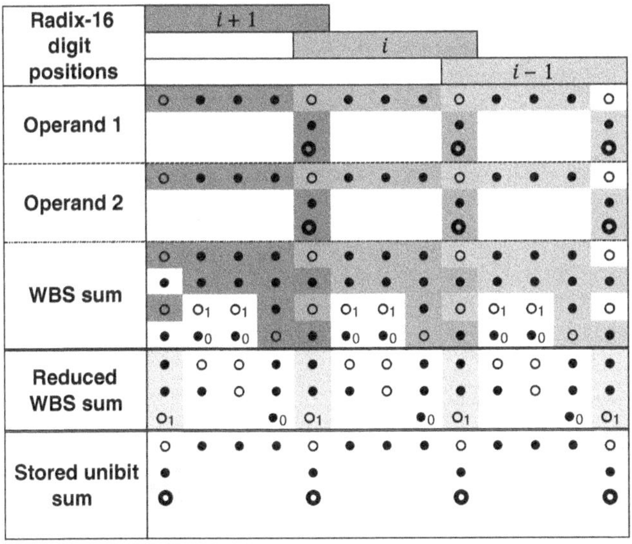

Figure 8.17 Three digit-slice of addition of two stored-unibit numbers.

in Table 8.9. Note that this is a totally cost-free operation such that the posibit and the negabit are just logical copies of unibits. This transformation can be also expressed via Eq. (8.13), where x^u, y^u, x^p, and y^n represent arithmetic values of source unibits, target posibit and negabit, and x and y denote both the logical states of source unibits and those of target posibit and negabit, respectively. Note that since the addition scheme of Figure 8.17 does not directly follow Algorithm 8.1, no three-bit carry-digits are actually generated.

$$x^u + y^u = 2x - 1 + 2y - 1 = 2(x + (y - 1)) = 2(x^p + y^n) \tag{8.13}$$

Operand digits in Figure 8.17 are distinguished via shading strength. WBS sum represents a four-deep encoding of sum digits that is obviously obtained at no cost. Note that the arithmetic value of inserted one-indexed negabits [0-indexed posibits], with logical state 1 [0], are zero. These constant bits help in getting at the right encoding for final sum digits. This WBS sum is compressed to another one via four $_2^4C$s per digit. The resulted WBS sum is further reduced to final sum digits via three-bit adders ($(h-1)$-bit adders in case of radix-2^h), while the three shaded bits are reduced to a posibit/unibit pair, via an enforced carry FA (i.e., HA^{+1}) (based on Table 8.8). The light-shaded bit [s] in the rightmost [leftmost] column belong to digit position $i+2$ [$i-2$].

8.4.4.2 Stored Unibit Subtraction

Figure 8.18 depicts a digit-slice of carry-free subtraction for the same stored-unibit number system, where Operand 2 appears in negated form based on Figure 8.6, and the shade-free posibit and negabit in position 0 belong to the less significant digit. Note that reduction cells and adders are the same in both Figures 8.17 and 8.18, while value and polarity of constant bits are opposite. This calls for a unified circuitry for addition and subtraction, as is depicted by Figure 8.19, where $S = 1$ [0] denotes subtraction [addition], and critical delay path travels through a unification XOR gate, a $_2^4C$ (normally with three XORs latency) and an $(h-1)$-bit adder. Therefore, the total delay roughly amounts to that of one $(h+3)$-bit adder.

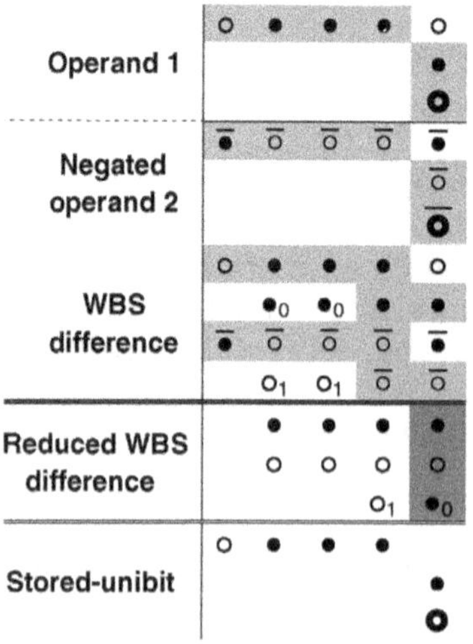

Figure 8.18 A digit-slice of subtraction of two stored-unibit numbers.

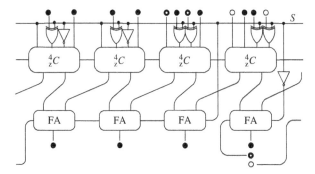

Figure 8.19 One radix-16 digit-slice of unified stored-unibit adder/subtractor.

8.5 Binary to Redundant Conversion and the Reverse

Conversion of n-bit binary numbers to any redundant representation can be generally done in constant time not depending on n (i.e., $O(1)$), while the reverse is a carry-propagate operation (i.e., at best $O(\log n)$). Conversion details and circuitry, greatly depends on the redundant digit set and encoding, which is used for representation of redundant digits. Therefore, we discuss conversion methods for the number systems that were presented in Sections 8.4.2 and 8.4.4, separately.

8.5.1 Binary to MRSD Conversion and the Reverse

Recall the radix-2^h maximally redundant signed digit number system with digit set $[-2^h + 1, \ 2^h - 1]$, where each digit is represented by an $(h + 1)$-bit NTC number. Figure 8.20 depicts the cost-free conversion of a 16-bit TC number to the equivalent four-digit (digits are distinguished with different shadings) radix-16 MRSD number (i.e., $h = 4$), where a one-indexed negabit (with arithmetic value of zero) is inserted in each 2^{4k}-weighted ($k \in \{1, \ 2, \ 3\}$) position.

To complete the forward conversion process, we sign extend the source, as illustrated by gray dots, where logical state of the corresponding target posibit is the same as the most significant bit of the source, and that of the negabit is the opposite. This process is justified as follows.

Let the logical state of the most significant source posibit (i.e., the gray dot in Figure 8.20), whose weight is -2^{15}, be denoted by x. Then we replace it with

Figure 8.20 2's complement to MRSD conversion.

Radix-16 o • • • • o • • • • o • • • • o • • • •
MRSD   • • • • • • • • • • • • • •

```
          o • • • • o • • • • o • • • • o • • • •
        •o o1 o1 o1 • o1 o1 o1 • o1 o1 o1 •
                                              o1
```

TC •

Figure 8.21 MRSD to nonredundant conversion.

a 2^{15}-weighted posibit with the same logical state and a 2^{16}-weighted negabit, whose logical state is $\bar{x}$. Arithmetic value of source is $-2^{15}x$, while that of the two target bits, collectively, is $2^{16}(\bar{x}-1)+2^{15}x = 2^{16}(-x)+2^{15}x = -2^{15}x$ (i.e., equal to source). Therefore, the conversion cost is only one inverter for the most significant source bit.

The procedure for reverse conversion is illustrated by Figure 8.21, where one-indexed negabits help to get at the desired TC output, which is achieved via inputting the shaded bits to a 13-bit adder of any nonredundant architecture (e.g., ripple carry or carry look-ahead). Note that the negabit carry out of this adder should be inverted to be valid as the most significant bit of the TC result.

8.5.2 Binary to Stored Unibit Conversion and the Reverse

The forward conversion is similar to what we did in Section 8.5.1. The only difference is that, as illustrated in Figure 8.22, we insert in each 2^{4k}-weighted ($k \in \{1, 2, 3\}$) position, a zero-indexed negabit and a one-indexed unibit with collective arithmetic value of 0. However, in position 0, where no negabit is required, we replace the source posibit with an inverted one and insert a unibit whose logical state is the same as the source posibit. Therefore, the conversion cost is only two inverters for the least and most significant source bits.

The reverse conversion, as is shown by Figure 8.23, is not as easy. We start with a cost free conversion of each equally weighted posibit/unibit pair to a negabit, and a doubly weighted posibit, whose logical states are the same as that of the source posibit and unibit, respectively. Then, it is easy to see that applying a word-wide adder, as in Section 8.5.1, does not lead to the desired TC result.

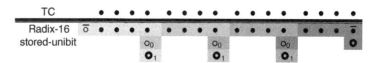

Figure 8.22 2's complement to stored-unibit conversion.

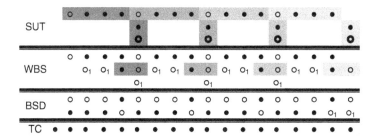

Figure 8.23 Stored-unibit to 2's complement conversion.

Therefore, we insert arithmetically zero-valued negabits, where necessary, and apply HAs and HA^{+1}s as appropriate to get at an equivalent BSD representation, which is just right for applying the word-wide adder, where again two constant bits are inserted. Note that the negabit carry out should be inverted to represent the sign bit of the TC result.

A general method that converts any two-deep WBS number to equivalent BSD number (quite similar to the above method) is described elsewhere [5]. Significance of BSD target encoding is that it can be directly converted to the equivalent TC number via any realization of a word-wide adder (e.g., ripple, carry look-ahead, and carry skip).

8.6 Special Arithmetic Circuits for Redundant Number Systems

Specialized circuits for multiplication and division by power-of-two constants are common in binary arithmetic that are usually implemented via left and right arithmetic shifts, respectively. The same shift operations are required for normalizing operands of floating point addition that are typically handled by parallel shifters. Similar shifts are needed in redundant digit arithmetic.

8.6.1 Radix-2^h MRSD Arithmetic Shifts

While h-position shifts are straightforward, smaller shifts need special treatment. For example, Figure 8.24 illustrates the situation for one binary left shift on radix-16 MRSD numbers, where equally shaded bits belong to same digit. Part II represents a WBS, where each bit is left shifted with respect to WBS of Part I. However, the radix-16 MRSD encoding is distorted. Indexed-one negabits are inserted prior to three-bit binary addition per radix-16 digit that leads to the desired original encoding. No adder is required for the least significant digit, where insertion of a one-indexed negabit and a zero-indexed posibit is

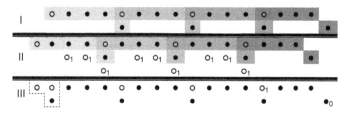

Figure 8.24 Binary left shift of 4-digit radix-16 MRSD number.

sufficient. The three dash-surrounded bits in Part III should be arithmetically reduced to one negabit, otherwise the ×2 operation overflows.

One bit right shift is easier and requires one HA^{+1} per radix-2^h digit, except for the most significant digit, where an HA will do. Figure 8.25 depicts this operation for a four-digit radix-16 MRSD number.

8.6.2 Stored Unibit Arithmetic Shifts

Figure 8.26 depicts one-bit right shift, where all bits of the original shifting operand, in Part I, are shifted one binary position to the right as shown in Part II. However, the rightmost posibit-and-unibit pair should not be discarded and need special treatment. Let this pair be denoted by x and y, respectively. Since, in case of $x = y = 1$, their collective value can be equal to 2, we insert a posibit $v = xy$ in the rightmost position in Part II. To transform this shifted operand to the one in Part III, which is in the same format as the original operand, we replace the shaded WBSs in Part II to arithmetically equivalent WBSs in part III that are distinguished by the same shading.

Let the two rightmost posibits in Part II be denoted by u and $v = xy$ and the rightmost posibit [unibit] in Part III as u' [v']. Therefore, $u + v = u' + 2v' - 1 \Rightarrow u + v + 1 = 2v' + u'$, which suggest the use of an HA^{+1} that, by receiving u and v as inputs, produces outputs u' (sum) and v' (carry). Other four-bit transformations are illustrated by Figure 8.27, where p, n, and u variables denote

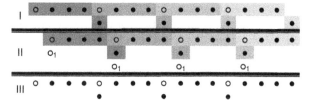

Figure 8.25 Binary right shift of 4-digit radix-16 MRSD number.

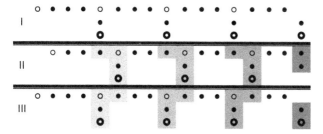

Figure 8.26 Right arithmetic shift of four-digit stored-unibit number.

posibits, negabits, and unibits, respectively, and $n'_1 = p_1$, $p'_0 = n_0 \oplus p_0$, $p'_1 = u_0 \oplus n_0 p_0$, and $u'_1 = u_0 \vee n_0 p_0$.

A conceptually simpler way to arrive at the desired right-shifted result is depicted by Figure 8.28, where the transformation from Part I to Part II is merely cost-free. Let n, p, and u denote logical states of the negabit, posibit, and unibit components of three-deep shaded WBSs of Part I, respectively. These are transformed to part II, as posibits p', p'', and negabit n', respectively, with the same logical states, and equally shaded. The reason is that equivalence of arithmetic value of the three-tuple source and target WBSs can be expressed as $n - 1 + p + 2u - 1 = 2(u - 1) + n + p = 2(n' - 1) + p' + p''$, and that of two-tuple light-shaded WBSs is $2u_0 - 1 + p_0 = 2u_0 + p_0 - 1 = 2p'_1 + n'_1 - 1$. Part III is achieved via a simple binary right shift of all bits. Applying HAs on the very light shaded bits, in Part III, leads to Part IV, where double posibits of the three-deep columns can turn to the desired posibit–unibit pair of Part V, exactly as was the case for the dark-shaded transformation of Figure 8.26. Note that in both cases (i.e., Figures 8.26 and 8.27) the required shifting logic per digit amounts to one HA and one HA^{+1}.

The left shift operation, as depicted by Figure 8.29, can start by one position left shift of all bits (see Part II). The same cost-free transformation that was applied to Part I of Figure 8.28, is now applied here, where a $\circ_1$ is inserted in position 1 for uniformity sake. This leads to Part III, where $(h - 2)$-bit adders ($h = 4$ in Figure 8.29) manipulate the dark-shaded bits. In Part IV,

Logical state		Arithmetic value		Arithmetic value		Logical state	
p_1	n_0	p_1	$n_0 - 1$	$p_1 - 1$	n_0	n'_1	p'_0
	p_0 $\Rightarrow$		p_0 $\Rightarrow$		p_0 $\Rightarrow$		p'_1
	u_0		$2u_0 - 1$		u_0		u'_1
Weight 2	1	2	1	2	1	2	1

Figure 8.27 Four-bit transformations of Figure 8.26.

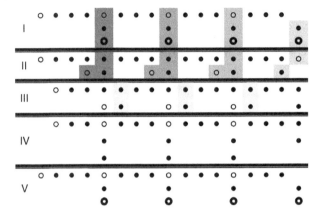

Figure 8.28 Alternative right- shift procedure.

$(h - 1)$-bit adders (that can start at the same time) manipulate the dark-shaded bits, which leads to Part V, where an HA^{+1} per digit terminates the shifting operation.

Figure 8.30 depicts the required circuitry, where the top digits represent binary positions of the corresponding fits, each radix-16 digit-slice is shaded the same as the corresponding digits in Part I of Figure 8.29, and the logic box receives the light-shaded fits of Part V to produce the most significant negabit of Part VI (i.e., U), or signal an overflow (i.e., v). The corresponding logical

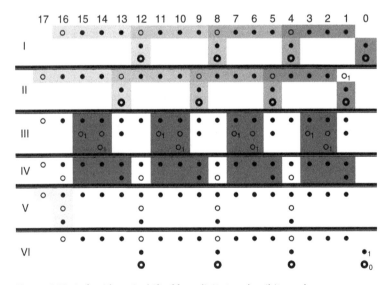

Figure 8.29 Left arithmetic shift of four-digit stored-unibit number.

expressions are described by Eq. (8.14), where inputs (outputs) are denoted by w, x, y, and z (v and u).

$$u = w(x \lor y \lor z) \lor xyz, v = w(yz \lor x(y \lor z)) \lor \bar{w}(\bar{y}\,\bar{z} \lor \bar{x}(\bar{y} \lor \bar{z})) \qquad (8.14)$$

8.6.3 Apparent Overflow

Overflow detection in nonredundant addition is easy and deterministic, where the carry out of the most significant position of sum serves as overflow indicator. However, a similar nonzero carry digit in redundant addition does not necessarily indicate an overflow, in the sense that the collective value of the obtained sum and carry digit may still be in the valid range of the RDNS. For example, a carry-out digit 1 in the addition scheme of Figure 8.10 can be propagated back to the most significant digit of the sum as 2^h. This can be absorbed if the attendant MSD sum digit is negative (i.e., in the range $\left[-2^h + 1, -1\right]$), which leads to a new sum digit in $\left[1, 2^h - 1\right]$. Therefore, we have an apparent overflow [26], which resolves to a real overflow if the most significant sum digit is positive, and is not an overflow, if it is negative. However, if it is 0, the back propagation can go on down to the first nonzero digit, leaving sum digits equal to $2^h - 1$. This process ends either by real overflow detection or sum correction. For example, in case of $h = 4$, a four-digit result with a carry-out 1, as $(1 - 15 - 15 - 15)_{16}$, is numerically equivalent to the four four-digit numbers $(1 - 15 - 15 - 15)_{16}$, $(0\ 1 - 15 - 15)_{16}$, $(0\ 0\ 1 - 15)_{16}$, and $(0\ 0\ 0\ 1)_{16}$.

8.7 Applications

Numbers that are represented in a redundant number system (RDNS) are not, by and large, readable. For instance, recall three alternative representations of 2000, in Example 8.7 (e.g., 1A00), that are hard to understand and commonly unacceptable to be included in any report. On the other hand, it is often the case that redundant representations require wider code words that are not usually supported by register files and memory words. Therefore, such redundant representations have to be converted to their nonredundant equivalent numbers before storage or being sent to an output device. This conversion requires, in the worst case, word-wide carry propagation that jeopardizes the speed gain that is achieved due to redundant representation, unless several arithmetic operations take place on redundant numbers before the need arises for a conversion. This is, therefore, the main characteristic of applications of RDNSs. In this section, we examine few such application examples.

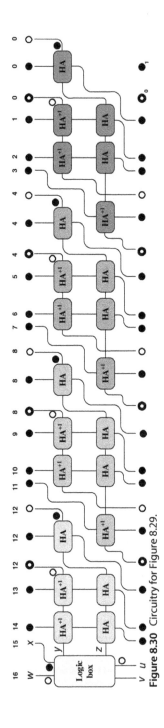

Figure 8.30 Circuitry for Figure 8.29.

8.7.1 Redundant Representation of Partial Products

Multiplication is generally performed in the following three operational phases:

- PPG: Partial product generation
- PPR [PPA]: Partial product reduction [accumulation] in parallel [sequential] realizations
- FPC: Final product computation

The bulk of multiplication effort is due to PPR or PPA that are special cases of multioperand addition, where multiplier designers can freely decide on the representation of intermediate results regardless of input/output representations. Given the large number of operands to be added in PPR or PPA (i.e., normally n operands for $n \times n$ multiplication), representing intermediate results of the required multioperand addition via an RDNS is quite justifiable, as is commonly practiced in most realizations. The CSAs, BSD adders, and $^4_2$Cs are commonly used as PPR cells, where partial products are represented in one of CS formats, per Definition 8.16.

Often PPG leads to nonredundant partial products, such that the first PPR level receives nonredundant operands to add them into a redundant intermediate result, and in sequential multipliers new nonredundant partial products are added to the redundant accumulated partial product. The final product is usually the outcome of redundant to nonredundant conversion.

8.7.2 Recoding the Multiplier to a Redundant Representation

To reduce the number of original partial products that are produced by PPG phase, one may recode the normal binary multiplier to a radix-2^h number and precompute all the required multiples (e.g., $[0, \ 2^h - 1]$) of the multiplicand to be selected by the multiplier's radix-2^h digits. For example for $h = 2$, multiples $0, X, 2X$, and $3X$ are required, where X denotes the n-bit multiplicand and one partial product is generated per two bits of the multiplicand. Unlike $2X$ that is attained via a simple left shift, multiple $3X$ is hard to arrive at, since it requires an n-bit adder to compute $2X + X$.

A common optimization, known as modified Booth recoding, is to recode the multiplier to a minimally redundant radix-4 signed-digit number, where digit set is $[-2, 2]$. While the hard multiple $3X$ is no longer needed, the recoding is merely a cost-free operation, as illustrated in Figure 8.31, where multiplier is an eight-bit TC number and logical states of negabits are equal to NOT of corresponding odd-indexed y-variables [5]. Note that equations corresponding to values of multiples are exactly the same as in conventional Booth recoding [1].

Weight	-2^7	2^6	2^5	2^4	2^3	2^2	2^1	2^0
I	y_7	y_6	y_5	y_4	y_3	y_2	y_1	y_0
II	y_7 y_5	y_6	$-y_5$	y_4 y_3	$-y_3$	y_2 y_1	$-y_1$	y_0
III	○ ●	●	○	●	○ ●	●	○	● $●_0$
Multiple	$-2y_7+y_6$ $+y_5$		$-2y_5+y_4$ $+y_3$		$-2y_3+y_2$ $+y_1$		$-2y_1+y_0$ $+0$	

Figure 8.31 Recoding of the multiplier.

Similar recodings are common in decimal multiplication, where multiplier's BCD digits are recoded to decimal signed digits (e.g., $[-5, 5]$ in Reference [27]). Also redundant digit sets are used for decimal partial product representation such as $[-5, 5]$ in Reference [27], and $[-7, 7]$ in Reference [12].

8.7.3 Use of Redundant Number Systems in Digit Recurrence Algorithms

Most circuit realizations of dividers and square rooters are based on digit recurrence algorithms, where one digit of result is obtained per cycle. The output digit selection is based on the value of partial remainders that are manipulated via add/subtract operations [28]. To shorten the cycle time, it is common to keep partial remainders in a redundant format to allow fast carry-free addition/subtraction [29].

On the other hand, use of intermediate signed digit sets in order to represent the output digits, is a common practice with the benefit of less costly and faster output digit selection. For example, the SRT [30] division algorithm primarily produces BSD quotient digits, and the minimally redundant decimal digit set $[-5, 5]$ is common in radix-10 division [29]. To save one cycle that would be normally required for conversion of redundant output to conventional nonredundant representation, on the fly conversion [31] is used to gradually produce final nonredundant result; one BCD digit per cycle.

8.7.4 Transcendental Functions and Redundant Number Systems

Another domain of RDNS applications is in realization of transcendental functions (e.g., trigonometric, logarithmic). The series-based (e.g., Taylor) implementations require considerable number of add and multiply operations to take place before the required precision is achieved. Therefore, an RDNS with the corresponding redundant add and multiply operations can be devised to represent and manipulate partial results, where the required multiplier with inputs and outputs in the RDNS is quite similar to conventional multipliers. However, the final product computation is also carry-free [12]. Nevertheless, note that such multiplications are not constant-time operations, since the number of partial products are proportional to operand's width and at best can be reduced to two in logarithmic time. Consequently, the whole latency savings due to use of RDNS come from carry-free additions, including that of the final product computation.

CORDIC algorithm [32] is also commonly used, where the main operations are addition, shift, and sign detection of partial results. Again representing partial results in an RDNS enhances the computation speed due to carry-free additions. However, sign detection of redundant numbers is by and large a carry-propagating operation. Nevertheless, constant-time sign estimation techniques, with possible correction operation in the subsequent recurrence, exist that lead to carry-free recurrence [29].

8.7.5 RDNS and Fused Multiply–Add Operation

The fused multiply–add (FMA) operation that is widely used in digital signal processors, is an important application for RDNS. An FMA can be described by $A + B \times C$, where A is the accumulated partial result and B and C represent the incoming input operands. Partial products can be maintained in the same RDNS as that of A, which is therefore, fused within the PPR process to lead to new accumulated partial result.

8.7.6 RDNS and Floating Point Arithmetic

Use of RDNS in floating point add/subtract architectures has been proposed in a couple of articles. One is due to Reference [16] that uses radix-16 MRSD number system, and another one proposes stored-unibit representation [33]. A common problem in RDNS floating point addition is to guarantee that any intermediate redundant result, if hypothetically converted to nonredundant format, as defined by IEEE 754 standard [34], should be exactly the same as the result that is obtainable via a nonredundant manipulation. This is a challenging requirement that has been dealt with in both of the aforementioned works.

A similar problem is overflow detection, where a nonzero carry out of the most significant position of addition process in nonredundant manipulation definitely signals an overflow. However, the same event can only signal an apparent overflow (see Section 8.6.3), since if the intermediate redundant result is converted to nonredundant format, the apparent overflow signal can be vanished by a possible signed carry out with opposite sign. A comprehensive study of apparent overflow correction is found elsewhere [35].

Another challenging problem is detection of the actual number of leading zero digits that is required by post normalization. For example, consider radix-16 maximally redundant four-digit number $(-1 \quad 15 \quad 15 \quad 2)_{16}$. It, apparently, has no leading zero digit. However, the most significant digit (i.e., -1) can be back propagated down to the least significant digit, where the equivalent four-digit numbers in the process of back propagating are: $(0 \ -1 \ 15 \ 2)_{16}$, $(0 \ 0 \ -1 \ 2)_{16}$, and $(0 \ 0 \ 0 \ -14)$. Therefore, the actual number of leading zero digits is three. However, only one digit back propagation is required in case of less redundant digit sets. For example, consider radix-16 digit set $[-14, 14]$, where the most significant digit -1 in $-1 \ 14 \ldots$ can be replaced by 0, while a compensating -2^h ($h = 4$) should be added to the next less significant digit. This leads to a new valid digit $-2^h + (2^h - 2) = -2$, where no new zero digit can be obtained.

8.7.7 RDNS and RNS Arithmetic

RDNS is ideal for addition and subtraction, where $O(1)$ latency is possible. However, RDNS multiplication is at best an $O(\log n)$-latency operation for n-bit operands. On the other hand, latency of an $n \times n$-bit RNS multiplication is on the order of $\lceil \log \frac{n}{k} \rceil$, where k is the number of modulus of the RNS in hand. Therefore, the RNS is more attractive than RDNS for multiplication.

Unfortunately, however, the lowest achievable latency for RNS addition and subtraction is $O(\lceil \log \frac{n}{k} \rceil)$. Consequently, representing the residual RNS operands in an RDNS leads to $O(1)$-latency addition and subtraction and $O(\lceil \log \frac{n}{k} \rceil)$ multiplication; hence, redundant RNS representations take advantage of both RNS and RDNS benefits.

8.8 Summary and Further Reading

Redundant number systems (RDNS) are useful for representing intermediate results in multioperand addition and/or subtraction, where an individual add/subtract operation is performed in constant time not depending on operand length (i.e., $O(1)$), while same operations on conventional nonredundant n-bit operands require $O(n)$ time to complete in the worst case, or $O(\log n)$ with carry acceleration logic.

Partial product reduction (PPR), as the bulk of multiplication of nonredundant operands, is the prime example of multioperand addition that usually benefits from RDNS. However, constant-time PPR and accordingly whole multiplication in $O(1)$ is not possible. The reason is that there are $O(n)$ partial products in an $n \times n$ multiplication, whose reduction to two partial products requires $O(n)$ or $O(\log n)$ operations in sequential or parallel modes, respectively. Therefore, RDNS cannot help in achieving multiplication with better performance than $O(\log n)$. The only improvement is that final product computation can be removed or reduced to an $O(1)$ operation. Nevertheless, this is useful and beneficial in fused multiply–add sequences.

Use of RDNS to maintain partial remainders, in digit-recurrence division or square rooting with nonredundant operands, shortens the cycle time, where one subtraction per output digit is required. Therefore, such use of RDNS reduces the latency of digit recurrence algorithms from $O(n \log n)$ to $O(n)$. Again, similar to multiplication, division with RDNS operands does not lead to any speedup, due to the digit recurrence nature of division algorithm.

Although latency of add and subtract operations is constant in RDNS, proper encoding of digits can further enhance the performance. WBS encoding and use of common arithmetic cells (i.e., FA, HA, $_2^4C$) to manipulate arbitrary mixes of posibits and negabits, help in reducing the number of $O(h)$ steps in execution of addition and subtraction on radix-2^h redundant operands.

There are several textbooks on computer arithmetic in general, which are mostly dealing with algorithms and hardware designs related to conventional number systems [1,28,36,37]. Specialized books on one of the subjects of computer arithmetic are not very common. Nevertheless, we have come across one on division and square root [38] and another one on RNS [2], but none on redundant number systems. However, few pages on signed digit numbers within the chapter on general number systems and a touch on carry save adders in multiplication and division chapters are common.

A separate 10-page chapter is dedicated to "signed digit number operations" in Reference [39]. None of tfhese books have discussed redundant number systems in details and as an independent subject. Therefore, interested readers are recommended to explore the articles listed in Reference section.

References

1 B. Parhami, *Computer Arithmetic, Algorithms and Hardware Designs,* Oxford University Press, 2010.

2 A. Omondi and B. Premkumar, *Residue Number Systems, Theory and Implementation,* Imperial College Press, 2007.

3 R. Conway and J. Nelson, Improved RNS FIR filter architectures, *IEEE Trans. Circuits Syst. II,* vol. 51, no. 1, 2004, pp. 26–28.

4 G. Jaberipur, B. Parhami, and M. Ghodsi, Weighted two-valued digit-set encodings: unifying efficient hardware representation schemes for redundant number systems, *IEEE Trans. Circuits Syst. I,* vol. 52, no. 7, 2005, pp. 1348–1357.

5 G. Jaberipur and B. Parhami, Efficient realization of arithmetic algorithms with weighted collections of posibits and negabits, *IET Comput. Digit. Tech.,* vol. 6, no. 5, 2012, pp. 259–268.

6 B. Parhami, Generalized signed-digit number systems: a unifying framework for redundant number representations, *IEEE Trans. Comput.,* vol. 39, no. 1, 1990, pp. 89–98.

7 G. Metze and J. E. Robertson, Elimination of carry propagation in digital computers, in *Proc. Int. Conf. Inform. Process., Paris*, 1959, pp. 389–396.

8 A. Avizienis, Signed-digit number representations for fast parallel arithmetic, *IRE Trans. Electron. Comput.*, vol. 10, 1961, pp. 389–400.

9 G. Jaberipur, B. Parhami, and M. Ghodsi, Weighted bit-set encodings for redundant digit sets: theory and applications, in *Proc. 36th Asilomar Conf. Signals Syst. Comput.*, November 2002, pp. 1629–1633.

10 G. Jaberipur and B. Parhami, Stored-transfer representations with weighted digit-set encodings for ultrahigh-speed arithmetic, *IET Circuits, Dev. Syst.*, vol. 1, no. 1, 2007, pp. 102–110.

11 R. D. Kenney and M. J. Schulte, High-speed multioper and decimal adders, *IEEE Trans. Comput.*, vol. 54, no. 8, 2005, pp. 953–963.

12 S. Gorgin and G. Jaberipur, Fully redundant decimal arithmetic, in *Proc. 19th IEEE Symp. Comput. Arith., Portland, USA*, June 2009, pp. 145–152.

13 A. Vazquez, E. Antelo, and J. D. Bruguera, Fast radix-10 multiplication using redundant BCD codes, *IEEE Trans. Comput.*, vol. 63, no. 8, 2014, pp. 1902–1914.

14 B. Shirazi, D. Y. Yun, and C. N. Zhang, RBCD: redundant binary coded decimal adder, *IEEE Proc. Comput. Digit. Tech. (CDT)*, vol. 136, no. 2, 1989, pp. 156–160.

15 S. Gorgin and G. Jaberipur, A fully redundant decimal adder and its application in parallel decimal multipliers, *Microelectronics J.*, vol. 40, no. 10, 2009, pp. 1471–1481.

16 H. Fahmy and M. J. Flynn, The case for a redundant format in floating point arithmetic, in *Proc. 16th IEEE Symp. Comput. Arith., Santiago de Compostela, Spain*, June 2003.

17 S. Gorgin and G. Jaberipur, A family of high radix signed digit adders, in *Proc. 20th IEEE Symp. Comput. Arith., Tubingen, Germany*, July 25–27, 2011, pp. 112–120.

18 V. G. Oklobdzija and D. Villeger, Improving multiplier design by using improved column compression tree and optimized final adder in CMOS technology, *IEEE Trans. Very Large Scale Integr. (VLSI) Syst.*, vol. 3, no. 2, 1995, pp. 292–301.

19 Mingyan Zhang, Jiangmin Gu, and Chip-Hong Chang, A novel hybrid pass logic with static CMOS output drive full-adder cell, *IEEE Proc. Int. Symp. Circuits Syst.*, vol. 5, 2003, pp. 317–320.

20 G. Jiangmin and Chip-Hong Chang, Ultra low voltage low power 4-2 compressor for high speed multiplications, *IEEE Int. Symp. Circuits Syst.*, vol. 5, 2003, pp. 321–324.

21 C. H. Chang, J. Gu, and M. Zhang, Ultra low-voltage low-power CMOS 4-2 and 5-2 compressors for fast arithmetic circuits. *IEEE Trans. Circuits Syst. I*, vol. 51, no. 10, 2004, pp. 1985–1997.

22 S. Veeramachaneni, K. M. Krishna, L. Avinash, S. R. Puppala, and M. B. Srinivas, Novel architectures for high-speed and low-power 3-2, 4-2 and 5-2 compressors, in *Proc. 20th Int. Conf. VLSI Des.*, 2007, pp. 324–329.

23 D. Baran, M. Aktan, and V. G. Oklobdzija, Energy efficient implementation of parallel CMOS multipliers with improved compressors, in *16th ACM/IEEE Int. Symp. Low-Power Electron. Des.*, 2010, pp. 147–152.

24 A. Pishvaie, G. Jaberipur, and A. Jahanian, Improved CMOS (4; 2) compressor designs for parallel multipliers, *Comput. Electr. Eng.*, vol. 38, 2012, pp. 1703–1716.

25 B. Jose and D. Radhakrishnan, Delay optimized redundant binary adders, in *Proc. 13th IEEE Int. Conf. Electron., Circuits Syst.*, 2006, pp. 514–517.

26 B. Parhami, On the implementation of arithmetic support functions for generalized signed-digit number systems, *IEEE Trans. Comput.*, vol. 42, no. 3, 1993, pp. 379–384.

27 M. A. Erle, E. M. Schwartz, and M. J. Schulte, Decimal multiplication with efficient partial product generation, in *Proc. 17th IEEE Symp. Comput. Arith.*, 2005, pp. 21–28.

28 M. D. Ercegovac and T. Lang, *Digital Arithmetic*, Morgan Kaufmann Publishers, 2004.

29 A. Kaivani and G. Jaberipur, Decimal CORDIC rotation based on selection by rounding, *Comput. J.*, vol. 54, no. 11, 2011, pp. 1798–1809.

30 D. E. Atkins, The theory and implementation of SRT division, Department of Computer Science, University of Illinois at Urbana-Champaign, Urbana, IL, Tech. Rep. UIUCDCS-R-67-230, 1967.

31 M. D. Ercegovac and T. Lang, On-the-fly rounding, *IEEE Trans. Comput.*, vol. 41, no. 12, 1992, pp. 1497–1503.

32 J. Volder, The CORDIC trigonometric computing technique, *IRE Trans. Electron. Comput.*, vol. EC-8, no. 3, 1959, pp. 330–334.

33 G. Jaberipur, B. Parhami, and S. Gorgin, Redundant-digit floating-point addition scheme based on a stored rounding value, *IEEE Trans. Comput.*, vol. 59, no. 5, 2010, pp. 694–706.

34 CAU IEEE, Inc., IEEE 754–2008 Standard for Floating-Point Arithmetic, August 2008.

35 G. Jaberipur, Frameworks for the Exploration and Implementation of Generalized Carry-Free Redundant Number Systems, Ph.D. Thesis, Sharif University of Technology, December 2004.

36 I. Koren, *Computer Arithmetic Algorithms*, EDN 2nd Edn, A K Press Ltd, 2002.

37 P. Kornerup and D. W. Matula, *Finite Presision Number System and Arithmetic*, Cambrige University Press, 2010.

38 M. D. Ercegovac and T. Lang, *Division and Square Root: Digit Recurrence Algorithms and Implementations*, Kluwer Academic Publishers, 1994.

39 L. Mi, *Arithmetic and Logic in Computer Systems*, John Wiley and Sons, Inc., 2004.

INDEX

Arithmetic Circuits for DSP Applications, First Edition. Edited by Pramod Kumar Meher and Thanos Stouraitis.
© 2017 by The Institute of Electrical and Electronics Engineers, Inc. Published 2017 by John Wiley & Sons, Inc.